Iterative Methods and Their Dynamics with Applications

A Contemporary Study

Iterative Methods and Their Dynamics with Applications

A Contemporary Study

Ioannis K. Argyros
Department of Mathematical Sciences
Cameron University
Lawton, Oklahoma, USA

and

Á. Alberto Magreñán
Universidad Internacional de La Rioja
Escuela Superior de Ingeniería y Tecnología
Logroño, La Rioja, Spain

CRC Press
Taylor & Francis Group
Boca Raton London New York

CRC Press is an imprint of the
Taylor & Francis Group, an **informa** business
A SCIENCE PUBLISHERS BOOK

CRC Press
Taylor & Francis Group
6000 Broken Sound Parkway NW, Suite 300
Boca Raton, FL 33487-2742

Printed on acid-free paper
Version Date: 20160826

International Standard Book Number-13: 978-1-4987-6360-8 (Hardback)

Library of Congress Cataloging-in-Publication Data

Names: Argyros, Ioannis K. | Magreñán, Aì. Alberto (Aìngel Alberto)
Title: Iterative methods and their dynamics with applications : a
contemporary study / Ioannis K. Argyros and Aì. Alberto Magreñán.
Description: Boca Raton, FL : CRC Press, [2016] | "A science publishers
book." | Includes bibliographical references and index.
Identifiers: LCCN 2016038689| ISBN 9781498763608 (hardback : alk. paper) |
ISBN 9781498763622 (e-book : alk. paper)
Subjects: LCSH: Iterative methods (Mathematics) | Numerical analysis.
Classification: LCC QA297.8 .A7364 2016 | DDC 518/.26--dc23
LC record available at https://lccn.loc.gov/2016038689

Visit the Taylor & Francis Web site at
http://www.taylorandfrancis.com

and the CRC Press Web site at
http://www.crcpress.com

Dedication

To my parents Mercedes and Alberto

my grandmother Ascensión

and my beloved Lara.

Preface

This book comprises a modern approach which combines recent and past results in numerical methods with applications in fields such us Engineering, Chemistry, etc. The book also provides a comparison between various investigations of recent years. Each chapter contains several new theoretical results and important applications. The applications appear in the form of examples or study cases or they are implied since our results improve earlier ones that have already been applied in concrete problems. Chapters and sections have been written independent of each other. Hence, the interested reader can go directly to a certain chapter or section and understand the material without having to go back and forth. Furthermore, several references to recent studies appear along the whole book.

This monograph presents recent works of authors on Convergence and Dynamics of iterative methods. It is the natural outgrowth of their related publications in these areas. Chapters are self-contained and can be read independently. Moreover, an extensive list of references is given in each chapter, in order to allow reader to use the previous ideas. For these reasons, we think that several advanced courses can be taught using this book.

The book's results are expected to find applications in many areas of applied mathematics, engineering, computer science and real problems. As such this monograph is suitable to researchers, graduate students and seminars in the above subjects, also to be in all science and engineering libraries.

The preparation of this book took place during the year 2016 in Lawton, Oklahoma, USA and Logroño, La Rioja, Spain.

October 2016

Ioannis K. Argyros
Á. Alberto Magreñán

Contents

List of Figures

List of Tables

Symbol Description

α	To solve the generator maintenance scheduling, in the past, several mathematical techniques have been applied.
σ^2	These include integer programming, integer linear programming, dynamic programming, branch and bound etc.
$\sum$	Several heuristic search algorithms have also been developed. In recent years expert systems,
abc	fuzzy approaches, simulated annealing and genetic algorithms have also been tested.
$\theta\sqrt{abc}$	This paper presents a survey of the literature
ζ	over the past fifteen years in the generator
∂	maintenance scheduling. The objective is to
sdf	present a clear picture of the available recent literature
ewq	of the problem, the constraints and the other aspects of
bvcn	the generator maintenance schedule.

Chapter 1

Halley's method

Ioannis K. Argyros

Department of Mathematical Sciences, Cameron University, Lawton, OK 73505, USA, Email: iargyros@cameron.edu

Á. Alberto Magreñán

Universidad Internacional de La Rioja, Escuela Superior de Ingeniería y Tecnología, 26002 Logroño, La Rioja, Spain, Email: alberto.magrenan@unir.net

CONTENTS

1.1 Introduction

In this chapter we are concerned with the problem of approximating a locally unique solution $x^\star$ of the nonlinear equation

$$F(x) = 0, \tag{1.1}$$

where F is twice Fréchet-differentiable operator defined on a nonempty open and convex subset of a Banach space X with values in a Banach space Y.

Many problems from computational sciences and other disciplines can be brought in a form similar to equation (1.1) using mathematical modelling [1,3, 5,13,19–21]. The solutions of these equations can be rarely be found in closed form. That is why most solution methods for these equations are iterative. The convergence matter is usually divided in local and semilocal study. The local study is based on the solution whereas the semilocal one is based on the initial point.

In the present chapter we provide a semilocal convergence analysis for Halley's method defined by

$$x_{n+1} = x_n - \Gamma_F(x_n)F(x_n), \quad for\ each\ n = 0,1,2,\ldots, \tag{1.2}$$

where, $\Gamma_F(x) = (I - L_F(x))^{-1}$ and $L_F(x) = \frac{1}{2}F'(x)^{-1}F''(x)F'(x)^{-1}F(x)$.

The convergence of Halley's method has been studied in many works such as [1–5,8,11,12]. The most used conditions for the study of the semilocal convergence of Halley's method are the following ones

(C_1) There exists $x_0 \in D$ such that $F'(x_0)^{-1} \in L(Y,X)$, the space of bounded linear operator from Y into X;

(C_2) $\|F'(x_0)^{-1}F(x_0)\| \leq \gamma$, $\gamma > 0$;

(C_3) $\|F'(x_0)^{-1}F''(x)\| \leq M$ for each x in D, $M > 0$;

(C_4) $\|F'(x_0)^{-1}[F''(x) - F''(y)]\| \leq K\|x - y\|$ for each x and y in D, $K > 0$.

The corresponding sufficient convergence condition is given by

$$\gamma \leq \frac{4K + M^2 - M\sqrt{M^2 + 2K}}{3K(M + \sqrt{M^2 + 2K})}, \tag{1.3}$$

where all the parameters comes from the above conditions.

Condition (C_4) is not satisfied in many cases as the following one.

Example 1.1 Let $X = Y = \mathbb{R}$, define $D = [0, +\infty)$ and let

$$F(x) = \frac{4}{15}x^2\sqrt{x} + x^2 + x - 3.$$

Then, we obtain

$$|F''(x) - F''(y)| = |\sqrt{x} - \sqrt{y}| = \frac{|x - y|}{\sqrt{x} + \sqrt{y}}.$$

Therefore, there is no constant K satisfying (C_4).

Ezquerro and Hernández in [12], using recurrence relations, expanded the applicability of Halley's method dropping condition (C_4) and replacing (1.3) by

$$M\gamma < \frac{4 - \sqrt{6}}{5}. \tag{1.4}$$

In the present chapter we show how to expand even further the applicability of Halley's method using (C_1), (C_2), (C_3) and the well-known center-Lipschitz condition, defined by

$$(C_5) \quad \|F'(x_0)^{-1}[F'(x)-F'(x_0)]\| \leq L\|x-x_0\| \quad for\ each\ x \in D, \quad L>0.$$

We have that

$$L \leq M \tag{1.5}$$

holds in general. Moreover, as it was stated in [4] and [5] $\frac{M}{L}$ can be arbitrarily large. In the literature (C_3) is used to obtain

$$\|F'(x)^{-1}F'(x_0)\| \leq \frac{1}{1-M\|x-x_0\|} \quad for\ each\ x \in D. \tag{1.6}$$

However, using condition (C_5) and if $L<M$ we obtain the tighter estimate

$$\|F'(x)^{-1}F'(x_0)\| \leq \frac{1}{1-L\|x-x_0\|} \quad for\ each\ x \in D. \tag{1.7}$$

This idea, used in the study of many methods [4, 5], allows to obtain:

- a tighter convergence analysis,
- weaker sufficient convergence conditions,
- more precise error bounds on the location of the solution.

As a consequence of the idea, we can use a different approach than recurrent relations in our analysis. The chapter is organized as follows: Section 1.2 contains the study of the semilocal convergence of Halley's method and the numerical examples are presented Section 1.3. Finally, some basins of attraction related to Halley's method are presented in the concluding Section 1.4.

1.2 Semilocal convergence of Halley's method

In order to present the semilocal convergence analysis of Halley's method, we shall first introduce an additional condition.

Denoting

$$a = \frac{M\gamma}{2} \quad and \quad \gamma_0 = \frac{\gamma}{1-a},$$

we can define condition

(C_6) Suppose that

$$(L+\frac{M}{2})\gamma < 1 \tag{1.8}$$

and there exists

$$b \in (0, 1 - L\gamma_0) \tag{1.9}$$

which is the smallest positive zero of function g defined on $(0, 1 - L\gamma_0)$ given by

$$g(t) \quad = \quad \frac{\dfrac{a}{1 - \dfrac{L\gamma_0}{1-t}} + \dfrac{a}{1-a}}{(1 - \dfrac{L\gamma_0}{1-t})(1 - \dfrac{a}{(1-a)(1 - \dfrac{L\gamma_0}{1-t})^2}(\dfrac{a}{1 - \dfrac{L\gamma_0}{1-t}} + \dfrac{a}{1-a}))} - t. \tag{1.10}$$

Then, we get

$$b = \frac{\dfrac{a}{1 - \dfrac{L\gamma_0}{1-b}} + \dfrac{a}{1-a}}{(1 - \dfrac{L\gamma_0}{1-b})(1 - \dfrac{a}{(1-a)(1 - \dfrac{L\gamma_0}{1-b})^2}(\dfrac{a}{1 - \dfrac{L\gamma_0}{1-b}} + \dfrac{a}{1-a}))}. \tag{1.11}$$

Now, define

$$R_0 = \frac{\gamma_0}{1-b} \quad \textit{and} \quad \beta = LR_0.$$

Then, we have that

$$a \in (0,1), \quad \beta \in (0,1) \quad \textit{and}$$
$$\frac{a}{(1-a)(1-\beta)^2}(\frac{a}{1-\beta} + \frac{a}{1-a}) = \frac{M^2\gamma_0}{4(1-\beta)^2}(\frac{\gamma}{1-\beta} + \gamma_0) < 1. \tag{1.12}$$

Rewriting (1.11) as

$$b = \frac{\dfrac{M}{2}(\dfrac{\gamma}{1-\beta} + \gamma_0)}{(1-\beta)(1 - \dfrac{M^2\gamma_0}{4(1-\beta)^2}(\dfrac{\gamma}{1-\beta} + \gamma_0))}.$$

Furthermore, before stating the main theorem, we must define the following condition.

(C_7) Suppose that

$$d = \frac{\dfrac{M}{2(1-\beta)}\gamma + \dfrac{M}{2}\gamma_0}{(1-\beta)(1 - \dfrac{M}{2(1-\beta)^2}\gamma)} < 1.$$

We refer to $(C_1) - (C_3)$, $(C_5) - (C_7)$ as the (C) conditions. Let $B(x,R)$, $\overline{B}(x,R)$

stand, respectively, for the open and closed balls in X with center x and radius $R > 0$.

Theorem 1.1

Let $F : D \subset X \to Y$ be continuously twice Fréchet differentiable, where X, Y are Banach spaces and D is an open and convex subset of X. Suppose that the (C) conditions hold and $\overline{B}(x_0, R_0) \subset D$. Then, the sequence $\{x_n\}$ generated by Halley's method (1.2) is well defined, remains in $B(x_0, R_0)$ for all $n \geq 0$ and converges to a unique solution $x^\star \in \overline{B}(x_0, R_0)$ of equation $F(x) = 0$. Moreover, the following estimate holds for each $n = 0, 1, 2, \ldots$

$$\|x_{n+2} - x_{n+1}\| \leq b\|x_{n+1} - x_n\|. \tag{1.13}$$

Furthermore, if there exists $R^\star \geq R_0$ such that

$$\overline{B}(x_0, R^\star) \subseteq D \tag{1.14}$$

and

$$\frac{L}{2}(R_0 + R^\star) < 1, \tag{1.15}$$

then, $x^\star$ is the only solution of equation $F(x)$ in $\overline{B}(x_0, R^\star)$.

Proof: By the definition of $(C_1) - (C_3)$ we get that

$$\begin{aligned} \|I - (I - L_F(x_0))\| &= \|L_F(x_0)\| = \frac{1}{2}\|F'(x_0)^{-1}F''(x_0)F'(x_0)^{-1}F(x_0)\| \\ &\leq \frac{1}{2}\|F'(x_0)^{-1}F''(x_0)\|\|F'(x_0)^{-1}F(x_0)\| \\ &\leq \frac{1}{2}M\gamma = a < 1. \end{aligned} \tag{1.16}$$

Moreover, it follows from the Banach lemma on invertible operators [2], [13] and (1.16) that $(I - L_F(x_0))^{-1}$ exists and $\|(I - L_F(x_0))^{-1}\|$ is bounded by

$$\|(I - L_F(x_0))^{-1}\| \leq \frac{1}{1 - \|L_F(x_0)\|} \leq \frac{1}{1 - a}.$$

Then, by (C_2), (1.2) and the above estimate we obtain that

$$\begin{aligned} \|x_1 - x_0\| &= \|(I - L_F(x_0))^{-1}F'(x_0)^{-1}F(x_0)\| \leq \|(I - L_F(x_0))^{-1}\| \\ &\quad \|F'(x_0)^{-1}F(x_0)\| \\ &\leq \frac{\gamma}{1 - a} = \gamma_0 < R_0. \end{aligned}$$

Now, using (C_5) we get that

$$\|I - F'(x_0)^{-1}F'(x_1)\| \leq L\|x_1 - x_0\| \leq L\gamma_0 < LR_0 = \beta < 1.$$

Hence, $F'(x_1)^{-1}$ exists and $\|F'(x_1)^{-1}F'(x_0)\|$ is bounded by

$$\|F'(x_1)^{-1}F'(x_0)\| < \frac{1}{1-\beta}.$$

By (1.2) we can write

$$[I - L_F(x_0)](x_1 - x_0) + F'(x_0)^{-1}F(x_0) = 0$$

or

$$F(x_0) + F'(x_0)(x_1 - x_0) = \frac{1}{2}F''(x_0)F'(x_0)^{-1}F(x_0)(x_1 - x_0)$$

or

$$\begin{aligned} F'(x_0)^{-1}F(x_1) &= F'(x_0)^{-1}[F(x_1) - F(x_0) - F'(x_0)(x_1 - x_0) + F(x_0) \\ &\quad + F'(x_0)(x_1 - x_0)] \\ &= \frac{1}{2}F'(x_0)^{-1}F''(x_0)F'(x_0)^{-1}F(x_0)(x_1 - x_0) \\ &\quad + \int_0^1 F'(x_0)^{-1}F''(x_0 + \theta(x_1 - x_0))(1-\theta)d\theta(x_1 - x_0)^2. \end{aligned}$$

Hence, we have that

$$\begin{aligned} \|F'(x_0)^{-1}F(x_1)\| &\leq \frac{1}{2}M\gamma\|x_1 - x_0\| + \frac{1}{2}M\|x_1 - x_0\|^2 \\ &\leq \frac{M}{2(1-\beta)}\gamma\|x_1 - x_0\| + \frac{1}{2}M\|x_1 - x_0\|^2 \\ &\leq \frac{M}{2}(\frac{\gamma}{1-\beta} + \gamma_0)\|x_1 - x_0\|. \end{aligned}$$

Moreover, we get that

$$\begin{aligned} \|L_F(x_1)\| &= \frac{1}{2}\|F'(x_1)^{-1}F'(x_0)F'(x_0)^{-1}F''(x_1)F'(x_1)^{-1} \\ &\quad F'(x_0)F'(x_0)^{-1}F(x_1)\| \\ &\leq \frac{1}{2}\|F'(x_1)^{-1}F'(x_0)\|^2\|F'(x_0)^{-1}F''(x_1)\|\|F'(x_0)^{-1}F(x_1)\| \\ &\leq \frac{M}{2(1-\beta)^2}(\frac{M}{2(1-\beta)}\gamma\|x_1 - x_0\| + \frac{1}{2}M\|x_1 - x_0\|^2) \\ &\leq \frac{M^2(\frac{\gamma}{1-\beta} + \gamma_0)\|x_1 - x_0\|}{4(1-\beta)^2} \leq \frac{M^2\gamma_0}{4(1-\beta)^2}(\frac{\gamma}{1-\beta} + \gamma_0) < 1. \end{aligned}$$

Then, $(I - L_F(x_1))^{-1}$ exists,

$$\begin{aligned} \|(I - L_F(x_1))^{-1}\| &\leq \frac{1}{1 - \|L_F(x_1)\|} \\ &\leq \frac{1}{1 - \frac{M}{2(1-\beta)^2}(\frac{M}{2(1-\beta)}\gamma\|x_1 - x_0\| + \frac{1}{2}M\|x_1 - x_0\|^2)} \\ &\leq \frac{1}{1 - \frac{M^2\gamma_0}{4(1-\beta)^2}(\frac{\gamma}{1-\beta} + \gamma_0)}. \end{aligned}$$

So, x_2 is well defined, and using (1.2) we get

$$\begin{aligned}
\|x_2 - x_1\| &\leq \frac{\|F'(x_1)^{-1}F'(x_0)\| \|F'(x_0)^{-1}F(x_1)\|}{1 - \|L_F(x_1)\|} \\
&\leq \frac{\frac{M}{2(1-\beta)}\gamma\|x_1 - x_0\| + \frac{1}{2}M\|x_1 - x_0\|^2}{(1-\beta)(1 - \frac{M}{2(1-\beta)^2}[\frac{M}{2(1-\beta)}\gamma\|x_1 - x_0\| + \frac{1}{2}M\|x_1 - x_0\|^2])} \\
&\leq \frac{\frac{M}{2}(\frac{\gamma}{1-\beta} + \gamma_0)}{(1-\beta)(1 - \frac{M^2\gamma_0}{4(1-\beta)^2}(\frac{\gamma}{1-\beta} + \gamma_0))}\|x_1 - x_0\| = b\|x_1 - x_0\|.
\end{aligned}$$

Therefore, we have that

$$\begin{aligned}
\|x_2 - x_0\| &\leq \|x_2 - x_1\| + \|x_1 - x_0\| \leq b\|x_1 - x_0\| + \|x_1 - x_0\| \\
&= (1+b)\|x_1 - x_0\| \\
&= \frac{1-b^2}{1-b}\|x_1 - x_0\| < \frac{\|x_1 - x_0\|}{1-b} \leq \frac{\gamma_0}{1-b} = R_0.
\end{aligned}$$

Hence, $x_2 \in B(x_0, R_0)$.

The above shows that for $n = 0$, the following statements hold

(i) $F'(x_{n+1})^{-1}$ exists and

$$\|F'(x_{n+1})^{-1}F'(x_0)\| < \frac{1}{1-\beta};$$

(ii)

$$\|F'(x_0)^{-1}F(x_{n+1})\| \leq d_{n+1};$$

(iii)

$$d_{n+1} \leq d d_n \leq d^{n+1} d_0;$$

(iv) $(I - L_F(x_{n+1}))^{-1}$ exists and

$$\|(I - L_F(x_{n+1}))^{-1}\| \leq \frac{1}{1 - \frac{M}{2(1-\beta)^2}d_{n+1}};$$

(v) x_{n+2} is well defined and

$$\|x_{n+2} - x_{n+1}\| \leq \frac{d_{n+1}}{(1-\beta)(1 - \frac{M}{2(1-\beta)^2}d_{n+1})} \leq b\|x_{n+1} - x_n\| \leq b^{n+1}\|x_1 - x_0\|;$$

(vi) $\|x_{n+2} - x_0\| < R_0$.

Here, sequence $\{d_n\}$ is defined by

$$\begin{aligned} d_0 &= \gamma, \\ d_{n+1} &= \frac{M}{2(1-\beta)} d_n \|x_{n+1}-x_n\| + \frac{M}{2}\|x_{n+1}-x_n\|^2 \quad n \geq 0. \end{aligned}$$

The rest of the proof will be obtained using mathematical induction. Assuming that (i)-(vi) are true for all natural integers $n \leq k$, where $k \geq 0$ is a fixed integer. Then $F'(x_{k+2})$ exists, since $x_{k+2} \in B(x_0, R_0)$ and

$$\|I - F'(x_0)^{-1}F'(x_{k+2})\| \quad \leq \quad L\|x_{k+2}-x_0\| < LR_0 = \beta < 1.$$

Hence, $F'(x_{k+2})^{-1}$ exists and

$$\|F'(x_{k+2})^{-1}F'(x_0)\| \leq \frac{1}{1 - L\|x_{k+2}-x_0\|} < \frac{1}{1-\beta}.$$

Next, we shall estimate $\|F'(x_0)^{-1}F(x_{k+2})\|$. We have by Halley's method that

$$\begin{aligned} F(x_{k+2}) &= F(x_{k+2}) - F(x_{k+1}) - F'(x_{k+1})(x_{k+2}-x_{k+1}) \\ &\quad + \frac{1}{2}F''(x_{k+1})F'(x_{k+1})^{-1}F(x_{k+1})(x_{k+2}-x_{k+1}) \end{aligned}$$

or

$$\begin{aligned} F'(x_0)^{-1}F(x_{k+2}) &= \frac{1}{2}F'(x_0)^{-1}F''(x_{k+1})F'(x_{k+1})^{-1}F'(x_0)F'(x_0)^{-1} \\ &\quad F(x_{k+1})(x_{k+2}-x_{k+1}) \\ &\quad + \int_0^1 F'(x_0)^{-1}F''(x_{k+1}+\theta(x_{k+2}-x_{k+1}))(1-\theta) \\ &\quad d\theta(x_{k+2}-x_{k+1})^2. \end{aligned}$$

Hence, we get that

$$\begin{aligned} \|F'(x_0)^{-1}F(x_{k+2})\| &\leq \frac{M}{2(1-\beta)} d_{k+1}\|x_{k+2}-x_{k+1}\| + \frac{M}{2}\|x_{k+2}-x_{k+1}\|^2 = d_{k+2} \\ &= \frac{M}{2}(\frac{d_{k+1}}{1-\beta} + \|x_{k+2}-x_{k+1}\|)\|x_{k+2}-x_{k+1}\| \\ &\leq \frac{M}{2}(\frac{d^{k+1}\gamma}{1-\beta} + b^{k+1}\|x_1-x_0\|)\|x_{k+2}-x_{k+1}\| \\ &\leq \frac{M}{2}(\frac{d\gamma}{1-\beta} + b\gamma_0)\|x_{k+2}-x_{k+1}\| \leq \frac{M}{2}(\frac{d\gamma}{1-\beta} + b\gamma_0)b^{k+1} \\ &\quad \|x_1-x_0\| \\ &\leq \frac{M}{2}(\frac{d\gamma}{1-\beta} + b\gamma_0)b\gamma_0, \end{aligned} \tag{1.17}$$

$$
\begin{aligned}
d_{k+2} &\le \frac{M}{2}(\frac{d\gamma}{1-\beta}+b\gamma_0)\|x_{k+2}-x_{k+1}\| \\
&\le \frac{M}{2}(\frac{d\gamma}{1-\beta}+b\gamma_0)\frac{d_{k+1}}{(1-\beta)(1-\frac{M}{2(1-\beta)^2}d^{k+1}\gamma)} \\
&\le \frac{M}{2}(\frac{d\gamma}{1-\beta}+b\gamma_0)\frac{d_{k+1}}{(1-\beta)(1-\frac{M}{2(1-\beta)^2}d\gamma)} \\
&\le \frac{M}{2}(\frac{\gamma}{1-\beta}+\gamma_0)\frac{d_{k+1}}{(1-\beta)(1-\frac{M}{2(1-\beta)^2}\gamma)} = dd_{k+1} \le d^{k+2}d_0
\end{aligned}
$$

and

$$
\begin{aligned}
\|L_F(x_{k+2})\| &\le \frac{1}{2}\|F'(x_{k+2})^{-1}F'(x_0)\|^2\|F'(x_0)^{-1}F(x_{k+2})\| \\
&\quad \|F'(x_0)^{-1}F''(x_{k+2})\| \\
&\le \frac{M}{2(1-\beta)^2}d_{k+2} \le \frac{M^2}{4(1-\beta)^2}(\frac{\gamma}{1-\beta}+\gamma_0)\gamma_0 < 1.
\end{aligned}
$$

As a consequence, $(I-L_F(x_{k+2}))^{-1}$ exists and

$$
\|(I-L_F(x_{k+2}))^{-1}\| \le \frac{1}{1-\frac{M}{2(1-\beta)^2}d_{k+2}} \le \frac{1}{1-\frac{M^2}{4(1-\beta)^2}(\frac{\gamma}{1-\beta}+\gamma_0)\gamma_0}.
$$

Therefore, x_{k+3} is well defined. Moreover, we obtain that

$$
\begin{aligned}
\|x_{k+3}-x_{k+2}\| &\le \|(I-L_F(x_{k+2}))^{-1}\|\|F'(x_{k+2})^{-1}F'(x_0)\|\|F'(x_0)^{-1}F(x_{k+2})\| \\
&\le \frac{d_{k+2}}{(1-\beta)(1-\frac{M}{2(1-\beta)^2}d_{k+2})} \\
&\le \frac{\frac{M}{2}(\frac{\gamma}{1-\beta}+\gamma_0)\|x_{k+2}-x_{k+1}\|}{(1-\beta)(1-\frac{M^2}{4(1-\beta)^2}(\frac{\gamma}{1-\beta}+\gamma_0)\gamma_0)} = b\|x_{k+2}-x_{k+1}\| \\
&\le b^{k+2}\|x_1-x_0\|.
\end{aligned}
$$

Furthermore, we have that

$$
\begin{aligned}
\|x_{k+3}-x_0\| &\le \|x_{k+3}-x_{k+2}\| + \|x_{k+2}-x_{k+1}\| + \cdots + \|x_1-x_0\| \\
&\le (b^{k+2}+b^{k+1}+\cdots+1)\|x_1-x_0\| \\
&= \frac{1-b^{k+3}}{1-b}\|x_1-x_0\| < \frac{\gamma_0}{1-b} = R_0.
\end{aligned}
$$

Hence, we deduce that $x_{k+3} \in B(x_0, R_0)$. The induction for (i)-(vi) is complete. Let m be a natural integer. Then, we have that

$$\begin{aligned}
\|x_{k+m} - x_k\| &\leq \|x_{k+m} - x_{k+m-1}\| + \|x_{k+m-1} - x_{k+m-2}\| + \cdots + \|x_{k+1} - x_k\| \\
&\leq (b^{m-1} + \cdots + b + 1)\|x_{k+1} - x_k\| \\
&\leq \frac{1-b^m}{1-b} b^k \|x_1 - x_0\|.
\end{aligned}$$

It follows that $\{x_k\}$ is complete sequence in a Banach space X and as such it converges to some $x^\star \in \overline{B}(x_0, R_0)$ (since $\overline{B}(x_0, R_0)$ is a closed set). By letting $k \to \infty$ in (1.17) we obtain $F(x^\star) = 0$. Moreover, we get

$$\|x^\star - x_k\| \leq \frac{b^k}{1-b}\|x_1 - x_0\|.$$

In order to show the uniqueness part, let $y^\star$ be a solution of $F(x) = 0$ in $\overline{B}(x_0, R^\star)$. Denoting $Q = \int_0^1 F'(x_0)^{-1} F'(x^\star + \theta(y^\star - x^\star))d\theta$. Using (C_5) we have in turn that

$$\begin{aligned}
\|I - Q\| &= \|\int_0^1 F'(x_0)^{-1}[F'(x^\star + \theta(y^\star - x^\star)) - F'(x_0)]d\theta\| \\
&\leq L\int_0^1 [(1-\theta)\|x^\star - x_0\| + \theta\|y^\star - x_0\|]d\theta \leq \frac{L}{2}(R_0 + R^\star) < 1.
\end{aligned}$$

It follows that Q^{-1} exists. Now, using

$$0 = F'(x_0)^{-1}(F(y^\star) - F(x^\star)) = F'(x_0)^{-1}T(y^\star - x^\star)$$

we deduce $y^\star = x^\star$. $\square$

1.3 Numerical examples

In order to show the applicability of the main result we will apply the conditions to two different examples.

Example 1.3.1 Consider the scalar function $F(x) = x^3 - 0.64$ on $D = (0, 1)$ and choose as initial point $x_0 = 0.85$. Then,

$$F'(x) = 3x^2, \quad F''(x) = 6x. \tag{1.18}$$

So, $F(x_0) = -0.025875$, $F'(x_0) = 2.1675$ and $\gamma = 0.011937716$.

Moreover, we have for any $x \in D$ that

$$|F'(x_0)^{-1}F''(x)| = \frac{2|x|}{x_0^2} \leq \frac{2}{x_0^2} = 2.76816609 \tag{1.19}$$

and

$$|F'(x_0)^{-1}[F'(x) - F'(x_0)]| = \frac{|x + x_0|}{x_0^2}|x - x_0| \leq 2.560553633|x - x_0|. \quad (1.20)$$

That is, we obtain $M = 2.76816609$, $L = 2.560553633$, $a = 0.016522791$, $\gamma_0 = 0.012138274$ and $1 - L\gamma_0 = 0.968919297$.

Furthermore, we have that $b = 0.035021656$ by using the secant method. Then we get $R_0 = 0.012578805$, $\beta = 0.032208705$ and $d = 0.035628901$. As a consequence, conditions (C_6) and (C_7) are satisfied.

On the other hand,

$$\overline{B}(x_0, R_0) = [0.837421195, 0.862578805] \subset D.$$

So, we can ensure the convergence of Halley's method by Theorem 1.1.

Now, we want to compare our uniqueness ball with the ones obtained in [11]. Using (1.4) we get that

$$\beta_0 = M\gamma = 0.03304558 < \frac{4 - \sqrt{6}}{5}.$$

So, conditions of [11] are also satisfied.

The uniqueness ball is $\overline{B}(x_0, r_0\gamma)$, where

$$r_0 = \frac{2 - 3\beta_0}{2(1 - 3\beta_0 + \beta_0^2)} = 1.053745860.$$

Then, we get that $R_1 = r_0\gamma = 0.012579319 > R_0$.

So, it is clear that we provide a better information on the location of the solution. Moreover by (1.14) and (1.15) we can set $R^\star = 0.15$, which extends the uniqueness ball from $\overline{B}(x_0, R_0)$ to $\overline{B}(x_0, R^\star)$.

Example 1.3.2 In this example we provide an application of our results to a special nonlinear Hammerstein integral equation of the second kind. In this example we will use the max-norm.

$$x(s) = 1 + \frac{4}{5}\int_0^1 G(s,t)x(t)^3 dt, \quad s \in [0,1], \quad (1.21)$$

where, G is the Green's kernel on $[0,1] \times [0,1]$ defined by

$$G(s,t) = \begin{cases} t(1-s), & t \leq s; \\ s(1-t), & s \leq t. \end{cases} \quad (1.22)$$

Let $X = Y = C[0,1]$ and D be a suitable open convex subset of $D := \{x \in X : x(s) > 0, s \in [0,1]\}$, which will be given below. Define $F : D \to Y$ by

$$[F(x)](s) = x(s) - 1 - \frac{4}{5}\int_0^1 G(s,t)x(t)^3 dt, \quad s \in [0,1]. \quad (1.23)$$

We get that

$$[F'(x)y](s) = y(s) - \frac{12}{5}\int_0^1 G(s,t)x(t)^2y(t)dt, \quad s \in [0,1], \tag{1.24}$$

and

$$[F''(x)yz](s) = -\frac{24}{5}\int_0^1 G(s,t)x(t)y(t)z(t)dt, \quad s \in [0,1]. \tag{1.25}$$

Let $x_0(s) = 1$ for all $s \in [0,1]$, then, for any $y \in D$, we have

$$[(I - F'(x_0))y](s) = \frac{12}{5}\int_0^1 G(s,t)y(t)dt, \quad s \in [0,1], \tag{1.26}$$

which means

$$\|I - F'(x_0)\| \leq \frac{12}{5}\max_{s\in[0,1]}\int_0^1 G(s,t)dt = \frac{12}{5\times 8} = \frac{3}{10} < 1. \tag{1.27}$$

It follows from the Banach lemma on invertible operators that $F'(x_0)^{-1}$ exists and

$$\|F'(x_0)^{-1}\| \leq \frac{1}{1-\dfrac{3}{10}} = \frac{10}{7}. \tag{1.28}$$

Moreover, from (1.23) we obtain $\|F(x_0)\| = \frac{4}{5}\max_{s\in[0,1]}\int_0^1 G(s,t)dt = \frac{1}{10}$.

Then, we get $\gamma = \frac{1}{7}$.

It is clear that $F''(x)$ is not bounded in X or its subset D. A solution $x^\star$ of $F(x) = 0$ must satisfy

$$\|x^\star\| - 1 - \frac{1}{10}\|x^\star\|^3 \leq 0, \tag{1.29}$$

i.e., $\|x^\star\| \leq \rho_1 = 1.153467305$ and $\|x^\star\| \geq \rho_2 = 2.423622140$, where ρ_1 and ρ_2 are the positive roots of the equation $z - 1 - \frac{1}{10}z^3 = 0$. Consequently, if we look for a solution such that $x^\star < \rho_1 \in X_1$, we can consider $D = \{x : x \in X_1 \textit{ and } \|x\| < r\}$, with $r \in (\rho_1, \rho_2)$, as a nonempty open convex subset of X. For example, choose $r = 1.7$.

Using (1.24) and (1.25), we have that for any $x, y, z \in D$

$$\begin{aligned} |[(F'(x) - F'(x_0))y](s)| &= \frac{12}{5}\left|\int_0^1 G(s,t)(x(t)^2 - x_0(t)^2)y(t)dt\right| \\ &\leq \frac{12}{5}\int_0^1 G(s,t)|x(t)+x_0(t)||x(t)-x_0(t)|y(t)dt \\ &\leq \frac{12}{5}\int_0^1 G(s,t)(r+1)|x(t)-x_0(t)|y(t)dt, \quad s \in [0,1] \end{aligned} \tag{1.30}$$

and

$$\|[F''(x)yz](s)\| \quad = \quad \frac{24}{5}\int_0^1 G(s,t)x(t)y(t)z(t)dt, \quad s \in [0,1]. \tag{1.31}$$

Then, we have

$$\|F'(x)-F'(x_0)\| \quad \leq \quad \frac{12}{5}\times\frac{1}{8}(r+1)\|x-x_0\| = \frac{81}{100}\|x-x_0\| \tag{1.32}$$

and

$$\|F''(x)\| \leq \frac{24}{5}\times\frac{r}{8} = \frac{51}{50}. \tag{1.33}$$

Choosing

$$\begin{aligned} &M=\frac{51}{35}, \quad L=\frac{81}{70}, \quad a=\frac{51}{490}, \quad \gamma_0=\frac{70}{439}, \quad (L+\frac{M}{2})\gamma=\frac{66}{245}, \\ &1-L\gamma_0=\frac{358}{439}, \\ &b=0.4266517573, \quad R_0=0.2781089940, \quad \beta=0.3218118359, \\ &d-0.5138822165. \end{aligned} \tag{1.34}$$

We get

$$\overline{B}(x_0,R_0)\subset D. \tag{1.35}$$

So, all conditions in Theorem are satisfied and we can state the convergence of Halley's method (1.2) starting in x_0. Moreover, by (1.14) and (1.15) we can set $R^\star = 0.7$, which extends the uniqueness ball from $\overline{B}(x_0,R_0)$ to $\overline{B}(x_0,R^\star)$.

1.4 Basins of attraction

In this section, we present some basins of attraction related to Halley's method applied to different polynomials using the techniques that appear in [6, 7, 9, 10, 14–18].

1.4.1 2 roots

Throughout this subsection, we have used cyan color is related to the convergence to $r_1 = -1$ and magenta is related to the convergence to $r_2 = 1$. Moreover, black color is related to that pairs for which there is no convergence to any of the roots. In Figures 1.1–1.3 some basins of attraction related to polynomials with two different roots with different multiplicities are shown.

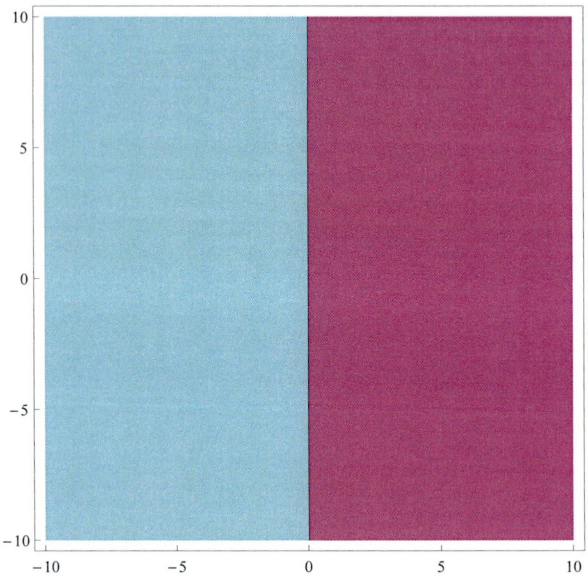

Figure 1.1: Basins of attraction associated to the Halley's method applied to $p(z) = z^2 - 1$.

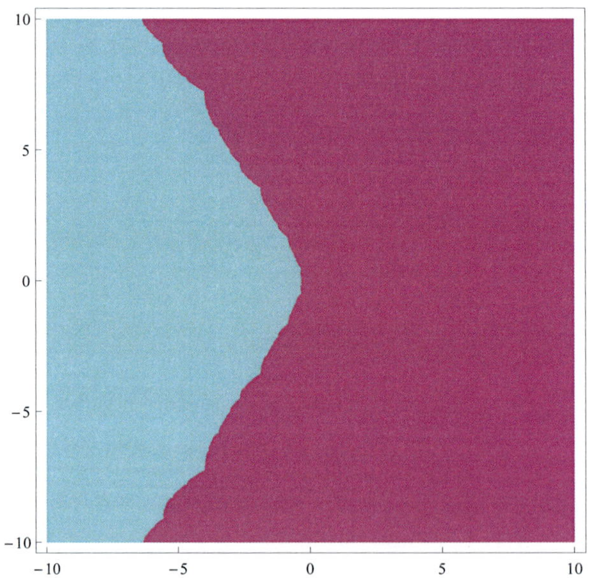

Figure 1.2: Basins of attraction associated to the Halley's method applied to $p(z) = (z^2 - 1)(z - 1)$.

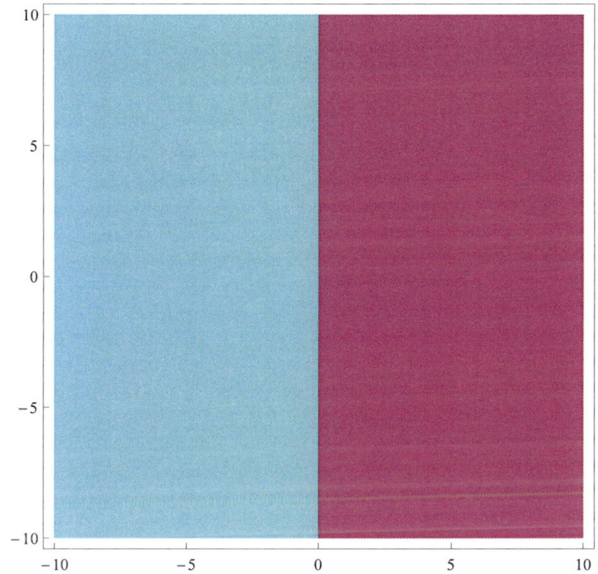

Figure 1.3: Basins of attraction associated to the Halley's method applied to $p(z) = (z^2 - 1)^2$.

1.4.2 3 roots

Throughout this subsection, we have used cyan color is related to the convergence to $r_1 = -1$, magenta is related to the convergence to $r_2 = -\frac{1}{2} - \frac{\sqrt{3}}{2}i$ and yellow to $r_3 = -\frac{1}{2} + \frac{\sqrt{3}}{2}i$. Moreover, black color is related to that pairs for which there is no convergence to any of the roots. In Figures 1.4–1.11 some basins of attraction related to polynomials with three different roots with different multiplicities are shown.

1.4.3 4 roots

Throughout this subsection, we have used cyan color is related to the convergence to $r_1 = -1$, magenta is related to the convergence to $r_2 = 1$, yellow to $r_3 = i$ and red to $r_4 = -i$. Moreover, black color is related to that pairs for which there is no convergence to any of the roots. In Figures 1.12–1.16 some basins of attraction related to polynomials with four different roots with different multiplicities are shown.

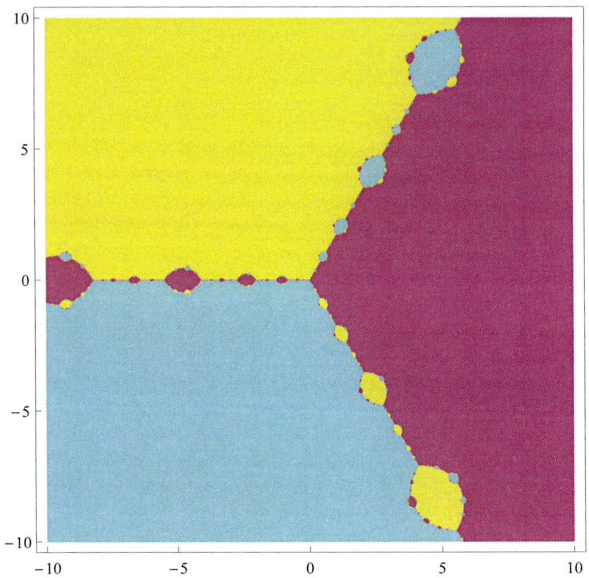

Figure 1.4: Basins of attraction associated to the Halley's method applied to $p(z) = z^3 - 1$.

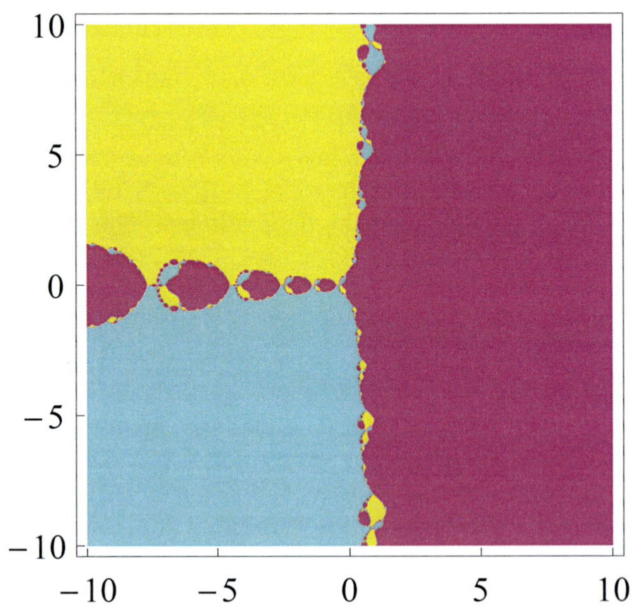

Figure 1.5: Basins of attraction associated to the Halley's method applied to $p(z) = (z^3 - 1)(z - 1)$.

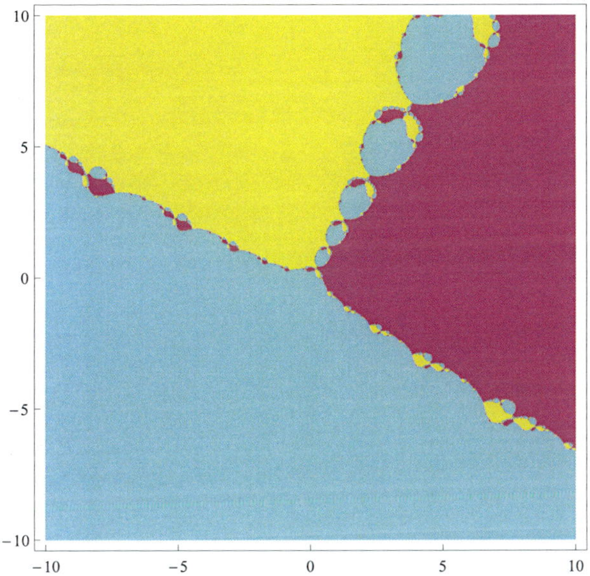

Figure 1.6: Basins of attraction associated to the Halley's method applied to $p(z) = (z^3 - 1)(z - (-\frac{1}{2} - \frac{\sqrt{3}}{2}i))$.

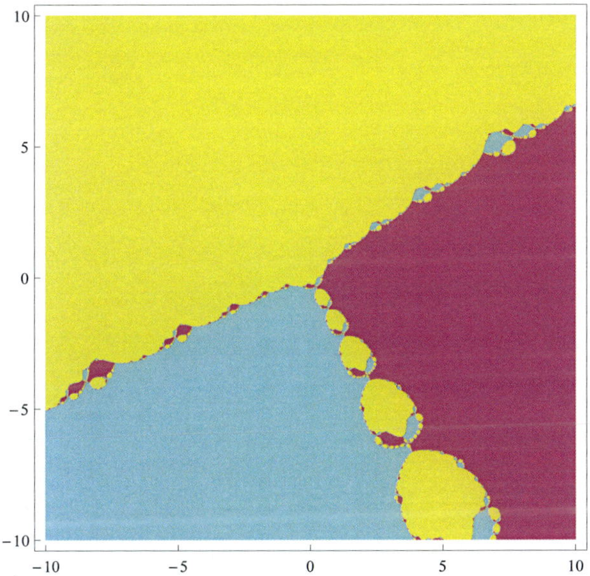

Figure 1.7: Basins of attraction associated to the Halley's method applied to $p(z) = (z^3 - 1)(z - (-\frac{1}{2} + \frac{\sqrt{3}}{2}i))$.

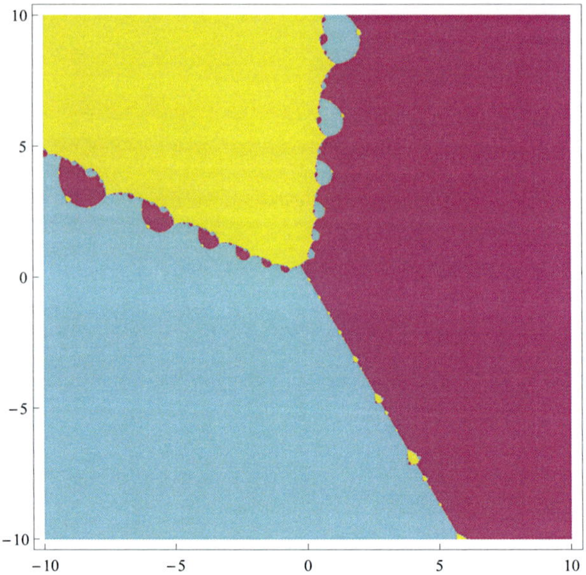

Figure 1.8: Basins of attraction associated to the Halley's method applied to $p(z) = (z^3 - 1)(z - (-\frac{1}{2} - \frac{\sqrt{3}}{2}i))(z - 1)$.

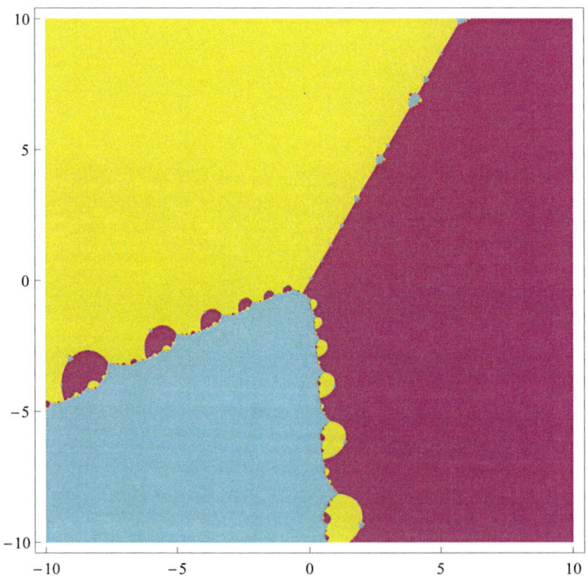

Figure 1.9: Basins of attraction associated to the Halley's method applied to $p(z) = (z^3 - 1)(z - (-\frac{1}{2} + \frac{\sqrt{3}}{2}i))(z - 1)$.

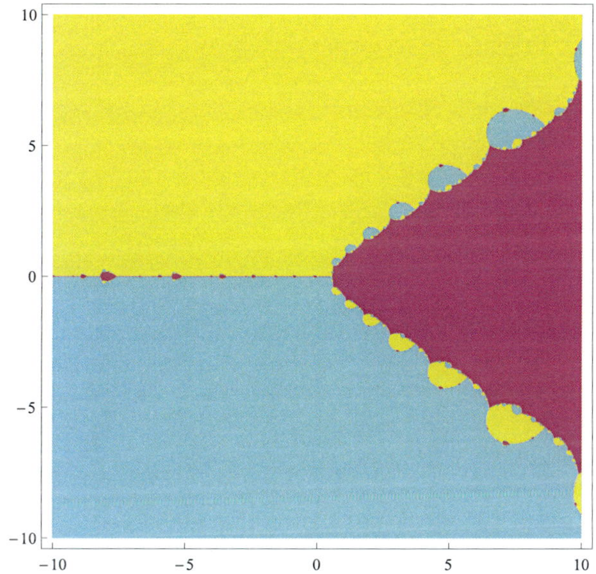

Figure 1.10: Basins of attraction associated to the Halley's method applied to $p(z) = (z^3 - 1)(z^2 + z + 1)$.

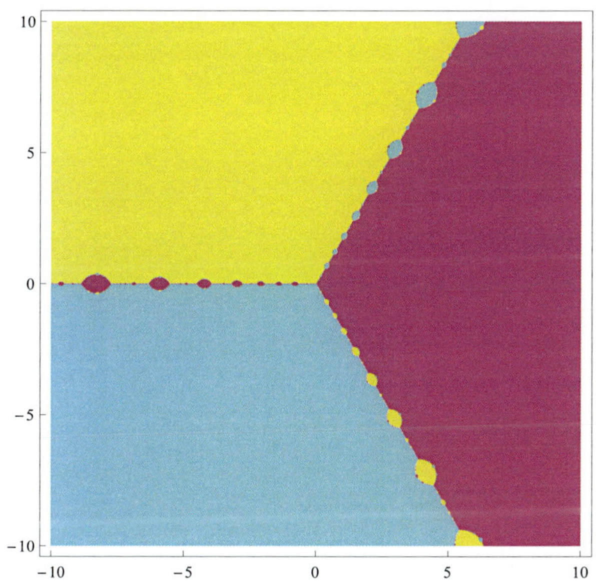

Figure 1.11: Basins of attraction associated to the Halley's method applied to $p(z) = (z^3 - 1)^2$.

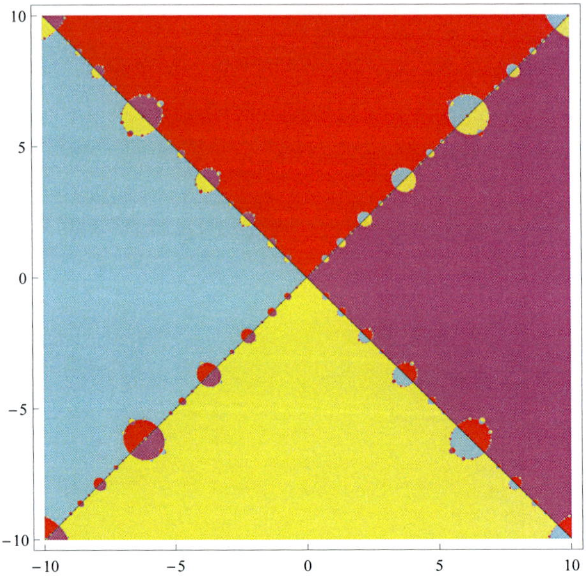

Figure 1.12: Basins of attraction associated to the Halley's method applied to $p(z) = z^4 - 1$.

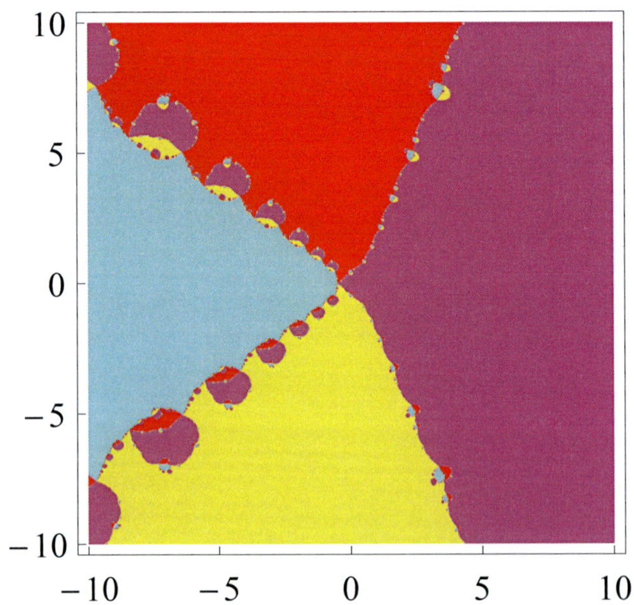

Figure 1.13: Basins of attraction associated to the Halley's method applied to $p(z) = (z^4 - 1)(z - 1)$.

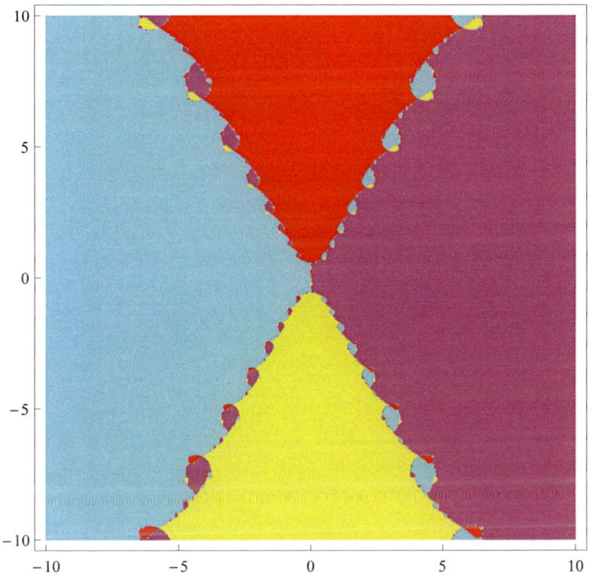

Figure 1.14: Basins of attraction associated to the Halley's method applied to $p(z) = (z^4 - 1)(z^2 - 1)$.

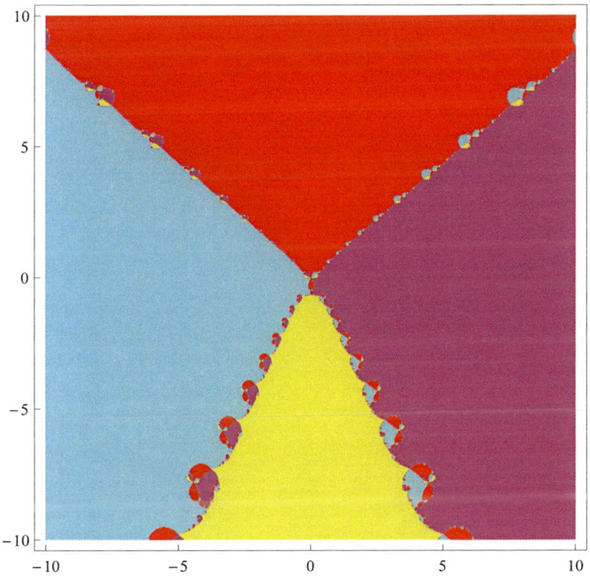

Figure 1.15: Basins of attraction associated to the Halley's method applied to $p(z) = (z^4 - 1)(z^2 - 1)(z - i)$.

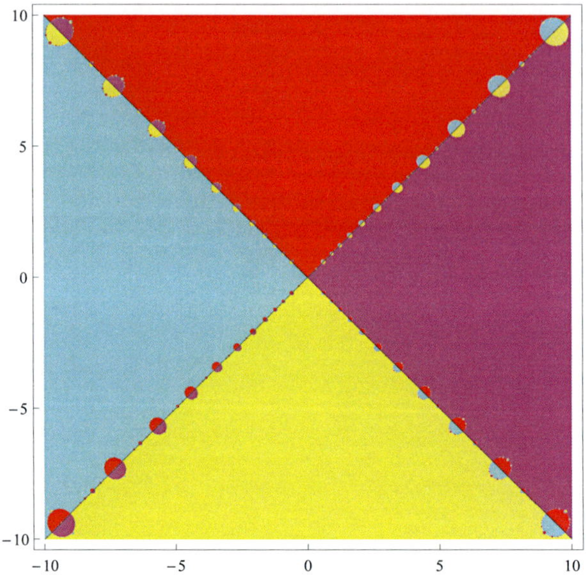

Figure 1.16: Basins of attraction associated to the Halley's method applied to $p(z) = (z^4 - 1)^2$.

References

[1] S. Amat, S. Busquier and J. M. Gutiérrez, Geometric constructions of iterative functions to solve nonlinear equations, *J. Comput. Appl. Math.*, 157, 197–205, 2003.

[2] I. K. Argyros, The convergence of Halley-Chebyshev type method under Newton-Kantorovich hypotheses, *Appl. Math. Lett.*, 6, 71–74, 1993.

[3] I. K. Argyros, Convergence and Application of Newton-type Iterations, Springer, 2008.

[4] I. K. Argyros and S. Hilout, Numerical methods in Nonlinear Analysis, World Scientific Publ. Comp. New Jersey, 2013.

[5] I. K. Argyros and H. M. Ren, Ball convergence theorems for Halley's method in Banach spaces, *J. Appl. Math. Computing*, 38, 453–465, 2012.

[6] I. K. Argyros and Á. A. Magreñán, On the convergence of an optimal fourth-order family of methods and its dynamics, *Appl. Math. Comput.* 252, 336–346, 2015.

[7] I. K. Argyros and Á. A. Magreñán, Local Convergence and the Dynamics of a Two-Step Newton-Like Method, *Int. J. Bifurcation and Chaos*, 26 (5), 1630012 2016.

[8] V. Candela and A. Marquina, Recurrence relations for rational cubic methods I: The Halley method, *Computing*, 44, 169–184, 1990.

[9] A. Cordero, J. M. Gutiérrez, Á. A. Magreñán and J. R. Torregrosa, Stability analysis of a parametric family of iterative methods for solving nonlinear models, *Appl. Math. Comput.*, 285, 26–40, 2016.

[10] A. Cordero, L. Feng, Á. A. Magreñán and J. R. Torregrosa, A new fourth-order family for solving nonlinear problems and its dynamics, *J. Math. Chemistry*, 53 (3), 893–910, 2015.

[11] P. Deuflhard, Newton Methods for Nonlinear Problems: Affine Invariance and Adaptive Algorithms, Springer-Verlag, Berlin, Heidelberg, 2004.

[12] J. A. Ezquerro and M. A. Hernández, New Kantorovich-type conditions for Halley's method, *Appl. Num. Anal. Comp. Math.*, 2, 70–77, 2005.

[13] A. Fraile, E. Larrodé, Á. A. Magreñán and J. A. Sicilia, Decision model for siting transport and logistic facilities in urban environments: A methodological approach. *J. Comput. Appl. Math.*, 291, 478–487, 2016.

[14] Y. H. Geum, Y. I., Kim and Á. A., Magreñán, A biparametric extension of King's fourth-order methods and their dynamics, *Appl. Math. Comput.*, 282, 254–275, 2016.

[15] T. Lotfi, Á. A., Magreñán, K. Mahdiani, K. and J. J. Rainer, A variant of Steffensen-King's type family with accelerated sixth-order convergence and high efficiency index: Dynamic study and approach, *Appl. Math. Comput.*, 252, 347–353, 2015.

[16] Á. A. Magreñán, Different anomalies in a Jarratt family of iterative root-finding methods, *Appl. Math. Comput.*, 233, 29–38, 2014.

[17] Á. A. Magreñán, A new tool to study real dynamics: The convergence plane, *Appl. Math. Comput.*, 248, 215–224, 2014.

[18] Á. A. Magreñán and I. K. Argyros, On the local convergence and the dynamics of Chebyshev-Halley methods with six and eight order of convergence,*J. Comput. Appl. Math.*, 298, 236–251, 2016.

[19] B. Royo, J. A. Sicilia, M. J. Oliveros and E. Larrodé, Solving a long-distance routing problem using ant colony optimization. *Appl. Math.*, 9 (2L), 415–421, 2015.

[20] J. A. Sicilia, C. Quemada, B. Royo and D. Escuín, An optimization algorithm for solving the rich vehicle routing problem based on variable neighborhood search and tabu search metaheuristics. *J. Comput. Appl. Math.*, 291, 468–477, 2016.

[21] J. A. Sicilia, D. Escuín, B. Royo, E. Larrodé and J. Medrano, A hybrid algorithm for solving the general vehicle routing problem in the case of the urban freight distribution. In Computer-based Modelling and Optimization in Transportation (pp. 463–475). Springer International Publishing, 2014.

Chapter 2

Newton's method for k-Fréchet differentiable operators

Ioannis K. Argyros

Department of Mathematical Sciences, Cameron University, Lawton, OK 73505, USA, Email: iargyros@cameron.edu

Á. Alberto Magreñán

Universidad Internacional de La Rioja, Escuela Superior de Ingeniería y Tecnología, 26002 Logroño, La Rioja, Spain, Email: alberto.magrenan@unir.net

CONTENTS

2.1 Introduction

We determine a solution x^* of the equation

$$F(x) = 0,$$

where X and Y are Banach spaces, $D \subseteq X$ a convex set and $F : D \to Y$ is a Fréchet-differentiable operator. In particular, we expand the applicability of the Newton's method defined by

$$x_{n+1} = x_n - [F'(x_n)]^{-1} F(x_n), \quad \text{for each} \quad n = 0, 1, 2, \ldots, \tag{2.1}$$

by considering weaker sufficient convergence criteria than in earlier studies [13]. We will denote along the chapter:

- $\mathcal{L}(X,Y)$ as the space of bounded linear operators from the Banach space setting X into the Banach space setting Y.
- $U(x,r)$ as the open ball centered at the point $x \in X$ of radius $r > 0$.
- $\overline{U}(x,r)$ as the open ball centered at $x \in X$ of radius $r > 0$.

The Newton-Kantorovich theorem [20] needed the following conditions to be proved:

(C_1) There exists $\Gamma_0 = [F'(x_0)]^{-1} \in \mathcal{L}(Y,X)$ for some $x_0 \in \mathcal{D}$ and $\|\Gamma_0 F(x_0)\| \leq \eta$,

(C_2) $\|\Gamma_0 F''(x)\| \leq l$, for $x \in \mathcal{D}$,

(C_3) $h = l\eta \leq \dfrac{1}{2}$ and $\overline{U}\left(x_0, \dfrac{1-\sqrt{1-2l\eta}}{l}\right) \subset \mathcal{D}$.

The corresponding majorizing sequence $\{u_n\}$ for $\{x_n\}$ is defined by

$$\begin{cases} u_0 &= 0, \\ u_{n+1} &= u_n - \dfrac{p(u_n)}{p'(u_n)}, \quad \text{for each} \quad n = 0, 1, 2, \ldots \end{cases} \tag{2.2}$$

where

$$p(t) = \frac{l}{2}t^2 - t + \eta. \tag{2.3}$$

There exists a plethora of results studies in which authors uses Lipschitz-type conditions [1, 2, 4–6, 8–11, 13, 14, 17, 20]. The main idea of these works is to extend the results of the Theorem by means of relaxing second condition. In particular, the condition weakening (C_2) is given by

$$\|\Gamma_0 F''(x)\| \leq \omega(\|x\|) \quad \text{for each} \quad x \in \mathcal{D},$$

where $\omega : \mathbb{R}_+ \cup \{0\} \to \mathbb{R}$ is non-decreasing continuous real function, this kind of condition receives the name as ω-conditions.

Recently, in [13] for $\mathcal{D}_0 = \overline{U}(x_0, r_0)$ and $f \in \mathcal{C}^2([0,s'])$ a new result was given using the following conditions:

(D_1) There exists $\Gamma_0 = [F'(x_0)]^{-1} \in \mathcal{L}(Y,X)$ for some $x_0 \in \mathcal{D}$ and $\|\Gamma_0 F(x_0)\| \leq -\dfrac{f(0)}{f'(0)}$,

(D_2) $\|\Gamma_0 F''(x)\| \leq f''(s)$ if $\|x - x_0\| \leq s \leq s'$,

(D_3) the equation $f(s) = 0$ has a solution in $[0, s']$.

The majorizing sequence $\{u_n\}$ for $\{x_n\}$ is defined as in (2.2) with f replacing p.

In this chapter, we present:

- a more flexible way of computing upper bounds on the inverses $\|[F'(x_n)]^{-1}F'(x_0)\|$ in order to avoid the problems presented in the previous conditions,
- more precise majorizing sequence,
- a tighter convergence analysis and weaker sufficient semilocal convergence conditions in many interesting cases.

The chapter is organized as follows: In Section 2.2, we present the study of the semilocal convergence of Newton's method. Then, the numerical examples are presented in concluding Section 2.3.

2.2 Semilocal convergence analysis for Newton's method

Conditions $(\widetilde{D_1})$–$(\widetilde{D_5})$ introduced below will be denoted by $(\widetilde{D})$.

Suppose there exist two scalar continuously differentiable and non-decreasing functions $f \in \mathcal{C}^{(k)}([0,t_1])$ and $g : [0,t_1] \to \mathbb{R}$, with $t_1 > 0$, $k \geq 3$, that satisfy the following conditions

$(\widetilde{D_1})$ there exists $\Gamma_0 = [F'(x_0)]^{-1} \in \mathcal{L}(Y,X)$ for some $x_0 \in \mathcal{D}$, $\|\Gamma_0 F(x_0)\| \leq -\dfrac{f(0)}{g(0)-1}$ and $\|\Gamma_0 F^{(i)}(x_0)\| \leq f^{(i)}(0)$, with $i = 2,3,\ldots,k-1$ and $k \geq 3$,

$(\widetilde{D_2})$ $\|\Gamma_0 F^{(k)}(x)\| \leq f^{(k)}(t)$, for $\|x - x_0\| \leq t$, $x \in \mathcal{D}$ and $t \in [0,t_1]$,

$(\widetilde{D_3})$ $\|\Gamma_0(F'(x) - F'(x_0))\| \leq g(\|x - x_0\|)$, for $\|x - x_0\| \leq t$, $x \in \mathcal{D}$ and $t \in [0,t_1]$,

$(\widetilde{D_4})$ $f \in \mathcal{C}^{(k)}([0,t_1])$ be with $t_1 \in \mathbb{R}$, $k \geq 3$, such that $f(0) > 0$, $f'(0) < 0$, $f^{(i)}(0) > 0$, for $i = 2,3,\ldots,k-1$, $f^{(k)}(t) > 0$.

The key of the idea is to use condition $(\widetilde{D_3})$ to find more precise upper bounds on the norms $\|F'(x_n)^{-1}F'(x_0)\|$ than using $(\widetilde{D_2})$ [13].

Suppose that $g(t) < 1$, $x \in U(x_0,t)$. Then, using $(\widetilde{D_3})$, we get that

$$\|\Gamma_0(F'(x) - F'(x_0))\| \leq g(\|x - x_0\|) \leq g(t) < 1.$$

Then, it follows from the Banach lemma on invertible operators [20] and from the preceding estimate that $F'(x)^{-1} \in \mathcal{L}(Y,X)$ and $\|F'(x)^{-1}F'(x_0)\|$ is bounded by

$$\|F'(x)^{-1}F'(x_0)\| \leq \frac{1}{1 - g(t)}.$$

However, if $(\widetilde{D_2})$ is used instead, we obtain

$$\|F'(x)^{-1}F'(x_0)\| \leq -\frac{1}{f'(t)}.$$

Hence, if $g(t) \leq 1 + f'(t)$, the new estimate is at least as tight as the old one, so it is easy to see the advantages of this idea.

Theorem 2.2.1 *Let $\{t_n\}$ be the scalar sequence given by*

$$\begin{cases} t_0 = 0, \\ t_n = N(t_{n-1}) = t_{n-1} - \dfrac{f(t_{n-1})}{g(t_{n-1}) - 1}, \quad n \in \mathbb{N}. \end{cases} \tag{2.4}$$

where $f \in C^{(1)}([0,t_1])$ and $t_1 \in \mathbb{R}$.

Suppose there exists a solution $\beta \in (0,t_1)$ of $f'(t) = 0$ such that $f(\beta) \leq 0$ and g is continuously differentiable and non-decreasing on $[0,t_1]$. Moreover, suppose that

$(\widetilde{D_5})$

$$g(t) < 1 \quad \text{for each} \quad t \in [0,\beta] \tag{2.5}$$

and

$$(1 - g(t))^2 + f'(t)(1 - g(t)) + g'(t)f(t) \geq 0 \quad \text{for each} \quad t \in [t_1,\beta]. \tag{2.6}$$

Then, (2.4) is a non-decreasing sequence that converges to t^*.

Proof: As $f(0) > 0$, then $t^* \geq 0$. Using the well-known mean value theorem, we get that

$$t_1 - t^* = N(0) - N(t^*) = N'(\theta_0)(t_0 - t^*) \quad \text{with} \quad \theta_0 \in (0,t^*),$$

that is $t_1 < t^*$, since $N'(t) > 0$ in the interval $[\theta_0, t^*)$ by (2.6). Moreover, by the definition of the sequence, we obtain that

$$t_1 = -\frac{f(0)}{g(0)-1} > 0,$$

since

$$g(0) < 1 \quad \text{(by (2.5))}$$

and $f(0) > 0$.

Finally, using mathematical induction on n, we obtain $t_n < t^*$ and $t_n - t_{n-1} \geq 0$, since $(t_{n-1}, t^*) \subset (0, t^*)$. As a consequence, we find that sequence (2.4) converges to $r \in [0, t^*]$. Furthermore, as t^* is the unique root of $f(t) = 0$ in $[0, t^*]$, it clearly follows that $r = t^*$. $\square$.

The next step, is to prove that sequence $\{t_n\}$ is a majorizing sequence of Newton's method (2.1). We present this fact in the following result.

Lemma 2.1

Suppose that the $(\widetilde{D})$ conditions are satisfied; $x_n \in \mathcal{D}$, for all $n \geq 0$, and $f(\beta) \leq 0$, where β is a solution of equation $f'(t) = 0$ in $(0, t_1)$, and $\{x_n\}$ is the sequence generated by Newton's method (2.1). Moreover, denote $\Gamma_n = [F'(x_n)]^{-1}$. Then, for all $n \geq 1$, we have

(i_n) $\|\Gamma_n F'(x_0)\| \leq -\dfrac{1}{g(t_n)-1}$,

(ii_n) $\|\Gamma_0 F^{(i)}(x_n)\| \leq f^{(i)}(t_n)$, *for* $i = 2, 3, \ldots, k-1$,

(iii_n) $\|\Gamma_0 F(x_n)\| \leq f(t_n)$,

(iv_n) $\|x_{n+1} - x_n\| \leq t_{n+1} - t_n$.

Proof: We will use again mathematical induction on n to proof this result. Case $n = 1$ is clear so we omit writing it. We suppose that (i_n)–(iv_n) are true for $n = 1, 2, \ldots, m-1$ and we will prove it for $n = m$.

From the induction hypothesis (iv_{m-1}) we obtain that

$$\begin{aligned}\|x_m - x_0\| &\leq \|x_m - x_{m-1}\| + \|x_{m-1} - x_{m-2}\| + \cdots + \|x_1 - x_0\| \\ &\leq (t_m - t_{m-1}) + (t_{m-1} - t_{m-2}) + \cdots + (t_1 - t_0) = t_m - t_0 = t_m,\end{aligned}$$

as $t_0 = 0$.

(i_m) Using $(\widetilde{D_3})$ and (2.5) we get that

$$\|\Gamma_0(F'(x_m) - F'(x_0))\| \leq g(\|x_m - x_0\|) \leq g(t_m) < 1. \tag{2.7}$$

It follows the Banach lemma on invertible operators [7, 10, 20] and from (2.7) that, Γ_m exists and $\|\Gamma_m F'(x_0)\| \leq -\dfrac{1}{g(t_m)-1}$.

(ii_m) For any $i \in \{2,3,\ldots,k-1\}$ and Taylor's series, and denoting $x = x_{m-1} + \tau(x_m - x_{m-1})$ we obtain

$$\begin{aligned} F^{(i)}(x_m) &= F^{(i)}(x_{m-1}) + F^{(i+1)}(x_{m-1})(x_m - x_{m-1}) \\ &+ \frac{1}{2!}F^{(i+2)}(x_{m-1})(x_m - x_{m-1})^2 \\ &+ \cdots + \frac{1}{(k-1-i)!}F^{(k-1)}(x_{m-1})(x_m - x_{m-1})^{k-1-i} \\ &+ \frac{1}{(k-1-i)!}\int_{x_{m-1}}^{x_m} F^{(k)}(x)(x_m - x)^{k-1-i}\,dx. \end{aligned}$$

Moreover, as $\|x - x_0\| \leq t$ and $\|x_m - x_{m-1}\| \leq t_m - t_{m-1}$, it follows that

$$\begin{aligned} \|\Gamma_0 F^{(i)}(x_m)\| &\leq \sum_{j=1}^{k-i} \frac{1}{(j-1)!}\|\Gamma_0 F^{(j+i-1)}(x_{m-1})\| \|x_m - x_{m-1}\|^{j-1} \\ &+ \frac{1}{(k-1-i)!}\int_0^1 \left\|\Gamma_0 F^{(k)}\Big(x_{m-1} + \tau(x_m - x_{m-1})\Big)\right\| \\ &\times (1-\tau)^{k-2}\|x_m - x_{m-1}\|^{k-i}\,d\tau \\ &\leq \sum_{j=1}^{k-i} \frac{1}{(j-1)!} f^{(j+i-1)}(t_{m-1})(t_m - t_{m-1})^{j-1} \\ &+ \frac{1}{(k-1-i)!}\int_{t_{m-1}}^{t_m} f^{(k)}(t)(t_m - t)^{k-i-1}\,dt \\ &= f^{(i)}(t_m). \end{aligned}$$

(iii_m) From Taylor's series we have

$$\begin{aligned} F(x_m) &= \sum_{j=2}^{k-1} \frac{1}{j!}F^{(j)}(x_{m-1})(x_m - x_{m-1})^j \\ &+ \frac{1}{(k-1)!}\int_0^1 F^{(k)}\Big(x_{m-1} + t(x_m - x_{m-1})\Big) \\ &\times (1-t)^{k-1}(x_m - x_{m-1})^k\,dt, \end{aligned}$$

since $F(x_{m-1})+F'(x_{m-1})(x_m-x_{m-1})=0$. As a consequence, we get

$$\begin{aligned}
\|\Gamma_0 F(x_m)\| &\leq \sum_{j=2}^{k-1}\frac{1}{j!}f^{(j)}(t_{m-1})(t_m-t_{m-1})^j \\
&\quad +\frac{1}{(k-1)!}\int_0^1 f^{(k)}\Big(t_{m-1}+t(t_m-t_{m-1})\Big) \\
&\quad \times(1-t)^{k-1}(t_m-t_{m-1})^k\,dt \\
&= f(t_m)-f(t_{m-1})-f'(t_{m-1})(t_m-t_{m-1}) \\
&\leq f(t_m),
\end{aligned}$$

since $f(s)+f'(s)(t-s)\geq 0$ for each $s,t\in[0,t_1]$ with $s\leq t$. Indeed, let $\varphi(u)=f(u)+f'(u)(t-u)$ for each fixed t and $u\in[s,t]$. Then $\varphi'(u)=f''(u)(t-u)\geq 0$, since $f''(u)\geq 0$ and $t\geq u$. Hence, function φ is increasing for $u\in[s,t]$ and $\varphi(u)>0$, since $\varphi(s)=f(s)>0$ (see also Lemma 2.2).

(iv_m) $\|x_{m+1}-x_m\|\leq\|\Gamma_m F'(x_0)\|\|\Gamma_0 F(x_m)\|\leq -\dfrac{f(t_m)}{g(t_m)-1}=t_{m+1}-t_m.$

And now the proof is complete. □

Then, the main result for Newton's method (2.1) is presented.

Theorem 2.2.2 *Let X and Y be two Banach spaces and $F:\mathcal{D}\subseteq X\to Y$ be a nonlinear k ($k\geq 3$) times continuously differentiable operator on a non-empty open convex domain $\mathcal{D}$ of the Banach space X, $f\in C^{(k)}([0,t_1])$ with $t_1\in\mathbb{R}$ and $g:[0,t_1]\to\mathbb{R}$ are continuous and non-decreasing functions. Suppose that conditions $(\widetilde{D})$ hold, there exists $\beta\in(0,t_1)$ such that $f'(\beta)=0$, $f(\beta)\leq 0$ and $\overline{U}(x_0,t^*)\subseteq\mathcal{D}$. Then, Newton's sequence $\{x_n\}$, given by (2.1), converges to a solution x^* of $F(x)=0$ starting at x_0. Moreover, $x_n,x^*\in\overline{U}(x_0,t^*)$ and*

$$\|x^*-x_n\|\leq t^*-t_n,\quad \textit{for all}\quad n=0,1,2,\ldots,$$

where $\{t_n\}$ is defined in (2.4).

Proof: From $(\widetilde{D_1})$ and $(\widetilde{D_2})$,we get that x_1 is well-defined and $\|x_1-x_0\|=t_1<t^*$, so that $x_1\in U(x_0,t^*)\subset\mathcal{D}$. We now suppose that x_n is well-defined and $x_n\in U(x_0,t^*)$ for $n=1,2,\ldots,m-1$. It follows from Lemma 2.1, the operator Γ_{m-1} exists and $\|\Gamma_{m-1}F'(x_0)\|\leq -\dfrac{1}{g(t_{m-1})}$. In addition, x_m is well-defined. Moreover, since $\|x_{n+1}-x_n\|\leq t_{n+1}-t_n$ for $n=1,2,\ldots,m-1$, we have

$$\|x_m-x_0\|\leq\sum_{i=0}^{m-1}\|x_{i+1}-x_i\|\leq\sum_{i=0}^{m-1}(t_{i+1}-t_i)=t_m<t^*,$$

so that $x_m\in U(x_0,t^*)$. Therefore, Newton's sequence $\{x_n\}$ is well-defined and

$x_n \in U(x_0,t^*)$ for all $n \geq 0$. By the last lemma, we also have that $\|F(x_n)\| \leq f(t_n)$ and

$$\|x_{n+1}-x_n\| \leq \|\Gamma_n F'(x_0)\|\|\Gamma_0 F(x_n)\| \leq -\frac{f(t_n)}{g(t_n)-1} = t_{n+1}-t_n,$$

for all $n \geq 0$, so that $\{t_n\}$ is a majorizing sequence of $\{x_n\}$ and is convergent. Moreover, as $\lim_{n\to+\infty} t_n = t^*$, if $x^* = \lim_{n\to+\infty} x_n$, then $\|x^*-x_n\| \leq t^*-t_n$, for all $n=0,1,2,\ldots$ Furthermore, as $\|\Gamma_0 F(x_n)\| \leq f(t_n)$, for all $n=0,1,2,\ldots$, by letting $n \to +\infty$, it follows $\Gamma_0 F(x^*)=0$. Finally, by the continuity of F we get that $F(x^*)=0$, since $\Gamma_0 \in \mathcal{L}(Y,X)$. □

2.2.1 Uniqueness of solution

In order to prove the uniqueness part we need the following auxiliary result whose proof can be found in [15] (Lemma 2.10).

Lemma 2.2

Suppose the conditions of the Theorem 2.2.2 hold. Then, the scalar function $f(t)$ has two real zeros t^* and t^{**} such that $0 < t^* \leq t^{**}$ and if $x \in \overline{U}(x_0,t^{**}) \cap \mathcal{D}$, then

$$\|\Gamma_0 F''(x)\| \leq f''(t), \quad \textit{for} \quad \|x-x_0\| \leq t.$$

On the other hand, if $f(t)$ has two real zeros t^* and t^{**} such that $0 < t^* \leq t^{**}$, then the uniqueness of solution follows from the next result.

Theorem 2.2.3 *Under the conditions of Theorem 2.2.2, the scalar function $f(t)$ has two real zeros t^* and t^{**} such that $0 < t^* \leq t^{**}$. Let $\mu^{**} = \dfrac{f''(t^{**})t^{**}}{2(1-g(t^{**}))}$. Suppose that $\mu^{**} \in (0,1)$. Then, the solution x^* is unique in $U(x_0,t^{**}) \cap \mathcal{D}$ if $t^* < t^{**}$ or in $\overline{U}(x_0,t^*)$ if $t^{**} = t^*$.*

Proof: Suppose that $t^* < t^{**}$ and y^* is a solution of $F(x)=0$ in $U(x_0,t^{**}) \cap \mathcal{D}$. We get

$$\begin{aligned}
\|y^*-x_{n+1}\| &= \|-\Gamma_n (F(y^*)-F(x_n)-F'(x_n)(y^*-x_n))\| \\
&= \left\| -\Gamma_n \int_0^1 F''(x_n+\tau(y^*-x_n))(1-\tau)(y^*-x_n)^2\, d\tau \right\|.
\end{aligned}$$

As $\|x_n+\tau(y^*-x_n)-x_0\| \leq t_n+\tau(t^{**}-t_n) \leq t^{**}$, it follows from Lemma 2.1

$$\begin{aligned}
\|y^*-x_{n+1}\| &\leq -\frac{\int_0^1 f''(t_n+\tau(t^{**}-t_n))\, d\tau}{g(t_n)-1}(1-\tau)\|y^*-x_n\|^2 \\
&\leq \mu^{**}\|y^*-x_n\|^2 < \|y^*-x_n\|.
\end{aligned}$$

Therefore, $\lim_{n\to+\infty} x_n = y^*$. That is $x^* = y^*$.

If $t^{**} = t^*$ and y^* is a solution of $F(x) = 0$ in $\overline{U}(x_0, t^{**})$, then similarly

$$\|y^* - x_{n+1}\| \leq \mu^{**}\|y^* - x_n\|^2 < \|y^* - x_n\|.$$

Then, the proof is complete. □

Another result using only function g is presented next.

Theorem 2.2.4 *Under conditions of Theorem 2.2.3, if the scalar function $f(t)$ has two real zeros t^* and t^{**} such that $0 < t^* \leq t^{**}$ and there exists $r_0 \geq t^{**}$ such that*

$$\int_0^1 g((1-\tau)r_0 + \tau\bar{t})\, d\tau < 1 \quad \text{if} \quad \bar{t} = t^{**}$$

or

$$\int_0^1 g((1-\tau)r_0 + \tau\bar{t})\, d\tau \leq 1 \quad \text{if} \quad \bar{t} = t^*$$

then, the solution x^* is unique in $U(x_0, r_0) \cap \mathcal{D}$ if $\bar{t} = t^{**}$ or in $U(x_0, r_0)$ if $r_0 = \bar{t} = t^*$.

Proof: Let $y^* \in U(x_0, r_0)$ be a solution of equation $F(x) = 0$. Let $A = \int_0^1 F'(y^* + \tau(x^* - y^*))\, d\tau$. Using $(\widetilde{D_3})$ in the first case, we obtain in turn

$$\begin{aligned}
\|\Gamma_0(A - F'(x_0))\| &\leq \int_0^1 g(\|y^* + \tau(x^* - y^*) - x_0\|)\, d\tau \\
&\leq \int_0^1 g((1-\tau)\|y^* - x_0\| + \tau\|x^* - x_0\|)\, d\tau \\
&\leq \int_0^1 g((1-\tau)r_0 + \tau\bar{t})\, d\tau < 1.
\end{aligned}$$

Moreover, if $r_0 = \bar{t} = t^*$ and $x \in U(x_0, r_0)$, we have that

$$\|\Gamma_0(A - F'(x_0))\| < \int_0^1 g((1-\tau)r_0 + \tau\bar{t})\, dt \leq 1.$$

In either case, it follows from the Banach lemma on invertible operators that $A^{-1} \in \mathcal{L}(Y, X)$. Using the identity

$$0 = F(x^*) - F(y^*) = A(x^* - y^*)$$

we deduce that $x^* = y^*$. □

2.2.2 Special choices for function g

In this section we present two choices for function g, whereas function f is as previously determined in [15]. We also provide conditions on f and g that guarantee tighter majorizing sequences.

2.2.2.1 Choice 1

We choose $g(t) = 1 + f'(t)$. Then, in this case $(\widetilde{D_3})$ can be dropped, since only $(\widetilde{D_2})$ is needed for the derivation of the upper bounds on $\|\Gamma_m F'(x_0)\|$. Moreover, our scalar sequence $\{t_n\}$ specializes to the sequence $\{v_n\}$ considered in [15] and defined by

$$\begin{cases} v_0 = 0, \\ v_n = \overline{N}(v_{n-1}) := v_{n-1} - \dfrac{f(v_{n-1})}{f'(v_{n-1})}. \end{cases} \tag{2.8}$$

Indeed, we have in turn from Taylor's series that

$$\begin{aligned} I - \Gamma_{m-1} F'(x_m) \quad = \quad & I - \Gamma_{m-1} \left(\sum_{i=1}^{k-1} \frac{1}{(i-1)!} F^{(i)}(x_{m-1})(x_m - x_{m-1})^{i-1} \right. \\ & \left. + \frac{1}{(k-2)!} \int_{x_{m-1}}^{x_m} F^{(k)}(x)(x_m - x)^{k-2} \, dx \right). \end{aligned}$$

If conditions $(\widetilde{D}_2)$ are satisfied denote $v = v_{m-1} + \tau(v_m - v_{m-1})$ with $\tau \in [0,1]$, then

$$\begin{aligned} \|x - x_0\| & \leq \tau \|x_m - x_{m-1}\| + \|x_{m-1} - x_{m-2}\| + \cdots + \|x_1 - x_0\| \\ & \leq \tau(v_m - v_{m-1}) + v_{m-1} - v_{m-2} + \cdots + v_1 = v. \end{aligned}$$

Consequently, we have in turn

$$\begin{aligned}
\|I-\Gamma_{m-1}F'(x_m)\| &\leq \|\Gamma_{m-1}\|\Bigg(\sum_{i=2}^{k-1}\frac{1}{(i-1)!}\|F^{(i)}(x_{m-1})\|\|x_m-x_{m-1}\|^{i-1}\\
&\quad+\frac{1}{(k-2)!}\int_0^1\left\|F^{(k)}\Big(x_{m-1}+\tau(x_m-x_{m-1})\Big)\right\|\\
&\quad\times(1-\tau)^{k-2}\|x_m-x_{m-1}\|^{k-1}\,d\tau\Bigg)\\
&= -\frac{1}{f'(v_{m-1})}\Bigg(\sum_{i=2}^{k-1}\frac{1}{(i-1)!}f^{(i)}(v_{m-1})(v_m-v_{m-1})^{i-1}\\
&\quad+\frac{1}{(k-2)!}\int_{v_{m-1}}^{v_m}f^{(k)}(v)(v_m-v)^{k-2}\,dv\Bigg)\\
&= 1-\frac{f'(v_m)}{f'(v_{m-1})}<1.
\end{aligned}$$

Next we justify this choice for function g and then provide a comparison between sequences $\{t_n\}$ and $\{v_n\}$. It follows from the Banach lemma that

$$\|\Gamma_m F'(x_0)\|\leq -\frac{1}{f'(v_m)}.$$

Notice also that (2.5) and (2.6) are satisfied and $\overline{N}'(t)\geq 0$. Hence, we arrive at the results obtained in [15] and the choice $g(t)=1+f'(t)$ for g is justified under the $(\widetilde{D})$ conditions. However, if

$$f'(t)+1\geq g(t)\quad\text{for each}\quad t\in[0,t_1] \tag{2.9}$$

and

$$f(s)(1-g(t))+f(t)f'(s)\leq 0\quad\text{for each}\quad s,t\in[0,t_1]\quad\text{with}\quad s\leq t \tag{2.10}$$

hold, then an inductive arguments shows that

$$t_n\geq v_n\quad\text{for each}\quad n=1,2,\ldots \tag{2.11}$$

and

$$t_{n+1}-t_n\leq v_{n+1}-v_n\quad\text{for each}\quad n=1,2,\ldots. \tag{2.12}$$

Indeed, using (2.9) we have that

$$t_1=-\frac{f(0)}{g(0)-1}\geq-\frac{f(0)}{f'(0)}=v_1\Rightarrow t_1\geq v_1,$$

$$v_2=\overline{N}(v_1)=v_1-\frac{f(v_1)}{f'(v_1)}\leq v_1-\frac{f(v_1)}{g(v_1)-1}=N(v_1)\leq$$

$$\leq t_1 - \frac{f(t_1)}{g(t_1)-1} = N(t_1) = t_2 \Rightarrow v_2 \leq t_2,$$

(since N is an increasing function) and by (2.10)

$$t_2 - t_1 = -\frac{f(t_1)}{g(t_1)-1} \leq -\frac{f(v_1)}{f'(v_1)} = v_2 - v_1 \Rightarrow t_2 - t_1 \leq v_2 - v_1.$$

Suppose that $t_m \geq v_m$ and $t_m - t_{m-1} \leq v_m - v_{m-1}$ for each $m = 1, 2, \ldots, n$. Then, in an analogous way we show that (2.11) and (2.12) hold for $m = n+1$. The induction is complete. Notice that (2.5), (2.6), (2.9) and (2.10) hold for $g(t) = f'(t)+1$. Hence, the class of functions g that satisfy (2.5), (2.6), (2.9) and (2.10) is non-empty. Moreover, if strict inequality holds in (2.9) or (2.10), then so does in (2.11) and (2.12).

2.2.2.2 *Choice 2*

Define function g as follows:

$$g(t) = \sum_{i=2}^{k-1} \frac{f^{(i)}(0)}{(i-1)!} t^{i-1} + \int_0^t \frac{f^{(k)}(u)}{(k-2)!} (t-u)^{k-2}\, du.$$

Let us also define iteration $\{s_n\}$ as $\{t_n\}$ but using function g as above. Then, as in Choice 1 we have in turn that

$$\begin{aligned} I - \Gamma_0 F'(x_m) &= I - \Gamma_0 \left(\sum_{i=1}^{k-1} \frac{1}{(i-1)!} F^{(i)}(x_0)(x_m - x_0)^{i-1} \right. \\ &\quad \left. + \frac{1}{(k-2)!} \int_{x_0}^{x_m} F^{(k)}(x)(x_m - x)^{k-2}\, dx \right). \end{aligned}$$

Then, for $s = s_0 + \tau(s_m - s_0)$, $\tau \in [0,1]$,

$$\begin{aligned} \|I - \Gamma_0 F'(x_m)\| &\leq \sum_{i=2}^{k-1} \frac{1}{(i-1)!} \|\Gamma_0 F^{(i)}(x_0)\| \|x_m - x_0\|^{i-1} \\ &\quad + \frac{1}{(k-2)!} \int_0^1 \left\| \Gamma_0 F^{(k)}\left(x_0 + \tau(x_m - x_0)\right) \right\| \\ &\quad \times (1-\tau)^{k-2} \|x_m - x_0\|^{k-1}\, d\tau \\ &= \sum_{i=2}^{k-1} \frac{1}{(i-1)!} f^{(i)}(0)(s_m - s_0)^{i-1} \\ &\quad + \frac{1}{(k-2)!} \int_0^{s_m} f^{(k)}(s)(s_m - s)^{k-2}\, ds \\ &= g(s_m) < 1. \end{aligned}$$

Then, we have again that

$$\|\Gamma_m \Gamma_0\| \leq -\frac{1}{g(s_m)-1}.$$

In this, since $f^{(i)}(0) \leq f^{(i)}(t)$ for each $i=2,3,\ldots,k-1$, $t \in [0,t_1]$ we have that (2.5) is satisfied. Moreover, notice that $f^{(i)}(0) \leq f^{(i)}(t)$ for each $n=2,3,\ldots$ Then again, if (2.9) and (2.10) hold we have as in Choice 2.2.2.1 that

$$s_n \leq v_n \tag{2.13}$$

and

$$s_{n+1}-s_n \leq v_{n+1}-v_n \tag{2.14}$$

for each $n=1,2,3,\ldots$

Then, again (2.13) and (2.14) hold as strict inequalities if (2.9) or (2.10) hold as strict inequalities. Other choices of function g are possible such as g is any polynomial of degree $m \in \mathbb{N}$ satisfying $(\widetilde{D_3})$, (2.5), (2.6) (preferably as strict inequalities),(2.9) and (2.10). This way sequence $\{t_n\}$ will be tighter than $\{v_n\}$ as it was already shown in Choices 2.2.2.1 and 2.2.2.2.

Remark 2.2.5 *It follows from the proof of Lemma 2.1 and Theorem 2.2.2 that scalar sequence $\{r_n\}$ defined for each $n=0,1,2,\ldots$ by*

$$\begin{cases} r_0 &= 0, \quad r_1 = -\dfrac{f(0)}{g(0)-1}, \\ r_{n+2} &= r_{n+1} - \beta_{n+1}(r_{n+1}-r_n)^2, \end{cases}$$

where

$$\beta_{n+1} = \frac{\beta^1_{n+1}}{g(r_{n+1})-1},$$

$$\begin{aligned} \beta^1_{n+1} &= \sum_{j=2}^{k-1} \frac{f^{(j)}(r_n)}{j!}(r_{n+1}-r_n)^{j-2} \\ &+ \frac{1}{(k-1)!}\int_0^1 f^{(k)}(r_n+t(r_{n+1}-r_n))(1-t)^{k-1}(r_{n+1}-r_n)^k\,dt, \end{aligned}$$

is the tightest majorizing for $\{x_n\}$ (from the ones mentioned above in this paper) which is obviously converging under the hypotheses of theorem 2.2.2. These hypotheses depend on function f having a positive zero. However this may no be the case (see also the numerical examples). Then, one can simply directly study the convergence of $\{r_n\}$ hoping to obtain weaker sufficient convergence conditions. For example, if

$$0 \leq -\beta_{n+1}(r_{n+1}-r_n) \leq \gamma < 1 \tag{2.15}$$

for each $n = 0, 1, 2, \ldots$ and some $\gamma \in [0, 1)$ then, we obtain that sequence $\{r_n\}$ is non-decreasing and converges to some $r^* \geq r_1$. Hypotheses (2.15) can then replace the existence of the zero β of equation $f(t) = 0$. Note that we have provided sufficient convergence conditions which on the one hand imply (2.15); not necessarily imply the existence of β and can be weaker [4, 5, 10]. For simplicity we will only present the case when $k = 2$. Then, we have for $g(t) = l_0 t$ and $f(t) = p(t) = \frac{l}{2}t^2 - t + \eta$ that scalar sequence $\{\overline{u}_n\}$ defined by

$$\left\{ \begin{array}{rcl} \overline{u}_0 & = & 0, \quad \overline{u}_1 = \eta, \quad \overline{u}_2 = \overline{u}_1 - \dfrac{l_0(\overline{u}_1 - \overline{u}_0)^2}{2(l_0\overline{u}_1 - 1)}, \\ \overline{u}_{n+1} & = & \overline{u}_n - \dfrac{l(\overline{u}_n - \overline{u}_{n-1})^2}{2(l_0\overline{u}_n - 1)}, \quad \text{for each} \quad n = 2, 3, \ldots \end{array} \right. \tag{2.16}$$

is also a majorizing sequence for $\{x_n\}$ [10]. Note that

$$l_0 \leq l$$

holds in general and $\frac{l}{l_0}$ can be arbitrarily large [6, 9, 11]. Then, $\{\overline{u}_n\}$ converges provided that

$$h_1 = L\eta \leq \frac{1}{2}, \tag{2.17}$$

where

$$L = \frac{1}{8}\left(4l_0 + \sqrt{l_0 l} + \sqrt{l_0 l + 8l_0^2}\right).$$

Note that

$$h \leq \frac{1}{2} \Rightarrow h_1 \leq \frac{1}{2}$$

but not necessarily vice versa unless if $l_0 = l$ and $\frac{h_1}{h} \to 0$ as $\frac{l_0}{l} \to 0$. The sequence $\{u_n\}$ used in [13] (see also (2.2)) is given by

$$\left\{ \begin{array}{rcl} u_0 & = & 0, \\ u_n & = & u_{n-1} - \dfrac{f(u_{n-1})}{f'(u_{n-1})}, \quad \text{for each} \quad n =, 1, 2, \ldots \end{array} \right.$$

It can then easily be seen that (2.5) and (2.6) are satisfied if $2l_0 \leq l$.

We also have that

$$g(t) \leq 1 + f'(t) \quad \text{for each} \quad t \geq 0. \tag{2.18}$$

Moreover, strict inequality holds in (2.18) if $l_0 < l$.

2.3 Numerical examples

Example 2.3.1 Let $X = Y = \mathbb{R}$, $x_0 = 1$ and $\mathcal{D} = U(1, 1-\xi)$ for $\xi \in \left(0, \frac{1}{2}\right)$. Define the scalar function F on $\mathcal{D}$ by

$$F(x) = x^3 - \xi. \tag{2.19}$$

Then, we get

$$\eta = \frac{1}{3}(1-\xi), \quad l = 2(2-\xi) \quad \text{and} \quad l_0 = 3-\xi.$$

Then, the Kantorovich hypothesis is not satisfied

$$h = l\eta = \frac{2}{3}(2-\xi)(1-\xi) > \frac{1}{2} \quad \text{for all} \quad \xi \in \left(0, \frac{1}{2}\right).$$

Notice also that $f(t) = p(t) = (2-\xi)t^2 - t + \frac{1}{3}(1-\xi)$, $g(t) = l_0 t = (3-\xi)t$ and that function f has no real zeros. Hence, there is no guarantee that Newton's method starting from $x_0 = 1$ converges to $x = \sqrt[3]{\xi}$. However, the new conditions are satisfied for $\xi \in \left[0.4271907643\ldots, \frac{1}{2}\right]$. Notice also that γ used in [13] is given by [10, 11]

$$\gamma = \frac{2l}{l + \sqrt{l^2 + 8 l_0 l}}.$$

In order for us to compare the error bounds, let $\xi = 0.6$. Then, $h < \frac{1}{2}$ and $h_1 < \frac{1}{2}$. Then, we can construct sequences $\{u_n\}$ and $\{\overline{u}_n\}$ using (2.2) and (2.8) respectively. Using Theorem 2.2.2 we see that these sequences converge. Besides, we obtain the a priori error bounds for these sequences and we show that $\overline{u}_n < u_n$ and $|\overline{u}_{n+1} - \overline{u}_n| < |u_{n+1} - u_n|$, as we can see in the Table 2.1. In addition, $\overline{u}^* = 0.166997\ldots < u^* = 0.177385\ldots$.

Table 2.1: Sequences $\overline{u}_n$ and u_n and a priori error bounds.

Iteration	$\overline{u}_n$	u_n	$\lvert\overline{u}_{n+1} - \overline{u}_n\rvert$	$\lvert u_{n+1} - u_n\rvert$
1	0.133333...	0.133333...	0.002278...	0.039716...
2	0.164706...	0.173050...	0.000012...	0.004284...
3	0.166985...	0.177334...	$3.4382\ldots \times 10^{-10}$	0.000051...

Example 2.3.2 In this example we provide an application of our results to a special nonlinear Hammerstein integral equation of the second kind. In this example we will use the max-norm.

$$x(s) = \frac{1}{2} + \int_0^1 G(s,t)x(t)^3\,dt, \tag{2.20}$$

where $x \in \mathcal{C}([0,1])$, $t \in [0,1]$ and the kernel G is the Green's function

$$G(s,t) = \begin{cases} (1-s)t, & t \le s, \\ s(1-t), & s \le t. \end{cases} \tag{2.21}$$

Solving (2.20) is equivalent to solve $F(x) = 0$, where $F : \Omega \subseteq \mathcal{C}([0,1]) \to \mathcal{C}([0,1])$,

$$[F(x)](s) = x(s) - \frac{1}{2} - \int_0^1 G(s,t)x(t)^3\,dt, \quad s \in [0,1],$$

and Ω is a suitable non-empty open convex domain. Moreover,

$$[F'(x)y](s) = y(s) - 3\int_0^1 G(s,t)x(t)^2 y(t)\,dt,$$

$$[F''(x)(yz)](s) = -6\int_0^1 G(s,t)x(t)z(t)y(t)\,dt,$$

$$[F'''(x)(yzw)](s) = -6\int_0^1 G(s,t)w(t)z(t)y(t)\,dt.$$

Observe that $\|F''(x)\|$ is not bounded in a general domain Ω. Taking into account that a solution $x^*(s)$ of (2.20) in $\mathcal{C}([0,1])$ must satisfy:

$$\|x^*\| - \frac{1}{2} - \frac{1}{8}\|x^*\|^3 \le 0,$$

it follows that $\|x^*\| \le \sigma_1 = 0.5173\ldots$ or $\|x^*\| \ge \sigma_2 = 2.5340\ldots$, where σ_1 and σ_2 are the two real positive roots of $\frac{t^3}{8} - t + \frac{1}{2} = 0$. Thus, from Kantorovich's theory, we can only approximate one solution $x^*(s)$ by Newton's method, which satisfies $\|x^*\| \in [0, \sigma_1]$, since we can consider $\Omega = U(0,\sigma)$, with $\sigma \in (\sigma_1, \sigma_2)$, where $\|F''(x)\|$ is bounded, and takes $x_0 \in U(0,\sigma)$ as starting point. We see in this example that we improve the a priori error bounds obtained by Kantorovich.

We study the case where $\Omega = U(0,\sigma)$ with $\sigma \in (\sigma_1, \sigma_2)$. For example, we take $\sigma = 2$ and choose, as it is usually done [16], the starting point $x_0(s) = u(s) = \frac{1}{2}$. Since $\|I - F'(x_0)\| \le \frac{3}{32} < 1$, the operator Γ_0 is well defined and $\|\Gamma_0 F(x_0)\| \le \frac{1}{58} = \eta$. Moreover, $\|\Gamma_0 F''(x)\| \le \frac{48}{29} = l$ and $l\eta = \frac{24}{841} = 0.0285\ldots \le \frac{1}{2}$. Hence, Kantorovich's theory can be applied, so that $p(s) = \frac{24}{29}s^2 - s + \frac{1}{58}$,

$s^* = 0.0174\ldots$, and $U(x_0, s^*) \subseteq \Omega = U(0, \sigma)$. Consequently, Newton's method can be applied to approximate a solution $x^*(s)$ in $\Omega = U(0, \sigma)$.

For Theorem 2.2.1, if we choose $t_0 = 0$, we obtain $\|F''(x_0)\| \leq \frac{3}{8}$, $\|F'''(x)\| \leq \frac{3}{4}$, $f(t) = \frac{4}{29}t^3 + \frac{6}{29}t^2 - t + \frac{1}{58}$, $\beta = 1.1329\ldots$ and $f(q) = -0.6495\ldots \leq 0$, so that the conditions of Theorem 2.2.1 hold. On the other hand, we can get the function $g(t) = \frac{12}{29}(1+t)t$ and $l_0 = \frac{36}{29}$. We construct sequences $\{u_n\}$, $\{t_n\} = \{v_n\}$ and $\{\overline{u}_n\}$ by (2.2), (2.4), (2.8) and (2.16) respectively. Once again, these sequences converge because conditions of Theorem 2.2.2 hold.

After that, we apply Newton's method for approximating a solution with the features mentioned above. For this, we use a discretization process. Thus, we approximate the integral of (2.20) by the following Gauss-Legendre quadrature formula with 8 nodes:

$$\int_0^1 \phi(t)\,dt \simeq \sum_{j-1}^{8} \omega_j \phi(t_j),$$

where the nodes t_j and the weights ω_j are known. Moreover, we denote $x(t_i)$ by x_i, $i = 1, 2, \ldots, 8$, so that equation (2.20) is now transformed into the following system of nonlinear equations:

$$x_i = \frac{1}{2} + \sum_{j=1}^{8} a_{ij} x_j^3,$$

where

$$a_{ij} = \begin{cases} \omega_j t_j (1 - t_i) & \text{if } j \leq i, \\ \omega_j t_i (1 - t_j) & \text{if } j > i. \end{cases}$$

Then, we write the above system in the following matrix form:

$$F(\overline{x}) = \overline{x} - \overline{v} - A\overline{w} = 0, \tag{2.22}$$

where $\overline{x} = (x_1, x_2, \ldots, x_8)^T$, $\overline{v} = \left(\frac{1}{2}, \frac{1}{2}, \ldots, \frac{1}{2}\right)^T$, $A = (a_{ij})_{i,j=1}^8$ and $\overline{w} = (x_1^3, x_2^3, \ldots, x_8^3)^T$. Besides,

$$F'(\overline{x}) = I - 3A\,\mathrm{diag}\{x_1^2, x_2^2, \ldots, x_8^2\}.$$

Since $x_0(s) = \frac{1}{2}$ has been chosen as starting point for the theoretical study, a reasonable choice of initial approximation for Newton's method seems to be the vector $\overline{x}_0 = \left(\frac{1}{2}, \frac{1}{2}, \ldots, \frac{1}{2}\right)^T$.

We observe in Table 2.2 that $t_n < \overline{u}_n < u_n$ for the majorizing sequences given

Table 2.2: Sequences $\{t_n\}$, $\{\overline{u}_n\}$ and $\{u_n\}$.

n	t_n	$\overline{u}_n$	u_n
2	0.0173040441616297...	0.0174299241630703...	0.0174946186637403...
3	0.0173040450083055...	0.0174299542337089...	0.0174946733196340...
4	0.0173040450083055...	0.0174299542337097...	0.0174946733196365...

by (2.4), (2.16) and (2.8) respectively. Besides, the corresponding majorizing sequences $\{\overline{u}_n\}$ and $\{t_n\}$ provide better a priori error estimates than the majorizing sequence $\{u_n\}$ obtained by Kantorovich or in [13] using the polynomial $p(s)$. The a priori error estimates are shown in Table 2.3. Observe the improvement obtained from the majorizing sequences $\{t_n\}$ and $\{\overline{u}_n\}$ constructed in this work.

Table 2.3: A priori error estimates.

n	$\lvert\overline{u}_{n+1}-\overline{u}_n\rvert$	$\lvert t_{n+1}-t_n\rvert$	$\lvert u_{n+1}-u_n\rvert$
2	$7.6488\ldots\times 10^{-16}$	$8.4667\ldots\times 10^{-10}$	$5.4655\ldots\times 10^{-8}$
3	$4.9489\ldots\times 10^{-31}$	$1.5457\ldots\times 10^{-19}$	$2.5459\ldots\times 10^{-15}$
4	$2.0717\ldots\times 10^{-61}$	$5.1520\ldots\times 10^{-39}$	$5.5242\ldots\times 10^{-30}$

References

[1] S. Amat, C. Bermúdez, S. Busquier and D. Mestiri, A family of Halley-Chebyshev iterative schemes for non-Fréchet differentiable operators. *J. Comput. Appl. Math.*, 228 (1), 486–493, 2009.

[2] S. Amat and S. Busquier, Third-order iterative methods under Kantorovich conditions. *J. Math. Anal. Appl.*, 336 (1), 243–261, 2007.

[3] S. Amat, S. Busquier and J. M. Gutiérrez, Third-order iterative methods with applications to Hammerstein equations: a unified approach. *J. Comput. Appl. Math.*, 235, 2936–2943, 2011.

[4] I. K. Argyros, On the convergence of Newton's method for polynomial equations and applications in radiative transfer. Monatshefte für Mathematik, 127 (4), 265–276, 1999.

[5] I. K. Argyros, A Newton-Kantorovich theorem for equations involving m-Fréchet differentiable operators. *J. Comput. Appl. Math.*, 131, 149–159, 2001.

[6] I. K. Argyros, On the Newton-Kantorovich hypothesis for solving equations. *J. Comput. Appl. Math.*, 169 (2), 315–332, 2004.

[7] I. K. Argyros, Computational theory of iterative methods, Series: Studies in Computational Mathematics 15. Editors: C. K. Chui and L. Wuytack. Elsevier Publications, New York, USA, 2007.

[8] I. K. Argyros, A unified approach for the convergence of a certain numerical algorithm using recurrent functions. *Computing*,90 (3), 131–164, 2010.

[9] I. K. Argyros, On Newton's method using recurrent functions and hypotheses on the first and second Fréchet derivatives. Atti del seminario

Matemático e Físico Dell'Universitá de Modena, Regio Emilia, 57, 1–18, 2010.

[10] I. K. Argyros, Y. J. Cho and S. Hilout, Numerical methods for equations and its applications. CRC Press/Taylor and Francis, New York, 2012.

[11] I. K. Argyros and S. Hilout, Weaker conditions for the convergence of Newton's method. *Journal of Complexity*, 28, 364–387, 2012.

[12] K. E. Atkinson, The numerical solution of a nonlinear boundary integral equation on smooth surfaces. *IMA Journal of Numerical Analysis*, 14, 461–483, 1994.

[13] J. A. Ezquerro, D. González and M. A. Hernández, Majorizing sequences for Newton's method from initial value problems. *J. Comput. Appl. Math.*, 236, 2246–2258, 2012.

[14] J. A. Ezquerro, D. González and M. A. Hernández, A modification of the classic conditions of Newton-Kantorovich for Newton's method. *Math. Comput. Modelling*, 57 (3-4), 584–594, 2013.

[15] J. A. Ezquerro, D. González and M. A. Hernández, A semilocal convergence result for Newton's method under generalized conditions of Kantorovich, *J. Complexity*, 30 (3), 309–324, 2014.

[16] J. A. Ezquerro and M. A. Hernández, Generalized differentiability conditions for Newton's method. *IMA Journal of Numerical Analysis*, 22 (2), 187–205, 2002.

[17] J. A. Ezquerro and M. A. Hernández, Halley's method for operators with unbounded second derivative. *Appl. Numer. Math.*, 57 (3), 354–360, 2007.

[18] A. Fraile, E. Larrodé, Á. A. Magreñán and J. A. Sicilia, Decision model for siting transport and logistic facilities in urban environments: A methodological approach. *J. Comput. Appl. Math.*, 291, 478–487, 2016.

[19] J. M. Gutiérrez, A new semilocal convergence theorem for Newton's method. *J. Comput. Appl. Math.*, 79, 131–145, 1997.

[20] L. V. Kantorovich and G. P. Akilov, Functional analysis. Pergamon Press, Oxford, 1982.

[21] A. D. Polyanin and A. V. Manzhirov, Handbook of integral equations. CRC Press, Boca Raton, 1998.

[22] B. Royo, J. A. Sicilia, M. J. Oliveros and E. Larrodé, Solving a long-distance routing problem using ant colony optimization. *Appl. Math.*, 9 (2L), 415–421, 2015.

[23] J. A. Sicilia, C. Quemada, B. Royo and D. Escuín, An optimization algorithm for solving the rich vehicle routing problem based on variable neighborhood search and tabu search metaheuristics. *J. Comput. Appl. Math.*, 291, 468–477, 2016.

[24] J. A. Sicilia, D. Escuín, B. Royo, E. Larrodé and J. Medrano, A hybrid algorithm for solving the general vehicle routing problem in the case of the urban freight distribution. In Computer-based Modelling and Optimization in Transportation (pp. 463–475). Springer International Publishing, 2014.

Chapter 3

Nonlinear Ill-posed equations

Ioannis K. Argyros

Department of Mathematical Sciences, Cameron University, Lawton, OK 73505, USA, Email: iargyros@cameron.edu

Á. Alberto Magreñán

Universidad Internacional de La Rioja, Escuela Superior de Ingeniería y Tecnología, 26002 Logroño, La Rioja, Spain, Email: alberto.magrenan@unir.net

CONTENTS

3.1 Introduction

In this chapter we provide an extended the analysis of the Lavrentiev regularization for nonlinear ill-posed problems

$$F(x) = y, \tag{3.1}$$

where $F : D(F) \subseteq X \rightarrow X$ is a nonlinear monotone operator considered in [22].

First of all, let us recall that F is a monotone operator if, for all $x_1, x_2 \in D(F)$ it satisfies the relation

$$\langle F(x_1) - F(x_2), x_1 - x_2 \rangle \geq 0. \tag{3.2}$$

Throughout all the chapter we will denote by X as a real Hilbert space with inner product $\langle .,. \rangle$ and the norm $\|.\|$. Let $U(x,\rho)$ and $\overline{U(x,\rho)}$ stand respectively, for the open and closed ball in the Hilbert space X with center $x \in X$ and radius $\rho > 0$. We also denote by $L(X)$ the space of all bounded linear operators from X into X.

Our first assumption is that x^* is a solution. Moreover, we also assume that F has a locally uniformly bounded Fréchet derivative F' in a ball around the solution.

In application, usually only noisy data y^δ are available, such that

$$\|y - y^\delta\| \leq \delta. \tag{3.3}$$

Then the problem of recovery of x^* from noisy equation $F(x) = y^\delta$ is ill-posed.

In previous papers, we considered the iterative regularization method given by

$$x_{n+1,\alpha}^\delta = x_{n,\alpha}^\delta - (F'(x_0) + \alpha I)^{-1}(F(x_{n,\alpha}^\delta) - y^\delta + \alpha(x_{n,\alpha}^\delta - x_0)), \tag{3.4}$$

and proved that $(x_{n,\alpha}^\delta)$ converges linearly to the unique solution x_α^δ of

$$F(x) + \alpha(x - x_0) = y^\delta. \tag{3.5}$$

It is known, see Theorem 1.1 in [23], that the equation (3.5) has a unique solution x_α^δ for $\alpha > 0$, provided F is Fréchet differentiable and monotone in the ball $B_r(x^*) \subset D(F)$ with radius $r = \|x^* - x_0\| + \delta/\alpha$. However the regularized equation (3.5) remains nonlinear and it could be complicated to solve it numerically.

In the last years, several authors [8–11, 17] have considered iterative regularization methods for obtaining stable approximate solutions for equation (3.5).

As it appears in [18], an iterative method with iterations defined by

$$x_{k+1}^\delta = \Phi(x_0^\delta, x_1^\delta, \cdots, x_k^\delta; y^\delta),$$

where $x_0^\delta := x_0 \in D(F)$ is a known initial approximation of x^*, for a known function Φ together with a stopping rule which determines a stopping index $k_\delta \in \mathbb{N}$ is called an iterative regularization method if the condition $\|x_{k_\delta}^\delta - x^*\| \to 0$ as $\delta \to 0$ is satisfied.

Bakushinskii and Smirnova in [8] considered the iteratively regularized Lavrentiev method defined as

$$x_{k+1}^\delta = x_k^\delta - (A_k^\delta + \alpha_k I)^{-1}(F(x_k^\delta) - y^\delta + \alpha_k(x_k^\delta - x_0)), \quad k = 0, 1, 2, \cdots, \tag{3.6}$$

where $A_k^\delta := F'(x_k^\delta)$ and $\{\alpha_k\}$ is a sequence of positive real numbers such that $\lim_{k\to\infty}\alpha_k = 0$. Actually, the stopping index k_δ in [8] was chosen according to the discrepancy principle

$$\|F(x_{k_\delta}^\delta) - y^\delta\| \leq \tau\delta < \|F(x_k^\delta) - y^\delta\|, \qquad 0 \leq k < k_\delta$$

for some $\tau > 1$ and showed that $x_{k_\delta}^\delta \to x^*$ as $\delta \to 0$ under the following well-known assumptions:

- There exists $L > 0$ such that for all $x, y \in D(F)$
$$\|F'(x) - F'(y)\| \leq L\|x - y\|$$

- There exists $p > 0$ such that
$$\frac{\alpha_k - \alpha_{k+1}}{\alpha_k \alpha_{k+1}} \leq p, \quad \forall k \in \mathbb{N}, \tag{3.7}$$

- $\sqrt{(2+L\sigma)\|x_0 - x^*\|td} \leq \sigma - 2\|x_0 - x^*\|t \leq d\alpha_0$, where $\sigma := (\sqrt{\tau} - 1)^2$, $t := p\alpha_0 + 1$ and $d = 2(t\|x_0 - x^*\| + p\sigma)$.

However, authors didn't give error estimate for $\|x_{k_\delta}^\delta - x^*\|$. Later in [18], Mahale and Nair considered method (3.6) and obtained an error estimate for $\|x_{k_\delta}^\delta - x^*\|$ under a weaker condition than (3.7). Precisely they choose the stopping index k_δ as the first nonnegative integer such that x_k^δ in (3.7) is defined for each $k \in \{0, 1, 2, \cdots, k_\delta\}$ and

$$\|\alpha_{k_\delta}(A_{k_\delta}^\delta + \alpha_{k_\delta} I)^{-1}(F(x_{k_\delta}) - y^\delta)\| \leq c_0 \text{ with } c_0 > 4.$$

In [18], Mahale and Nair showed that $x_{k_\delta}^\delta \to x^*$ as $\delta \to 0$ and an optimal order error estimate for $\|x_{k_\delta}^\delta - x^*\|$ was obtained under the following assumptions:

Assumption 1 *There exists $r > 0$ such that $U(x^*, r) \subseteq D(F)$ and F is Fréchet differentiable at all $x \in U(x^*, r)$.*

Assumption 2 *(see Assumption 3 in [22]) There exists a constant $k_0 > 0$ such that for every $x, u \in U(x^*, r))$ and $v \in X$ there exists an element $\Phi(x, u, v) \in X$ such that $[F'(x) - F'(u)]v = F'(u)\Phi(x, u, v)$, $\|\Phi(x, u, v)\| \leq k_0\|v\|\|x - u\|$.*

Assumption 3 *There exists a continuous, strictly monotonically increasing function $\varphi : (0, a] \to (0, \infty)$ with $a \geq \|F'(x^*)\|$ satisfying*

1. $\lim_{\lambda\to 0}\varphi(\lambda) = 0$
2. *for* $\alpha \leq 1, \varphi(\alpha) \geq \alpha$
3. $\sup_{\lambda\geq 0}\frac{\alpha\varphi(\lambda)}{\lambda+\alpha} \leq c_\varphi\varphi(\alpha), \qquad \forall\lambda \in (0, a]$

4. *there exists $w \in X$ such that*

$$x_0 - x^* = \varphi(F'(x^*))w. \tag{3.8}$$

Assumption 4 *There exists a sequence $\{\alpha_k\}$ of positive real numbers such that $\lim_{k\to\infty} \alpha_k = 0$ and there exists $\mu > 1$ such that*

$$1 \leq \frac{\alpha_k}{\alpha_{k+1}} \leq \mu, \ \ \forall k \in \mathbb{N}. \tag{3.9}$$

It is worth noticing that (3.9) is weaker than (3.7).

In [14] motivated by iteratively regularized Lavrentiev method, we showed the quadratic convergence of the method defined by

$$x^\delta_{n+1,\alpha} = x^\delta_{n,\alpha} - (F'(x^\delta_{n,\alpha}) + \alpha I)^{-1}(F(x^\delta_{n,\alpha}) - y^\delta + \alpha(x^\delta_{n,\alpha} - x_0)), \tag{3.10}$$

where $x^\delta_{0,\alpha} := x_0$ is a starting point of the iteration. Let $R_\alpha(x) = F'(x) + \alpha I$ and

$$G(x) = x - R_\alpha(x)^{-1}(F(x^\delta_{n,\alpha}) - y^\delta + \alpha(x^\delta_{n,\alpha} - x_0)). \tag{3.11}$$

With the above notation $x^\delta_{n+1,\alpha} = G(x^\delta_{n,\alpha})$.

The alternative assumptions used instead of Assumptions 1–2 in [14] are, respectively

Assumption 5 *There exists $r > 0$ such that $U(x_0, r) \cup U(x^*, r)) \subseteq D(F)$ and F is Fréchet differentiable at all $x \in U(x_0, r) \cup U(x^*, r))$.*

Assumption 6 *There exists a constant $k_0 > 0$ such that for every $x, u \in U(x_0, r) \cup U(x^*, r))$ and $v \in X$ there exists an element $\Phi(x, u, v) \in X$ satisfying $[F'(x) - F'(u)]v = F'(u)\Phi(x, u, v), \|\Phi(x, u, v)\| \leq k_0\|v\|\|x - u\|$.*

The condition $[F'(x) - F'(u)]v = F'(u)\Phi(x, u, v), \|\Phi(x, u, v)\| \leq k_0\|v\|\|x - u\|$ in Assumption 6 is basically a Lipschitz-type condition (cf. [4, 16] and the references therein). However, it is in general very difficult to verify or may not even be satisfied [1]– [6]. In order for us to expand the applicability of the method, we consider the following weaker assumptions.

Assumption 7 *There exists $r > 0$ such that $U(x_0, r) \subseteq D(F)$ and F is Fréchet differentiable at all $x \in U(x_0, r)$.*

Assumption 8 *Let $x_0 \in X$ be fixed. There exists a constant $L > 0$ such that for every $x, u \in U(x_0, r) \subseteq D(F)$ and $v \in X$ there exists an element $\Phi(x, u, v) \in X$ satisfying $[F'(x) - F'(u)]v = F'(u)\Phi(x, u, v), \|\Phi(x, u, v)\| \leq L\|v\|(\|x - x_0\| + \|u - x_0\|)$.*

Note that from the estimate

$$\|x-u\| \leq \|x-x_0\| + \|x_0-u\|,$$

Assumption 6 clearly implies Assumption 8 with $k_0 = L$ but not necessarily vice versa. We will assume, throughout the chapter, that operator F satisfies Assumptions 7 and 8.

Remark 3.1.1 *If Assumptions 7 and 8 are fulfilled only for all $x,u \in U(x_0,r) \cap Q \neq \emptyset$, where Q is a convex closed a priori set for which $x^* \in Q$, then we can modify the method (3.10) as:*

$$x^{\delta}_{n+1,\alpha} = P_Q(G(x^{\delta}_{n,\alpha}))$$

to obtain the same estimates in this chapter; where P_Q is the metric projection onto the set Q.

The chapter is organized as follows. In Section 3.2, we provide a convergence result for the method. Then, in Section 3.3, we give error bounds. In Section 3.4, we deal with the starting point and the algorithm. Finally, some numerical examples are presented in Section 3.5.

3.2 Convergence analysis

We will use majorizing sequences in order to let us prove our results.

[1, Definition 1.3.11]: A nonnegative sequence $\{t_n\}$ is said to be a majorizing sequence of a sequence $\{x_n\}$ in X if

$$\|x_{n+1}-x_n\| \leq t_{n+1}-t_n, \quad \forall n \geq 0.$$

We will use the following reformulation of Lemma 1.3.12 in [1] in order to prove our results.

Lemma 3.2.1 *(cf. [13], Lemma 2.1) Let $\{t_n\}$ be a majorizing sequence for $\{x_n\}$ in X. If $\lim_{n\to\infty} t_n = t^*$, then $x^* = \lim_{n\to\infty} x_n$ exists and*

$$\|x^*-x_n\| \leq t^*-t_n, \quad \forall n \geq 0. \tag{3.12}$$

Moreover, we need the following auxillary result on majorizing sequences for method (3.10).

Lemma 3.2.2 *Suppose that there exist non-negative numbers L, η such that*

$$16L\eta \leq 1. \tag{3.13}$$

Let

$$q = \frac{1-\sqrt{1-16L\eta}}{2}. \tag{3.14}$$

Then, scalar sequence $\{t_n\}$ *given by*

$$t_0 = 0,\ t_1 = \eta,\ t_{n+1} = t_n + \frac{L}{2}(5t_n + 3t_{n-1})(t_n - t_{n-1}) \text{ for each } n = 1, 2, \cdots \quad (3.15)$$

is well defined, nondecreasing, bounded from above by t^{**} *given by*

$$t^{**} = \frac{\eta}{1-q} \quad (3.16)$$

and converges to its unique least upper bound t^* *which satisfies*

$$\eta \leq t^* \leq t^{**}. \quad (3.17)$$

Moreover the following estimates hold for each $n = 1, 2, \cdots$.

$$t_{n+1} - t_n \leq q(t_n - t_{n+1}) \quad (3.18)$$

and

$$t^* - t_n \leq \frac{q^n}{1-q}\eta. \quad (3.19)$$

Proof: It is clear that $q \in (0, 1)$. Now using induction we obtain

$$\frac{L}{2}(5t_m + 3t_{m-1}) \leq q. \quad (3.20)$$

Estimate is true for $m = 1$ by the definition of sequence $\{t_n\}$ and (3.13). Then, we have by (3.15) that $t_2 - t_1 \leq q(t_1 - t_0)$ and $t_2 \leq \eta + q\eta = (1+q)\eta = \frac{1-q^2}{1-q}\eta < \frac{\eta}{1-q} = t^{**}$. If we assume that condition (3.20) is satisfied for all integers $k \leq m$. As a consequence, we have

$$t_{m+1} - t_m \leq q(t_m - t_{m-1}). \quad (3.21)$$

and

$$t_{m+1} \leq \frac{1-q^{m+1}}{1-q}\eta. \quad (3.22)$$

Now, we must prove that (3.20) is satisfied for $m+1$. Using (3.21) and (3.22), (3.20) shall be true if

$$\frac{L}{2}[5(\frac{1-q^m}{1-q}) + 3(\frac{1-q^{m-1}}{1-q})]\eta \leq q. \quad (3.23)$$

Estimate (3.23) motivates us to define recurrent functions f_m on $[0, 1)$ by

$$f_m(t) = L[5(1 + t + \cdots + t^{m-1}) + 3(1 + t + \cdots + t^{m-2})]\eta - 2t. \quad (3.24)$$

We need n relationship between two consecutive functions f_m. Using (3.24) we get that

$$f_{m+1}(t) = f_m(t) + (5t+3)L\eta t^{m-1} \geq f_m(t). \quad (3.25)$$

Define function f_∞ on $[0,1)$ by

$$f_\infty(t) = \lim_{m\to\infty} f_m(t). \tag{3.26}$$

Then, using (3.24) we obtain

$$f_\infty(t) = \frac{8L}{1-t}\eta - 2t. \tag{3.27}$$

It is obvious that, (3.23) is true if

$$f_\infty(t) \leq 0, \tag{3.28}$$

since

$$f_m(q) \leq f_{m+1}(q) \leq \cdots \leq f_\infty(q). \tag{3.29}$$

But (3.28) is true by (3.23) and (3.24). The induction for (3.20) is complete. Hence, sequence $\{t_n\}$ is nondecreasing, bounded from above by t^{**} and such it converges to some t^* which satisfies (3.17). □

Now we must have to take into account the following Lemma.

Lemma 3.2.3 *([14], Lemma 2.3) For* $u,v \in B_{r_0}(x_0)$

$$F(u) - F(v) - F'(u)(u-v) = F'(u)\int_0^1 \Phi(v+t(u-v),u,u-v)dt.$$

From now on we assume that for $r > 0, L > 0$ and q given in (3.14):

$$\frac{\delta}{\alpha} < \eta \leq min\{\frac{1}{16L}, r(1-q)\} \tag{3.30}$$

and $\|x^* - x_0\| \leq \rho$, where

$$\rho \leq \frac{1}{L}(\sqrt{1+2L(\eta - \delta/\alpha)} - 1) \tag{3.31}$$

The following remark is also interesting for our study.

Remark 3.2.4 *Note that (3.30) and (3.31) imply*

$$\frac{L}{2}\rho^2 + \rho + \frac{\delta}{\alpha} \leq \eta \leq \min\{\frac{1}{16L}, r(1-q)\}. \tag{3.32}$$

Now, we present the main result for this chapter.

Theorem 3.2.5 *Suppose Assumption 8 is satisfied. Let the assumptions in Lemma 3.2.2 hold with* η *as in (3.32). Then the sequence* $\{x^\delta_{n,\alpha}\}$ *defined in (3.11) is well defined and* $x^\delta_{n,\alpha} \in U(x_0,t^*)$ *for all* $n \geq 0$*. Further* $\{x^\delta_{n,\alpha}\}$ *is a Cauchy* *sequence in* $U(x_0,t^*)$ *and hence converges to* $x_\alpha^\delta \in \overline{U(x_0,t^*)} \subset U(x_0,t^{**})$ *and* $F(x_\alpha^\delta) = y^\delta + \alpha(x_0 - x_\alpha^\delta)$.

Moreover, the following estimates hold for all $n \geq 0$:

$$\|x_{n+1,\alpha}^\delta - x_{n,\alpha}^\delta\| \leq t_{n+1} - t_n, \tag{3.33}$$

$$\|x_{n,\alpha}^\delta - x_\alpha^\delta\| \leq t^* - t_n \leq \frac{q^n \eta}{1-q} \tag{3.34}$$

and

$$\|x_{n+1,\alpha}^\delta - x_{n,\alpha}^\delta\| \leq \frac{L}{2}[5\|x_{n,\alpha}^\delta - x_{0,\alpha}^\delta\| + 3\|x_{n-1,\alpha}^\delta - x_{0,\alpha}^\delta\|]\|x_{n,\alpha}^\delta - x_{n-1,\alpha}^\delta\|. \tag{3.35}$$

Proof: We begin by proving that

$$\|x_{n+1,\alpha}^\delta - x_{n,\alpha}^\delta\| \leq \frac{L}{2}[5\|x_{n,\alpha}^\delta - x_{0,\alpha}^\delta\| + 3\|x_{n-1,\alpha}^\delta - x_{0,\alpha}^\delta\|]\|x_{n,\alpha}^\delta - x_{n-1,\alpha}^\delta\|. \tag{3.36}$$

With G as in (3.11), we have for $u, v \in B_{t^*}(x_0)$,

$$\begin{aligned}
G(u) - G(v) &= u - v - R_\alpha(u)^{-1}[F(u) - y^\delta + \alpha(u - x_0)] + R_\alpha(v)^{-1} \\
&\quad \times [F(v) - y^\delta + \alpha(v - x_0)] \\
&= u - v - [R_\alpha(u)^{-1} - R_\alpha(v)^{-1}](F(v) - y^\delta + \alpha(v - x_0)) \\
&\quad - R_\alpha(u)^{-1}(F(u) - F(v) + \alpha(u - v)) \\
&= R_\alpha(u)^{-1}[F'(u)(u - v) - (F(u) - F(v))] \\
&\quad - R_\alpha(u)^{-1}[F'(v) - F'(u)]R_\alpha(v)^{-1}(F(v) - y^\delta + \alpha(v - x_0)) \\
&= R_\alpha(u)^{-1}[F'(u)(u - v) - (F(u) - F(v))] \\
&\quad - R_\alpha(u)^{-1}[F'(v) - F'(u)](v - G(v)) \\
&= R_\alpha(u)^{-1}[F'(u)(u - v) + \int_0^1 (F'(u + t(v - u))(v - u))dt] \\
&\quad - R_\alpha(u)^{-1}[F'(v) - F'(u)](v - G(v)) \\
&= \int_0^1 R_\alpha(u)^{-1}[(F'(u + t(v - u)) - F'(u))(v - u)dt] \\
&\quad - R_\alpha(u)^{-1}[F'(v) - F'(u)](v - G(v)).
\end{aligned}$$

The last, but one step follows from the Lemma 3.2.1. So by Assumption 8 and the estimate $\|R_\alpha(u)^{-1}F'(u)\| \leq 1$, we have

$$\begin{aligned}
\|G(u) - G(v)\| &\leq L\int_0^1 [\|u + t(v - u) - x_0\| + \|u - x_0\|]dt\|v - u\| \\
&\quad + L[\|v - x_0\| + \|u - x_0\|]\|v - u\| \\
&\leq \frac{L}{2}[3\|u - x_0\| + \|v - x_0\|] + L[\|v - x_0\| + \|u - x_0\|] \\
&\leq \frac{L}{2}[5\|u - x_0\| + 3\|v - x_0\|]\|v - G(u)\|.
\end{aligned} \tag{3.37}$$

Now by taking $u = x^\delta_{n,\alpha}$ and $v = x^\delta_{n-1,\alpha}$ in (3.37), we obtain (3.36).

Next we shall prove that the sequence (t_n) defined in Lemma 3.2.1 is a majorizing sequence of the sequence $(x^\delta_{n,\alpha})$. Note that $F(x^*) = y$, so by Lemma 3.2.2,

$$\begin{aligned}
\|x^\delta_{1,\alpha} - x_0\| &= \|R_\alpha(x_0)^{-1}(F(x_0) - y^\delta)\| \\
&= \|R_\alpha(x_0)^{-1}(F(x_0) - y + y - y^\delta)\| \\
&= \|R_\alpha(x_0)^{-1}(F(x_0) - F(x^*) - F'(x_0)(x_0 - x^*) \\
&\quad + F'(x_0)(x_0 - x^*) + y - y^\delta)\| \\
&\leq \|R_\alpha(x_0)^{-1}(F(x_0) - F(x^*) - F'(x_0)(x_0 - x^*))\| \\
&\quad + \|R_\alpha(x_0)^{-1}F'(x_0)(x_0 - x^*)\| + \|R_\alpha(x_0)^{-1}(y - y^\delta)\| \\
&\leq \|R_\alpha(x_0)^{-1}F'(x_0)\int_0^1 \Phi(x^* + t(x_0 - x^*), x_0, (x_0 - x^*))dt\| \\
&\quad + \|R_\alpha(x_0)^{-1}F'(x_0))(x_0 - x^*)\| + \frac{\delta}{\alpha} \\
&\leq \frac{L}{2}\|x_0 - x^*\|^2 + \|x_0 - x^*\| + \frac{\delta}{\alpha} \\
&\leq \frac{L}{2}\rho^2 + \rho + \frac{\delta}{\alpha} \\
&\leq \eta = t_1 - t_0.
\end{aligned}$$

Assume that $\|x^\delta_{i+1,\alpha} - x^\delta_{i,\alpha}\| \leq t_{i+1} - t_i$ for all $i \leq k$ for some k. Then

$$\begin{aligned}
\|x^\delta_{k+1,\alpha} - x_0\| &\leq \|x^\delta_{k+1,\alpha} - x^\delta_{k,\alpha}\| + \|x^\delta_{k,\alpha} - x^\delta_{k-1,\alpha}\| + \cdots + \|x^\delta_{1,\alpha} - x_0\| \\
&\leq t_{k+1} - t_k + t_k - t_{k-1} + \cdots + t_1 - t_0 \\
&= t_{k+1} \leq t^*.
\end{aligned}$$

As a consequence, $x^\delta_{i+1,\alpha} \in B_{t^*}(x_0)$ for all $i \leq k$, and therefore, by (3.36),

$$\begin{aligned}
\|x^\delta_{k+2,\alpha} - x^\delta_{k+1,\alpha}\| &\leq \frac{L}{2}[5\|x^\delta_{n,\alpha} - x^\delta_{0,\alpha}\| + 3\|x^\delta_{n-1,\alpha} - x^\delta_{0,\alpha}\|]\|x^\delta_{n,\alpha} - x^\delta_{n-1,\alpha}\| \\
&\leq \frac{L}{2}(5t_{k+1} + 3t_{k-1}) = t_{k+2} - t_{k+1}.
\end{aligned}$$

Hence, by induction $\|x^\delta_{n+1,\alpha} - x^\delta_{n,\alpha}\| \leq t_{n+1} - t_n$ for all $n \geq 0$ and therefore $\{t_n\}, n \geq 0$ is a majorizing sequence of the sequence $\{x^\delta_{n,\alpha}\}$. In particular $\|x^\delta_{n,\alpha} - x_0\| \leq t_n \leq t^*$, i.e., $x^\delta_{n,\alpha} \in U(x_0, t^*)$, for all $n \geq 0$. So, $\{x^\delta_{n,\alpha}\}, n \geq 0$ is a Cauchy sequence and converges to some $x^\delta_\alpha \in \overline{U(x_0, t^*)} \subset U(x_0, t^{**})$ and by Lemma 3.2.3

$$\|x^\delta_\alpha - x^\delta_{n,\alpha}\| \leq t^* - t_n \leq \frac{q^n\eta}{1-q}.$$

In order to prove (3.35), we observe that $G(x^\delta_\alpha) = x^\delta_\alpha$, so (3.35) follows from

(3.37), by taking $u = x_{n,\alpha}^{\delta}$ and $v = x_{\alpha}^{\delta}$ in (3.37). Now letting $n \to \infty$ in (3.10) we obtain $F(x_{\alpha}^{\delta}) = y^{\delta} + \alpha(x_0 - x_{\alpha}^{\delta})$. □

Remark 3.2.6 *The convergence of the method is quadratic as it has been shown in [14], under Assumption 6. Since the error bounds shown in Theorem 3.2.5 are too pessimistic we will use the famous computational order of convergence (COC) ([5]) defined by*

$$\rho \approx \ln\left(\frac{\|x_{n+1} - x_{\alpha}^{\delta}\|}{\|x_n - x_{\alpha}^{\delta}\|}\right) / \ln\left(\frac{\|x_n - x_{\alpha}^{\delta}\|}{\|x_{n-1} - x_{\alpha}^{\delta}\|}\right).$$

3.3 Error bounds

We will provide an error estimate for $\|x_{n,\alpha}^{\delta} \quad x^*\|$ under the assumption

Assumption 9 *There exists a continuous, strictly monotonically increasing function $\varphi : (0,a] \to (0,\infty)$ with $a \geq \|F'(x_0)\|$ satisfying;*

(i) $lim_{\lambda \to 0}\varphi(\lambda) = 0,$

(ii) $sup_{\lambda \geq 0}\frac{\alpha\varphi(\lambda)}{\lambda+\alpha} \leq \varphi(\alpha) \qquad \forall \lambda \in (0,a]$ *and*

(iii) there exists $v \in X$ with $\|v\| \leq 1$ (cf. [20]) such that

$$x_0 - x^* = \varphi(F'(x_0))v.$$

Proposition 3.3.1 *Let $F : D(F) \subseteq X \to X$ be a monotone operator in X. Let x_{α}^{δ} be the solution of (3.5) and $x_{\alpha} := x_{\alpha}^{0}$. Then*

$$\|x_{\alpha}^{\delta} - x_{\alpha}\| \leq \frac{\delta}{\alpha}.$$

Proof: The result follows from the monotonicity of F and the following relation

$$F(x_{\alpha}^{\delta}) - F(x_{\alpha}) + \alpha(x_{\alpha}^{\delta} - x_{\alpha}) \quad = \quad y^{\delta} - y.$$

Theorem 3.3.2 *([22, Proposition 4.1], [23, Theorem 3.3]) Suppose that Assumptions 7, 8 and hypotheses of Proposition 3.3.1 hold. Let $x^* \in D(F)$ be a solution of (3.1). Then, the following assertion holds*

$$\|x_{\alpha} - x^*\| \leq (Lr+1)\varphi(\alpha).$$

Theorem 3.3.3 *Suppose hypotheses of Theorem 3.2.5 and Theorem 3.3.2 hold. Then, the following assertion are satisfied*

$$\|x_{n,\alpha}^{\delta} - x^*\| \leq \frac{q^n\eta}{1-q} + c_1(\varphi(\alpha) + \frac{\delta}{\alpha})$$

where $c_1 = \max\{1, (L_0 r + 1)\}$.

Let

$$\bar{c} := max\{\frac{\eta}{1-q} + 1, (L_0 r + 2)\}, \tag{3.38}$$

and let

$$n_\delta := min\{n : q^n \leq \frac{\delta}{\alpha}\}. \tag{3.39}$$

Theorem 3.3.4 *Let $\bar{c}$ and n_δ be as in (3.38) and (3.39) respectively. Suppose that hypotheses of Theorem 3.3.3 hold. Then, the following assertions hold*

$$\|x^\delta_{n_\delta,\alpha} - x^*\| \leq \bar{c}(\varphi(\alpha) + \frac{\delta}{\alpha}). \tag{3.40}$$

Note that the error estimate $\varphi(\alpha) + \frac{\delta}{\alpha}$ in (3.32) is of optimal order if $\alpha := \alpha_\delta$ satisfies, $\varphi(\alpha_\delta)\alpha_\delta = \delta$.

Now using the function $\psi(\lambda) := \lambda\varphi^{-1}(\lambda), 0 < \lambda \leq a$ we have $\delta = \alpha_\delta\varphi(\alpha_\delta) = \psi(\varphi(\alpha_\delta))$, so that $\alpha_\delta = \varphi^{-1}(\psi^{-1}(\delta))$.

Now, we present the following result.

Theorem 3.3.5 *Let $\psi(\lambda) := \lambda\varphi^{-1}(\lambda)$ for $0 < \lambda \leq a$, and the assumptions in Theorem 3.3.4 hold. For $\delta > 0$, let $\alpha := \alpha_\delta = \varphi^{-1}(\psi^{-1}(\delta))$ and let n_δ be as in (3.39). Then*

$$\|x^\delta_{n_\delta,\alpha} - x^*\| = O(\psi^{-1}(\delta)).$$

Now, we present a parameter choice rule based on the famous balancing principle studied in [12, 19, 21]. In this method, the regularization parameter α is chosen from some finite set

$$D_M(\alpha) := \{\alpha_i = \mu^i\alpha_0, i = 0, 1, \cdots, M\}$$

where $\mu > 1$, $\alpha_0 > 0$ and let

$$n_i := min\{n : e^{-\gamma_0 n} \leq \frac{\delta}{\alpha_i}\}.$$

Then for $i = 0, 1, \cdots, M$, we have

$$\|x^\delta_{n_i,\alpha_i} - x^\delta_{\alpha_i}\| \leq c\frac{\delta}{\alpha_i}, \quad \forall i = 0, 1, \cdots M.$$

Let $x_i := x^\delta_{n_i,\alpha_i}$. The parameter choice strategy that we are going to consider in this chapter, we select $\alpha = \alpha_i$ from $D_M(\alpha)$ and operate only with corresponding $x_i, \quad i = 0, 1, \cdots, M$.

Theorem 3.3.6 *([22, Theorem 3.1], see also the proof) Assume that there exists $i \in \{0, 1, 2, \cdots, M\}$ such that $\varphi(\alpha_i) \leq \frac{\delta}{\alpha_i}$. Suppose the hypotheses of Theorem 3.3.3 and Theorem 3.3.4 hold and let*

$$l := max\{i : \varphi(\alpha_i) \leq \frac{\delta}{\alpha_i}\} < M,$$

$$k := max\{i : \|x_i - x_j\| \leq 4\bar{c}\frac{\delta}{\alpha_j}, \qquad j = 0, 1, 2, \cdots, i\}.$$

Then $l \leq k$ and

$$\|x^* - x_k\| \leq c\psi^{-1}(\delta)$$

where $c = 6\bar{c}\mu$.

3.4 Implementation of adaptive choice rule

In this section we provide a starting point for the iteration approximating the unique solution x_α^δ of (3.5) and a determination of the parameter algorithm fulfilling the aforementioned balancing principle.

First of all, in order to select the starting point, we must consider the following steps:

- For $q = \frac{1-\sqrt{1-16L\eta}}{2}$ choose $0 < \alpha_0 < 1$ and $\mu > 1$.
- Choose η such that η satisfies (3.13).
- Choose ρ such that ρ satisfies (3.31).
- Choose $x_0 \in D(F)$ such that $\|x_0 - x^*\| \leq \rho$.
- Choose the parameter $\alpha_M = \mu^M \alpha_0$ big enough with $\mu > 1$, not too large.
- Choose n_i such that $n_i = min\{n : q^n \leq \frac{\delta}{\alpha_i}\}$.

On the other hand, the adaptive algorithm associated with the choice of the parameter specified in Theorem 3.3.6 involves the following steps:

3.4.1 Algorithm

- Set $i \leftarrow 0$.
- Solve $x_i := x_{n_i,\alpha_i}^\delta$ by using the iteration (3.10).
- If $\|x_i - x_j\| > 4c\frac{\sqrt{\delta}}{\mu^j}, \qquad j \leq i$, then take $k = i - 1$.
- Set $i = i + 1$ and return to Step 2.

3.5 Numerical examples

In this section we present the numerical examples.

Example 3.5.1 *Let $X = Y = \mathbb{R}$, $D = [0,\infty), x_0 = 1$ and define function F on D by*

$$F(x) = \frac{x^{1+\frac{1}{i}}}{1+\frac{1}{i}} + c_1 x + c_2, \tag{3.41}$$

where c_1, c_2 are real parameters and $i > 2$ an integer. Then $F'(x) = x^{1/i} + c_1$ is not Lipschitz on D. However central Lipschitz condition $(C2)'$ holds for $L_0 = 1$.

In fact, we have

$$\begin{aligned} \|F'(x) - F'(x_0)\| &= |x^{1/i} - x_0^{1/i}| \\ &= \frac{|x - x_0|}{x_0^{\frac{i-1}{i}} + \cdots + x^{\frac{i-1}{i}}} \end{aligned}$$

so

$$\|F'(x) - F'(x_0)\| \le L_0 |x - x_0|.$$

Example 3.5.2 *([7, Example 7.5]) We consider the integral equations*

$$u(s) = f(s) + \lambda \int_a^b G(s,t) u(t)^{1+1/n} dt, \quad n \in \mathbb{N}. \tag{3.42}$$

Here, f is a given continuous function satisfying $f(s) > 0, s \in [a,b], \lambda$ is a real number, and the kernel G is continuous and positive in $[a,b] \times [a,b]$.

Equation of the form (3.42) generalize equations of the form

$$u(s) = \int_a^b G(s,t) u(t)^n dt \tag{3.43}$$

studied in [1]– [6]. Instead of (3.42) we can try to solve the equation $F(u) = 0$ where

$$F : \Omega \subseteq C[a,b] \to C[a,b], \Omega = \{u \in C[a,b] : u(s) \ge 0, s \in [a,b]\},$$

and

$$F(u)(s) = u(s) - f(s) - \lambda \int_a^b G(s,t) u(t)^{1+1/n} dt.$$

We consider the max-norm.

The derivative F' is given by

$$F'(u)v(s) = v(s) - \lambda (1 + \frac{1}{n}) \int_a^b G(s,t) u(t)^{1/n} v(t) dt, \quad v \in \Omega.$$

It is worth noticing that F' does not satisfy a Lipschitz-type condition in Ω. Let us choose $[a,b] = [0,1], G(s,t) = 1$ and $y(t) = 0$. Then $F'(y)v(s) = v(s)$ and

$$\|F'(x) - F'(y)\| = |\lambda| (1 + \frac{1}{n}) \int_a^b x(t)^{1/n} dt.$$

If F' were a Lipschitz function, then

$$\|F'(x) - F'(y)\| \leq L_1 \|x - y\|,$$

or, equivalently, the inequality

$$\int_0^1 x(t)^{1/n} dt \leq L_2 \max_{x \in [0,1]} x(s), \tag{3.44}$$

would hold for all $x \in \Omega$ and for a constant L_2. But this is not true. Consider, for example, the functions

$$x_j(t) = \frac{t}{j}, \quad j \geq 1, \quad t \in [0,1].$$

If these are substituted into (3.44)

$$\frac{1}{j^{1/n}(1+1/n)} \leq \frac{L_2}{j} \Leftrightarrow j^{1-1/n} \leq L_2(1+1/n), \quad \forall j \geq 1.$$

This inequality is not true when $j \to \infty$.

Therefore, Assumption 6 is not satisfied in this case. However, Assumption 8 holds. To show this, let $x_0(t) = f(t)$ and $\gamma = \min_{s \in [a,b]} f(s), \alpha > 0$. Then for $v \in \Omega$,

$$\begin{aligned} \|[F'(x) - F'(x_0)]v\| &= |\lambda|(1+\frac{1}{n}) \max_{s \in [a,b]} |\int_a^b G(s,t)(x(t)^{1/n} - f(t)^{1/n})v(t)dt| \\ &\leq |\lambda|(1+\frac{1}{n}) \max_{s \in [a,b]} G_n(s,t) \end{aligned}$$

where $G_n(s,t) = \frac{G(s,t)|x(t)-f(t)|}{x(t)^{(n-1)/n} + x(t)^{(n-2)/n} f(t)^{1/n} + \cdots + f(t)^{(n-1)/n}}$.

Hence,

$$\begin{aligned} \|[F'(x) - F'(x_0)]v\| &= \frac{|\lambda|(1+1/n)}{\gamma^{(n-1)/n}} \max_{s \in [a,b]} \int_a^b G(s,t)v(t)dt \|x - x_0\| \\ &\leq L_0 \|v\| \|x - x_0\|, \end{aligned}$$

where $L_0 = \frac{|\lambda|(1+1/n)}{\gamma^{(n-1)/n}} N$ and $N = \max_{s \in [a,b]} \int_a^b G(s,t)dt$. Then by the following inequality;

$$\|[F'(x) - F'(u)]v\| \leq \|[F'(x) - F'(x_0)]v\| + \|[F'(x_0) - F'(u)]v\| \leq L_0 \|v\| (\|x - x_0\| + \|u - x_0\|)$$

Assumption 8 holds for sufficiently small λ.

References

[1] I. K. Argyros, Convergence and Applications of Newton-type Iterations, Springer, New York, 2008.

[2] I. K. Argyros, Approximating solutions of equations using Newton's method with a modified Newton's method iterate as a starting point. *Rev. Anal. Numer. Theor. Approx.*, 36, 123–138, 2005.

[3] I. K. Argyros, A Semilocal convergence for directional Newton methods, *Math. Comput. (AMS)*, 80, 327–343, 2011.

[4] I. K. Argyros, D. González and Á. A. Magreñán, A Semilocal Convergence for a Uniparametric Family of Efficient secant-Like Methods, *J. Funct. Spaces*, Volume 2014, Article ID 467980, 10 pages.http://dx.doi.org/10.1155/2014/467980.

[5] I. K. Argyros and S. Hilout, Weaker conditions for the convergence of Newton's method, *J. Complexity*, 28, 364–387, 2012.

[6] I. K. Argyros, Y. J. Cho and S. Hilout, Numerical methods for equations and its applications, CRC Press, Taylor and Francis, New York, 2012.

[7] I. K. Argyros, Y. J. Cho and S. George, Expanding the applicability of Lavrentiev regularization method for ill-posed problems, *Boundary Value Problems*, 2013:114, 2013.

[8] A. Bakushinsky and A. Smirnova, On application of generalized discrepancy principle to iterative methods for nonlinear ill-posed problems, *Numer. Funct. Anal. Optim.*, 26, 35–48, 2005.

[9] A. B. Bakushinskii, The problem of convergence of the iteratively regularized Gauss-Newton method, *Comput. Math. Phys.*, 32, 1353–1359, 1992.

[10] A. B. Bakushinskii, Iterative methods without saturation for solving degenerate nonlinear operator equations, *Dokl. Akad. Nauk.*, 344, 7–8, 1995.

[11] A. B. Bakushinsky and M. Yu. Kokurin, Iterative methods for approximate solution of inverse problems, volume 577 of Mathematics and Its Applications (New York). Springer, Dordrecht, 2004.

[12] F. Bauer and T. Hohage, A Lepskij-type stopping rule for regularized Newton methods. *Inverse Problems*, 21(6), 1975–1991, 2005.

[13] S. George and A. I. Elmahdy, An analysis of Lavrentiev regularization for nonlinear ill-posed problems using an iterative regularization method, *Int. J. Comput. Appl. Math.*, 5 (3), 369–381, 2010.

[14] S. George and A. I. Elmahdy, A quadratic convergence yielding iterative method for nonlinear ill-posed operator equations, *Comput. Methods Appl. Math.*, 12(1), 32–45, 2012.

[15] C. W. Groetsch, J. T. King and D. Murio, Asymptotic analysis of a finite element method for Fredholm equations of the first kind. In Treatment of Integral Equations by Numerical Methods, Eds.: C. T. H. Baker and G. F. Miller, Academic Press, London, 1–11, 1982.

[16] J. M. Gutiérrez, Á. A. Magreñán and N. Romero, On the semilocal convergence of Newton-Kantorovich method under center-Lipschitz conditions, *Appl. Math. Comput.*, 221, 79–88, 2013.

[17] B. Kaltenbacher, A. Neubauer and O. Scherzer, Iterative regularization methods for nonlinear ill-posed problems, volume 6 of Radon Series on Computational and Applied Mathematics. Walter de Gruyter GmbH & Co. KG, Berlin, 2008.

[18] P. Mahale and M. T. Nair, Iterated Lavrentiev regularization for nonlinear ill-posed problems. *ANZIAM Journal*, 51, 191–217, 2009.

[19] P. Mathe and S. V. Perverzev, Geometry of linear ill-posed problems in variable Hilbert scales, *Inverse Problems*, 19 (3), 789–803, 2003.

[20] M. T. Nair and P. Ravishankar, Regularized versions of continuous Newton's method and continuous modified Newton's method under general source conditions, *Numer. Funct. Anal. Optim.*, 29 (9-10), 1140–1165, 2008.

[21] S. V. Perverzev and E. Schock, On the adaptive selection of the parameter in regularization of ill-posed problems, *SIAM J. Numer. Anal.*, 43, 2060–2076, 2005.

[22] E. V. Semenova, Lavrentiev regularization and balancing principle for solving ill-posed problems with monotone operators, *Comput. Methods Appl. Math.*, 4, 444–454, 2010.

[23] U. Tautanhahn, On the method of Lavrentiev regularization for nonlinear ill-posed problems, *Inverse Problems*, 18, 191–207, 2002.

Chapter 4

Sixth-order iterative methods

Ioannis K. Argyros

Department of Mathematical Sciences, Cameron University, Lawton, OK 73505, USA, Email: iargyros@cameron.edu

Á. Alberto Magreñán

Universidad Internacional de La Rioja, Escuela Superior de Ingeniería y Tecnología, 26002 Logroño, La Rioja, Spain, Email: alberto.magrenan@unir.net

CONTENTS

4.1 Introduction

We develop sixth-order iterative methods in order to approximate zeros x^* of the function f defined on the real line. This method can be used to solve many problems from computational sciences and other disciplines [14, 17, 25, 34, 36, 37].

Sharma and Guha [35], through a modification of the well-known the Ostrowski method [32], developed the following sixth-order convergent method (**SG**)

$$\left\{\begin{array}{lcl} y_n & = & x_n - \dfrac{f(x_n)}{f'(x_n)}, \\ z_n & = & y_n - \dfrac{f(x_n)}{f(x_n)-2f(y_n)}\dfrac{f(y_n)}{f'(x_n)}, \\ x_{n+1} & = & z_n - \dfrac{f(x_n)+af(y_n)}{f(x_n)+(a-2)f(y_n)}\dfrac{f(z_n)}{f'(x_n)}, \end{array}\right. \tag{4.1}$$

where $a \in \mathbb{R}$.

Neta proposed a three-step uniparametric family of sixth-order (**N**)

$$\left\{\begin{array}{lcl} y_n & = & x_n - \dfrac{f(x_n)}{f'(x_n)}, \\ z_n & = & y_n - \dfrac{f(x_n)+\alpha f(y_n)}{f(x_n)+(\alpha-2)f(y_n)}\dfrac{f(y_n)}{f'(x_n)}, \\ x_{n+1} & = & z_n - \dfrac{f(x_n)-f(y_n)}{f(x_n)-3f(y_n)}\dfrac{f(z_n)}{f'(x_n)}, \end{array}\right. \tag{4.2}$$

[31].

- If $\alpha = 0$ the second steps of both methods are the same.
- If $a = -1$ the third steps of both methods are the same.

Grau and Díaz in [18] developed another sixth-order variant of the Ostrowski's method **GD**

$$\left\{\begin{array}{lcl} y_n & = & x_n - \dfrac{f(x_n)}{f'(x_n)}, \\ z_n & = & y_n - \dfrac{f(x_n)}{f(x_n)-2f(y_n)}\dfrac{f(y_n)}{f'(x_n)}, \\ x_{n+1} & = & z_n - \dfrac{f(x_n)}{f(x_n)-2f(y_n)}\dfrac{f(z_n)}{f'(x_n)}. \end{array}\right. \tag{4.3}$$

Chun and Ham [9] derived the following family of sixth-order variant of the Ostrowski's method **CH**

$$\left\{\begin{array}{lcl} y_n & = & x_n - \dfrac{f(x_n)}{f'(x_n)}, \\ z_n & = & y_n - \dfrac{f(x_n)}{f(x_n)-2f(y_n)}\dfrac{f(y_n)}{f'(x_n)}, \\ x_{n+1} & = & z_n - \mathcal{H}(u_n)\dfrac{f(z_n)}{f'(x_n)}, \end{array}\right. \tag{4.4}$$

where $u_n = f(y_n)/f(x_n)$ and $\mathcal{H}(t)$ represents a real valued function satisfying: $\mathcal{H}(0) = 1$ and $\mathcal{H}'(0) = 2$.

- If $\mathcal{H}(t) = \dfrac{1+\beta t}{1+(\beta-2)t}$, then **(CH)=(SG)**
- If $\mathcal{H}(t) = \dfrac{1}{1-2t}$, then **(CH)=(GD)**

Chun and Neta [10] proposed the following sixth-order iterative method based on a modification of the Kung and Traub method [24] **CN**

$$\left\{\begin{array}{lcl} y_n & = & x_n - \dfrac{f(x_n)}{f'(x_n)}, \\ z_n & = & y_n - \dfrac{f(y_n)}{f'(x_n)} \dfrac{1}{\left[1-\frac{f(y_n)}{f(x_n)}\right]^2}, \\ x_{n+1} & = & z_n - \dfrac{f(z_n)}{f'(x_n)} \dfrac{1}{\left[1-\frac{f(y_n)}{f(x_n)}-\frac{f(z_n)}{f(x_n)}\right]^2}. \end{array}\right. \tag{4.5}$$

In this chapter, we propose a three-step unification scheme of the sixth-order iterative methods. The rest of the chapter is organised as follows: Section 4.2 presents the scheme. In Section 4.3, through the scheme we generate various well-known sixth-order iterative methods. In the Section 4.4, numerical and dynamical comparisons of various methods is shown. Finally in Section 4.5 we present the conclusions drawn to this chapter.

4.2 Scheme for constructing sixth-order iterative methods

Our main purpose is to develop a unifying scheme of sixth-order methods presented by other authors in literature. For this purpose, we consider the following three-step iterative scheme

$$\left\{\begin{array}{lcl} y_n & = & x_n - \dfrac{f(x_n)}{f'(x_n)}, \\ z_n & = & y_n - \dfrac{f(y_n)}{f'(x_n)} \left(1+\sum_{j=1}^{m} a_j \left(\dfrac{f(y_n)}{f(x_n)}\right)^j\right), \\ x_{n+1} & = & z_n - \dfrac{f(z_n)}{f'(x_n)} \left(1+\sum_{k=1}^{l} b_k \left(\dfrac{\mu_1 f(y_n)+\mu_2 f(z_n)}{f(x_n)}\right)^k\right). \end{array}\right. \tag{4.6}$$

Here, a_j, b_k, μ_1, μ_2, m and l are independent parameters. The parameters a_j, b_k, μ_1, $\mu_2 \in \mathbb{R}$ while the parameters $m, l \in \mathbb{N}$.

In order to establish the values of the parameters we have the following result.

Theorem 4.2.1 *Let x^* be a simple zero of a sufficiently differentiable function $f\colon \mathbf{D}\subset\mathbb{R}\mapsto\mathbb{R}$ in an open interval* **D**. *If the starting point x_0 is sufficiently close to the solutions x^*, then the scheme* (4.6) *defines six order iterative methods if and only if $a_1 = 2$ and $b_1 = 2/\mu_1$. And, the error equation for the family of methods is given as*

$$e_{n+1} = \frac{c_2\left(c_3c_1 - 5c_2^2 + a_2c_2^2\right)\left(-6c_2^2 + c_2^2b_2\mu_1^2 + c_3c_1\right)}{c_1^5}e_n^6 + O\left({e_n}^7\right), \tag{4.7}$$

where $e_n = x_n - x^*$ and $c_m = \frac{f^m(x^*)}{m!}$ with $m \geq 1$.

Proof

Expanding $f(x_n)$, around the solution x^*, through the Taylor series

$$f(x_n) = c_1e_n + c_2e_n^2 + c_3e_n^3 + c_4e_n^4 + O\left(e_n^5\right), \tag{4.8}$$

From the preceding equation we have

$$f'(x_n) = c_1 + 2c_2e_n + 3c_3e_n^2 + 4c_4e_n^3 + O\left(e_n^4\right) \tag{4.9}$$

Now, we consider the following division

$$\frac{f(x)}{f'(x)} = e_n - \frac{c_2}{c_1}{e_n}^2 - 2\frac{c_3c_1 - {c_2}^2}{{c_1}^2}{e_n}^3 - \frac{3c_4{c_1}^2 - 7c_2c_3c_1 + 4{c_2}^3}{{c_1}^3}{e_n}^4 + O\left({e_n}^5\right). \tag{4.10}$$

From the first step, we get

$$y_n - x^* = e_n - \frac{f(x_n)}{f'(x_n)}, \tag{4.11}$$

now substituting, we obtain

$$y_n - x^* = \frac{c_2}{c_1}{e_n}^2 + 2\frac{c_3c_1 - {c_2}^2}{{c_1}^2}{e_n}^3 + \frac{3c_4{c_1}^2 - 7c_2c_3c_1 + 4{c_2}^3}{{c_1}^3}{e_n}^4 + O\left({e_n}^5\right). \tag{4.12}$$

Using Taylor expansion on $f(y_n)$ around the solution x^* we obtain the following and using $f(x^*) = 0$ we get

$$f(y_n) = \sum_{j=1}^{\infty} c_j\left(y_n - x^*\right),$$

again, substituting into the above equation, we get

$$f(y_n) = c_2 e_n{}^2 + 2\frac{c_3 c_1 - c_2{}^2}{c_1} e_n{}^3 + \frac{3c_4 c_1{}^2 - 7c_2 c_3 c_1 + 5c_2{}^3}{c_1{}^2} e_n{}^4 + O\left(e_n{}^5\right) \tag{4.13}$$

And considering the division

$$\frac{f(y_n)}{f(x_n)} = \frac{c_2}{c_1} e_n + \frac{2c_3 c_1 - 3c_2{}^2}{c_1{}^2} e_n{}^2 + \frac{3c_4 c_1{}^2 - 10c_2 c_3 c_1 + 8c_2{}^3}{c_1{}^3} e_n{}^3 + O\left(e_n{}^4\right) \tag{4.14}$$

From the second step of our scheme (4.6), we have

$$z_n = x_n - \frac{f(x_n)}{f'(x_n)} \left[1 + \frac{f(y_n)}{f(x_n)} \left(1 + a_1 \left(\frac{f(y_n)}{f(x_n)} \right) + a_2 \left(\frac{f(y_n)}{f(x_n)} \right)^2 + \cdots \right) \right], \tag{4.15}$$

substituting $f(x_n)/f'(x_n)$ from the equation (4.10), and $f(y_n)/f(x_n)$ from the equation (4.14), into the above equation we obtain

$$z_n - x^* = x^* - \frac{c_2{}^2 (a_1 - 2)}{c_1{}^2} e_n{}^3 - \frac{c_2 \left(4a_1 c_3 c_1 - 7a_1 c_2{}^2 + a_2 c_2{}^2 - 7c_3 c_1 + 9c_2{}^2\right)}{c_1{}^3} e_n{}^4 + O\left(e_n{}^5\right). \tag{4.16}$$

Expanding $f(z_n)$, around the solution x^*, through the Taylor series

$$f(z_n) = \sum_{k=1}^{\infty} c_k (z_n - x^*)^k, \tag{4.17}$$

and substituting, into the above equation we get

$$f(z_n) = -\frac{c_2{}^2 (a_1 - 2)}{c_1} e_n{}^3 - \frac{c_2 \left(4a_1 c_3 c_1 - 7a_1 c_2{}^2 + a_2 c_2{}^2 - 7c_3 c_1 + 9c_2{}^2\right)}{c_1{}^2} e_n{}^4 + O\left(e_n{}^5\right) \tag{4.18}$$

From the third step of our scheme (4.6), we obtain

$$x_{n+1} = z_n - \frac{f(z_n)}{f'(x_n)} \left[1 + b_1 \left(\mu_1 \frac{f(y_n)}{f(x_n)} + \mu_2 \frac{f(z_n)}{f(x_n)} \right) + b_2 \left(\mu_1 \frac{f(y_n)}{f(x_n)} + \mu_2 \frac{f(z_n)}{f(x_n)} \right)^2 + \cdots \right],$$

substituting from the equations (4.9), (4.14), (4.18) into the above equation we obtain the error relation

$$x_{n+1} = x^* + \frac{c_2^3(-2+b_1\mu_1)(a_1-2)e_n^4}{{c_1}^3} + \frac{c_2^2}{c_1^4}((-11a_1 - 11b_1\mu_1 + 6b_1a_1\mu_1 + 20)$$
$$c_1c_3 + ((-2+a_1){\mu_1}^2b_2 + ((-12a_1 + a_2 + 19)\mu_1 + (-4+4a_1 - {a_1}^2)\mu_2)b_1$$
$$-2a_2 + 18a_1 - 26)c_2^2)e_n^5 + \frac{c_2}{c_1^5}((-16a_1 - 16b_1\mu_1 + 28 + 9a_1b_1\mu_1){c_1}^2c_2c_4$$
$$+(12a_1b_1\mu_1 - 20b_1\mu_1 + 33 - 20a_1){c_1}^2{c_3}^2 + (131b_1\mu_1 - 89a_1b_1\mu_1 - 28b_1\mu_2$$
$$-15b_2{\mu_1}^2 + 8a_1b_2{\mu_1}^2 + 8b_1\mu_1a_2 + 30b_1\mu_2a_1 - 8b_1\mu_2{a_1}^2 - 15a_2 + 125a_1$$
$$-167)c_1{c_2}^2c_3 + (86a_1b_1\mu_1 - 15b_1\mu_1a_2 - 58b_1\mu_2a_1 + 17b_1\mu_2{a_1}^2 - 15a_1b_2{\mu_1}^2$$
$$+4a_2b_1\mu_2 + a_2b_2{\mu_1}^2 - 8b_2\mu_1\mu_2 + a_1b_3{\mu_1}^3 + b_1\mu_1a_3 - 2a_2b_1\mu_2a_1$$
$$+8b_2\mu_1\mu_2a_1 - 2b_2\mu_1\mu_2{a_1}^2 - {a_1}^2 - 98a_1 + 24a_2 - 2a_3 + 108 + 48b_1\mu_2$$
$$-111b_1\mu_1 + 25b_2{\mu_1}^2 - 2b_3{\mu_1}^3){c_2}^4)e_n^6 + O(e_n^7). \tag{4.19}$$

Three-step scheme (4.6) define a family of sixth order methods if the following three equations are satisfied simultaneously

$$\begin{cases} (a_1-2)(b_1\mu_1-2) & = & 0, \\ -11a_1 - 11b_1\mu_1 + 6b_1a_1\mu_1 + 20 & = & 0, \\ (-2+a_1){\mu_1}^2b_2 + ((-12a_1 + a_2 + 19)\mu_1 + (-4+4a_1-{a_1}^2)\mu_2) & & \\ b_1 - 2a_2 + 18a_1 - 26 & = & 0. \end{cases}$$

It is easy to get

$$a_1 = 2 \quad \text{and} \quad b_1 = \frac{2}{\mu_1}.$$

Substituting $a_1 = 2$ and $b_1 = 2/\mu_1$ in the equation (4.19) produces the required error equation (4.7).

☒

Therefore, we present the following three-step sixth-order iterative scheme **USS**

$$\begin{cases} y_n & = & x_n - \dfrac{f(x_n)}{f'(x_n)}, \\ z_n & = & y_n - \dfrac{f(y_n)}{f'(x_n)}\left(1 + 2\left(\dfrac{f(y_n)}{f(x_n)}\right) + \sum_{j=2}^m a_j\left(\dfrac{f(y_n)}{f(x_n)}\right)^j\right), \\ x_{n+1} & = & z_n - \dfrac{f(z_n)}{f'(x_n)}\left(1 + \dfrac{2}{\mu_1}\left(\dfrac{\mu_1 f(y_n) + \mu_2 f(z_n)}{f(x_n)}\right)\right. \\ & & \left. + \sum_{k=2}^l b_k\left(\dfrac{\mu_1 f(y_n) + \mu_2 f(z_n)}{f(x_n)}\right)^k\right). \end{cases} \tag{4.20}$$

It is worth noticing that parameters a_j, b_k (with $j \geq 2$ and $k \geq 2$) and μ_m (with $m = 1, 2$) are free to choose.

In the next sections we will present some values of the parameter which we obtain well-known methods.

4.3 Sixth-order iterative methods contained in family USS

We begin with Sharma et al. method ([35]). First of all, let us consider

$$\begin{array}{llll} a_j = 2^j & \text{for} \quad j \geq 2, & m = \infty, & \mu_1 = 1, \\ b_k = 2(2-\beta)^{k-1} & \text{for} \quad k \geq 2, & l = \infty, & \mu_2 = 0. \end{array}$$

Where $\beta \in \mathbb{R}$. Substituting these values in scheme **USS**, we obtain

$$\left\{ \begin{array}{lll} y_n & = & x_n - \dfrac{f(x_n)}{f'(x_n)}, \\ z_n & = & y_n - \dfrac{f(y_n)}{f'(x_n)} \left(1 + 2\left(\dfrac{f(y_n)}{f(x_n)}\right) + \sum_{j=2}^{\infty} 2^j \left(\dfrac{f(y_n)}{f(x_n)}\right)^j \right), \\ x_{n+1} & = & z_n - \dfrac{f(z_n)}{f'(x_n)} \left(1 + 2\left(\dfrac{f(y_n)}{f(x_n)}\right) + \sum_{k=2}^{\infty} 2(2-\beta)^{k-1} \left(\dfrac{f(y_n)}{f(x_n)}\right)^k \right). \end{array} \right.$$

Using the equivalence $1 + r + r^2 + r^3 + \cdots = 1/(1-r)$ for $|r| < 1$ in the second and third steps, we obtain have **SG**.

The next one to be studied is Chun et al. method [9]. We begin considering

$$\begin{array}{llll} a_j = 2^j & \text{for} \quad j \geq 2, & m = \infty, & \mu_1 = 1, \\ b_k = \omega_k & \text{for} \quad k \geq 2, & l = \infty, & \mu_2 = 0. \end{array}$$

Where $a \in \mathbb{R}$. Substituting in **USS**, we get

$$\left\{ \begin{array}{lll} y_n & = & x_n - \dfrac{f(x_n)}{f'(x_n)}, \\ z_n & = & y_n - \dfrac{f(y_n)}{f'(x_n)} \left(1 + 2\left(\dfrac{f(y_n)}{f(x_n)}\right) + \sum_{j=2}^{\infty} 2^j \left(\dfrac{f(y_n)}{f(x_n)}\right)^j \right), \\ x_{n+1} & = & z_n - \dfrac{f(z_n)}{f'(x_n)} \left(1 + 2\left(\dfrac{f(y_n)}{f(x_n)}\right) + \sum_{k=2}^{\infty} \omega_k \left(\dfrac{f(y_n)}{f(x_n)}\right)^k \right). \end{array} \right. \tag{4.21}$$

using again the equivalence $1 + r + r^2 + r^3 + \cdots = 1/(1-r)$ for $|r| < 1$ in the second step, we get the second step of method **CH**. The third step of method **CH** is given by $x_{n+1} = z_n - f(z_n)/f'(x_n)\mathcal{H}(u_n)$ (see the equation (4.4)). Here

$u_n = f(y_n)/f(x_n)$ and $\mathcal{H}(t)$ is a real valued function satisfying $\mathcal{H}(0) = 1$ and $\mathcal{H}'(0) = 2$. Moreover, notice that function $\mathcal{H}(u_n)$ can be expressed as

$$\mathcal{H}\left(\frac{f(y_n)}{f(x_n)}\right) = 1 + 2\left(\frac{f(y_n)}{f(x_n)}\right) + \sum_{k=2}^{\infty} \omega_k \left(\frac{f(y_n)}{f(x_n)}\right)^k.$$

Hence, method (4.21), that comes from scheme (**USS**) is method **CH**. Method (**GC**) can be obtained using

$$\begin{aligned} a_j &= 2^j \quad \text{for} \quad j \geq 2, \quad m = \infty, \quad \mu_1 = 1, \\ b_k &= 2^k \quad \text{for} \quad k \geq 2, \quad l = \infty, \quad \mu_2 = 0. \end{aligned}$$

in the second and third step of scheme **USS**.

Next, to derive the family (**N**), we consider

$$\begin{aligned} a_j &= 2(-(a-2))^{j-1} \quad \text{for} \quad j \geq 2, \quad m = \infty, \quad \mu_1 = 1, \\ b_k &= 2(-(-3))^{k-1} \quad \text{for} \quad k \geq 2, \quad l = \infty, \quad \mu_2 = 0. \end{aligned}$$

Where, $a \in \mathbb{R}$.

Finally, to obtain **CN**, parameters are

$$\begin{aligned} a_j &= (j+1) \quad \text{for} \quad j \geq 2, \quad m = \infty, \quad \mu_1 = 1, \\ b_k &= (k+1) \quad \text{for} \quad k \geq 2, \quad l = \infty, \quad \mu_2 = 1. \end{aligned}$$

Substituting in **USS**, we get

$$\left\{ \begin{array}{lcl} y_n & = & x_n - \dfrac{f(x_n)}{f'(x_n)}, \\ z_n & = & y_n - \dfrac{f(y_n)}{f'(x_n)} \left(1 + 2\left(\dfrac{f(y_n)}{f(x_n)}\right) + \sum_{j=2}^{\infty} 2^j \left(\dfrac{f(y_n)}{f(x_n)}\right)^j\right), \\ x_{n+1} & = & z_n - \dfrac{f(z_n)}{f'(x_n)} \left(1 + 2\left(\dfrac{f(y_n)}{f(x_n)}\right) + \sum_{k=2}^{\infty} \omega_k \left(\dfrac{f(y_n)}{f(x_n)}\right)^k\right). \end{array} \right. \tag{4.22}$$

Using the equivalence $1 + 2r + 3r^2 + 4r^3 + \cdots = 1/(1-r)^2$ for $|r| < 1$ in the second and third steps of preceding method we get method (**CN**).

4.4 Numerical work

Let $\{x_n\}_{n=0}^{\infty}$ be a sequence, generated by an iterative method, converging to x^* and $e_n = x_n - x^*$. If there exists a real number $\xi \in [1, \infty)$ and a nonzero constant C such that

$$\lim_{n \to \infty} |e_{n+1}|/|e_n|^{\xi} = C,$$

then ξ is called the convergence order of the sequence and the constant C is called the asymptotic error constant. From the preceding relation, the computational order of convergence (COC) is approximated as follows

$$\rho \approx \frac{\log |(x_{n+1}-x^*)/(x_n-x^*)|}{\log |(x_n-x^*)/(x_{n-1}-x^*)|}.$$

All the computations are performed in the programming language C^{++}. For numerical precision, the C^{++} library ARPREC [1] is being used. For convergence of the method, it is required that the distance of two consecutive iterates $(|x_{n+1}-x_n|)$ and the absolute value of the function $(|f(x_n)|)$, also referred to as residual, be less than 10^{-300}. The maximum allowed iterations are 200.

4.4.1 Solving nonlinear equations

The methods are tested for the functions

$$\begin{aligned}
f_1(x) &= x^3+4x^2-10, & x^* &\approx 1.365.\\
f_2(x) &= x^2-\exp(x)-3x+2, & x^* &\approx 0.257.\\
f_3(x) &= x\exp(x^2)-\sin^2(x)+3\cos(x)+5, & x^* &\approx -1.207.\\
f_4(x) &= x^4+\sin\frac{\pi}{x^2}-5, & x^* &= \sqrt{2}.\\
f_5(x) &= e^x\sin x+\log(1+x^2), & x^* &= 0.\\
f_6(x) &= \sqrt{2+x^2}\sin\frac{\pi}{x^2}+\frac{1}{x^4+1}-\frac{17\sqrt{3}+1}{17}, & x^* &= -2.
\end{aligned}$$

Table 4.1: The (number of function evaluations, COC) for various sixth order iterative methods.

$f(x)$	x_0	SG	N	GD	CH	CN
$f_1(x)$	100	(36,6)	(36,6)	**(28,6)**	(36,6)	(28,6)
$f_2(x)$	100	(172,6)	(884,6)	(160,6)	(172,6)	(108,6)
$f_3(x)$	−1	(20,6)	**(20,6)**	**(20,6)**	**(20,6)**	(32,6)
$f_4(x)$	1	(20,6)	**(20,6)**	**(20,6)**	**(20,6)**	(20,6)
$f_5(x)$	1	(24,6)	**(24,6)**	**(24,6)**	**(24,6)**	(24,6)
$f_5(x)$	3	(24,6)	**(28,6)**	**(20,6)**	**(24,6)**	(20,6)
$f_6(x)$	−1.5	(24,6)	**(28,6)**	**(20,6)**	**(24,6)**	(20,6)

Various free parameters are choosen as, in the method **SG**: $a = 2$, in the method **N**: $\alpha = 5$ and in the method **CH**: $\beta = 3$. Outcome of numerical experimentation is presented in the Table 4.1. The Table 4.1 reports (number of function evaluations, COC during the second last iterative step). COC is rounded to the nearest significant digits. In the Table 4.1, we see that the methods **GD** and **CN** are showing better results.

4.5 Dynamics for method SG

In this Section we will use the dynamical tools presented in related works as [7,8,11,12,16,27–30].

First of all, let us recall some basic concepts related to dynamics. Given a rational function $R:\hat{\mathbb{C}} \to \hat{\mathbb{C}}$, where $\hat{\mathbb{C}}$ is the Riemann sphere, the *orbit of a point* $z_0 \in \hat{\mathbb{C}}$ is defined by

$$\{z_0, R(z_0), R^2(z_0), ..., R^n(z_0), ...\}.$$

A point $z_0 \in \bar{C}$, is called a *fixed point* of $R(z)$ if it verifies that $R(z) = z$. Moreover, z_0 is called a *periodic point* of period $p > 1$ if it is a point such that $R^p(z_0) = z_0$ but $R^k(z_0) \neq z_0$, for each $k < p$. Moreover, a point z_0 is called *pre-periodic* if it is not periodic but there exists a $k > 0$ such that $R^k(z_0)$ is periodic.

There exist different types of fixed points depending on its associated multiplier $|R'(z_0)|$. Taking the associated multiplier into account a fixed point z_0 is called:

- *superattractor* if $|R'(z_0)| = 0$
- *attractor* if $|R'(z_0)| < 1$
- *repulsor* if $|R'(z_0)| > 1$
- and *parabolic* if $|R'(z_0)| = 1$.

The fixed points that do not correspond to the roots of the polynomial $p(z)$ are called *strange fixed points*. On the other hand, a *critical point* z_0 is a point which satisfies that, $R'(z_0) = 0$. Moreover, we call *free critical point* those critical points which are no roots of the polynomial $p(z)$.

The basin of attraction of an attractor α is defined as

$$\mathcal{A}(\alpha) = \{z_0 \in \hat{\mathbb{C}} \,:\, R^n(z_0) \to \alpha,\ n \to \infty\}.$$

The *Fatou set* of the rational function R, $\mathcal{F}(R)$, is the set of points $z \in \hat{\mathbb{C}}$ whose orbits tend to an attractor (fixed point, periodic orbit or infinity). Its complement in $\hat{\mathbb{C}}$ is the *Julia set*, $\mathcal{J}(R)$. That means that the basin of attraction of any fixed point belongs to the Fatou set and the boundaries of these basins of attraction belong to the Julia set.

By applying this operator on a quadratic polynomial with two different roots A and B, $p(z) = (z-A)(z-B)$ and using the Möebius map $h(z) = \frac{z-A}{z-B}$, which carries root A to 0, root B to ∞ and ∞ to 1, we obtain the rational operator associated to the family of iterative schemes is finally

$$G(z,a)=\frac{z^6\left(az^3+2az^2+2az+2a+z^4+2z^3+3z^2+4z+2\right)}{2az^4+2az^3+2az^2+az+2z^4+4z^3+3z^2+2z+1}. \tag{4.23}$$

4.5.1 Study of the fixed points and their stability

It is clear that $z=0$ and $z=\infty$ are fixed points of $G(z,a)$. Moreover, there exist some strange fixed points which are:

- $z=1$ related to divergence to ∞
- The roots of

$$\begin{aligned} p(z) &= az^7+3az^6+5az^5+7az^4+5az^3+3az^2+az+z^8 \\ &\quad +3z^7+6z^6+10z^5+12z^4+10z^3+6z^2+3z+1 \end{aligned}$$

These solutions of this polynomial depend on the value of the parameter a.

In Figure 4.1 the bifurcation diagram of the fixed points is shown

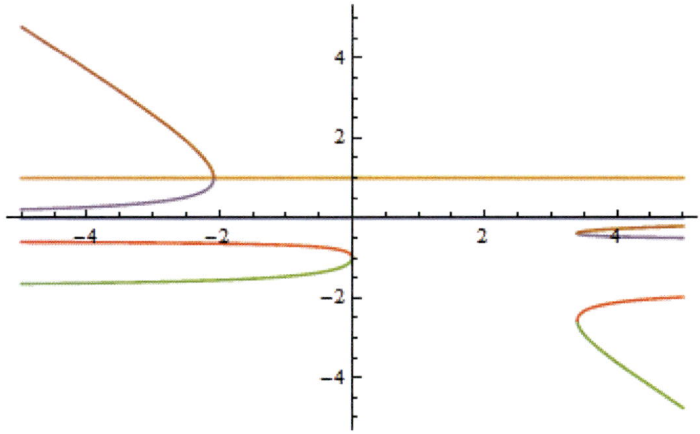

Figure 4.1: Bifurcation diagram of the fixed points.

Theorem 4.5.0

The character of the strange fixed point $z=1$ is:

i) If $-(52/25)<Re(a)<-\frac{76}{39}$ and $-\frac{\sqrt{-975\Re(a)^2-3928\Re(a)-3952}}{5\sqrt{39}}<\Im(a)<\frac{\sqrt{-975\Re(a)^2-3928\Re(a)-3952}}{5\sqrt{39}}$, then $z=1$ is an attractor. Moreover, if $a=-2$ it is superattractor.

ii) When $-(52/25) < Re(a) < -\frac{76}{39}$ *and* $\Im(a) = -\frac{\sqrt{-975\Re(a)^2-3928\Re(a)-3952}}{5\sqrt{39}} \vee \Im(a) = \frac{\sqrt{-975\Re(a)^2-3928\Re(a)-3952}}{5\sqrt{39}}$ $z = 1$ *is a parabolic point.*

iii) In other case, $z = 1$ *is a repulsor.*

Proof It follows from the expression

$$G'(z,1) = \Im(a) = -\frac{\sqrt{-975\Re(a)^2 - 3928\Re(a) - 3952}}{5\sqrt{39}} \vee \Im(a) = \frac{\sqrt{-975\Re(a)^2 - 3928\Re(a) - 3952}}{5\sqrt{39}}$$

⊠

In Figure 4.2 the region in which $z = 1$ is attractor is shown.

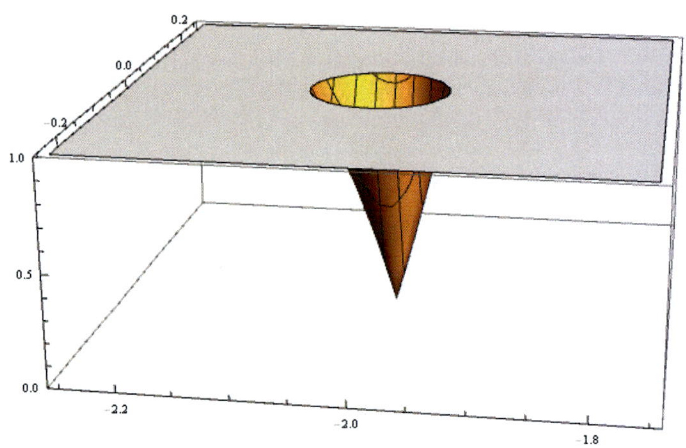

Figure 4.2: Region of stability of $z = 1$.

4.5.2 Study of the critical points and parameter spaces

It is a well-known fact that there is at least one critical point associated with each invariant Fatou component [13, 19]. The critical points of the family are the solutions of is $G'(z,a) = 0$, where

$$G'(z,a) = \frac{2z^5(z+1)^2(z^2+1)\left(5a^2z^3+4a^2z^2+5a^2z+6az^4+10az^3+20az^2+10az+6a+6z^4+12z^3+12z^2+12z+6\right)}{(2az^4+2az^3+2az^2+az+2z^4+4z^3+3z^2+2z+1)^2}$$

It is clear that $z=0$ and $z=\infty$ are critical points. Furthermore, the free critical points are the roots of the polynomial:

$$\begin{aligned} q(z) &= 5a^2z^3+4a^2z^2+5a^2z+6az^4+10az^3 \\ &\quad +20az^2+10az+6a+6z^4+12z^3+12z^2+12z+6 \end{aligned}$$

which will be called $cr_i(a)$ for $i=1,2,3,4$ and

$$cr_5(a) = -1$$
$$cr_6(a) = i$$

It is easy to see that cr_5 and cr_6 are preimages of the extraneous fixed point $z=1$, so we will not consider them in the dynamical study, as the stability of this fixed point has been already analyzed.

In Figure 4.3 the bifurcation diagram of the critical points is shown

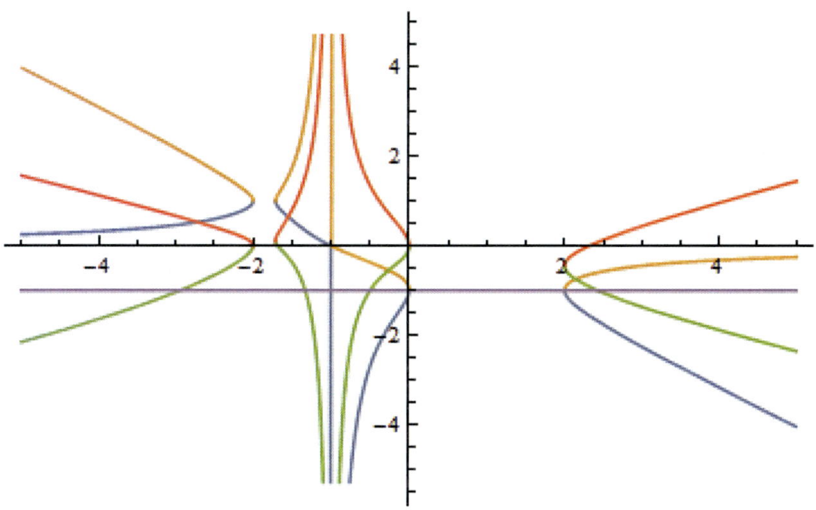

Figure 4.3: Bifurcation diagram of the critical points.

So, there are four independent free critical points in which we are interested, without loss of generality, we consider in this paper the free critical points $cr_1(a)$ and $cr_2(a)$, since its behavior is representative. Now, we are going to look for the best members of the family by means of using the parameter space associated to the free critical points.

The study of the orbits of the critical points gives rise about the dynamical behavior of an iterative method. In concrete, to determinate if there exists any attracting periodic orbit different to the roots of the polynomial $p(z)$, the following question must be answered: For which values of the parameter, the orbits of the free critical points are attracting periodic orbits? One way of giving response to that questions is drawing the parameter spaces associated to the free critical points. There will exist open regions of the parameter space for which the iteration of the free critical points does not converge to any of the roots of the polynomial $p(z)$, that is, in these regions the critical points converges to attracting cycles or to an strange fixed point or even they diverge.

In Figure 4.4, the parameter spaces associated to $cr_1(a)$ are shown in which we observe that there exists a zone with no convergence to the roots, this zone can be better seen in Figure 4.8, and in Figure 4.5 the parameter spaces associated to $cr_2(a)$ is shown. A point is painted in cyan if the iteration of the method starting in $z_0 = cr_1(a)$ converges to the fixed point 0 (related to root A), in magenta if it converges to ∞ (related to root B) and in yellow if the iteration converges to 1 (related to ∞). Moreover, it appears in red the convergence, after a maximum of 2000 iterations and with a tolerance of 10^{-6}, to any of the strange fixed points, in orange the convergence to 2-cycles, in light green the convergence to 3-cycles, in dark red to 4-cycles, in dark blue to 5-cycles, in dark green to 6-cycles, dark yellow to 7-cycles, and in white the convergence to 8-cycles. The regions in black correspond to zones of convergence to other cycles. As a consequence, every point of the plane which is neither cyan nor magenta is not a good choice of α in terms of numerical behavior.

Once the anomalies has been detected, the next step consist on describing them in the dynamical planes. In these dynamical planes the convergence to 0 will appear in magenta, in cyan it appears the convergence to ∞ and in black the zones with no convergence, after a maximum of 2000 iterations and with a tolerance of 10^{-6} to the roots.

Then, focussing the attention in the region shown in Figure 4.8 it is evident that there exist members of the family with complicated behavior. In Figure 4.9, the dynamical planes of a member of the family with regions of convergence to some of the strange fixed points is shown.

In Figures 4.10, 4.15 dynamical planes of members of the family with regions of convergence to an attracting 2-cycle is shown, those regions with convergence to 2-cycles are painted in black since there is no convergence to any of the roots. Both values of parameter correspond to red zones in the parameter planes.

On the other hand, in Figure 4.12, the dynamical plane of a member of the family with regions of convergence to $z = 1$, related to ∞ is presented, painted in black. This value of parameter a is located in a yellow zone of the parameter planes.

Other special cases are shown in Figures 4.13, 4.14, 4.15 and 4.16 where there are no convergence problems, as the only basins of attraction correspond to the roots of $p(z)$.

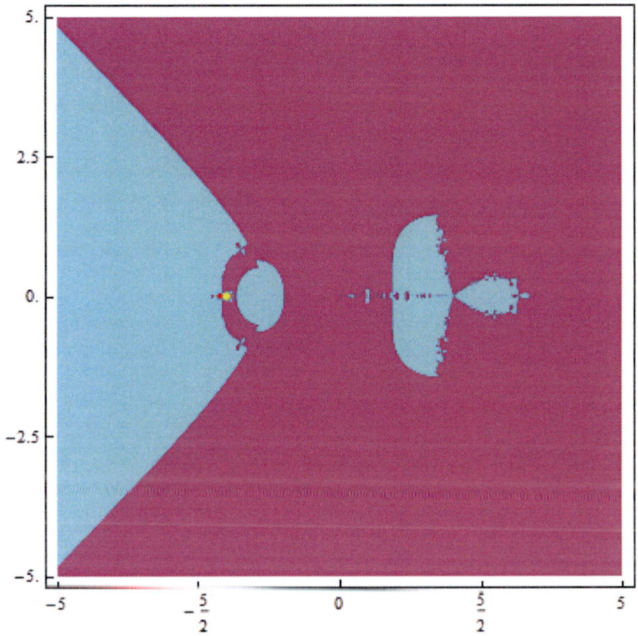

Figure 4.4: Parameter space associated to the free critical point $cr_1(a)$.

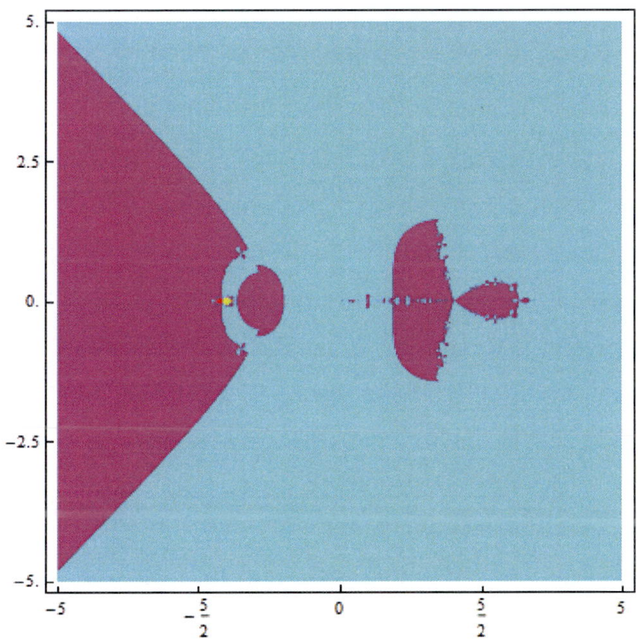

Figure 4.5: Parameter space associated to the free critical point $cr_2(a)$.

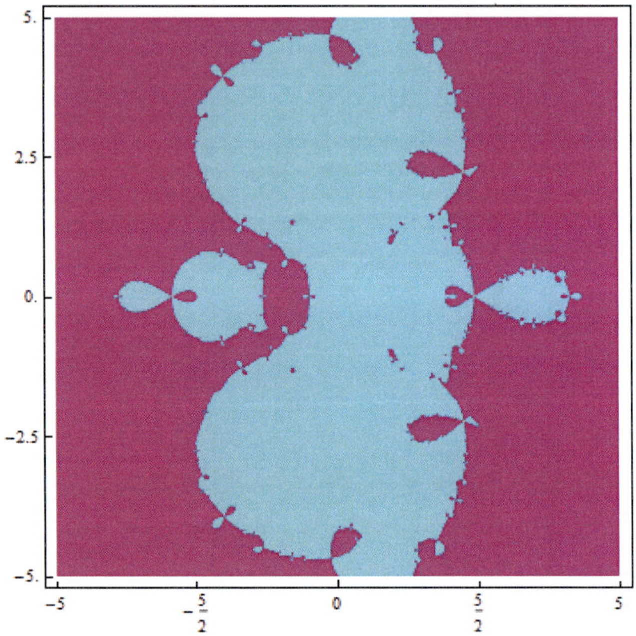

Figure 4.6: Parameter space associated to the free critical point $cr_3(a)$.

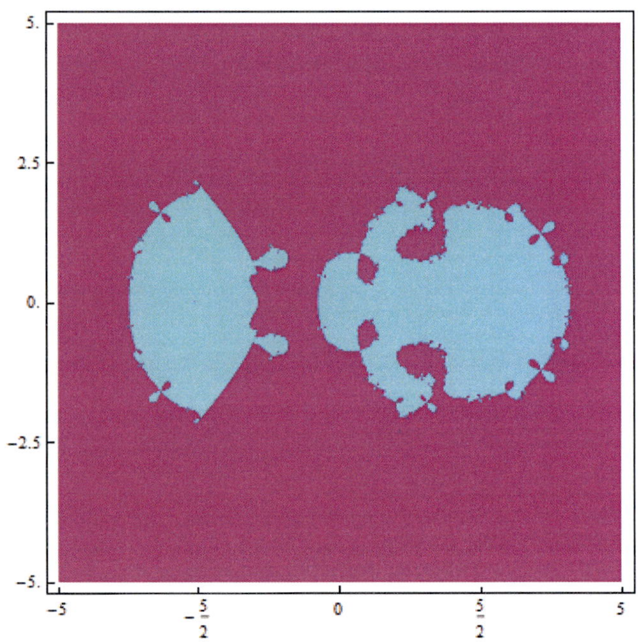

Figure 4.7: Parameter space associated to the free critical point $cr_4(a)$.

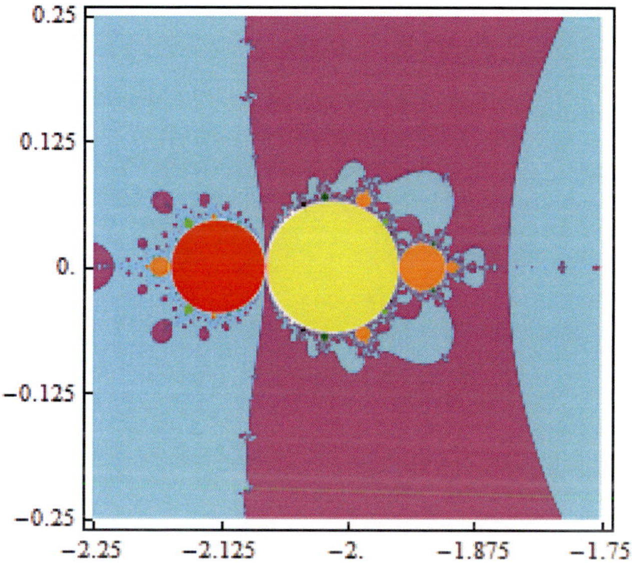

Figure 4.8: Detail of the parameter space associated to the free critical point $cr_1(a)$.

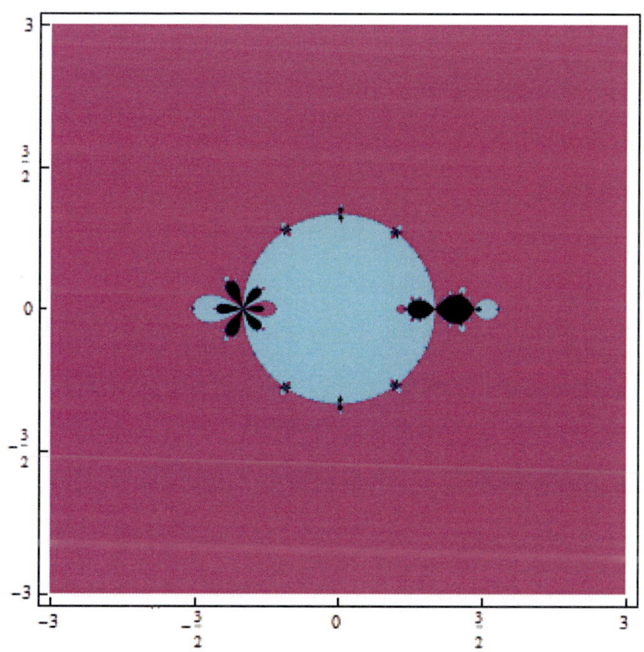

Figure 4.9: Basins of attraction associated to the method with $a = -2.125$.

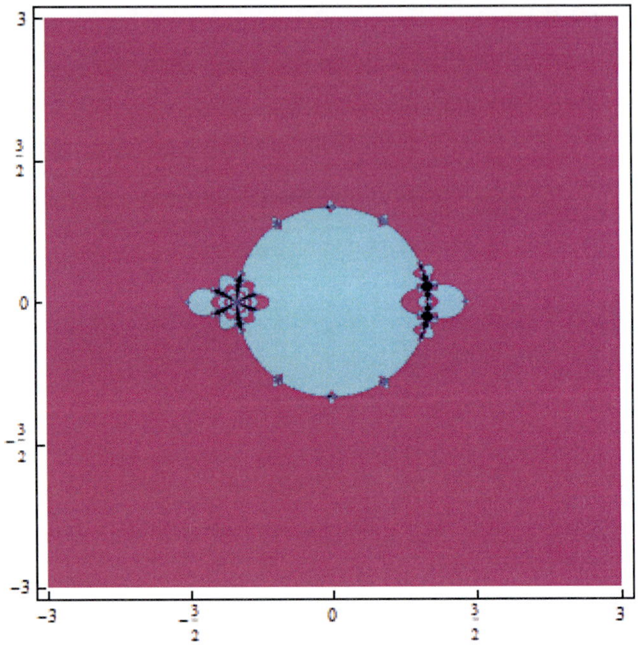

Figure 4.10: Basins of attraction associated to the method with $a = -1.9$.

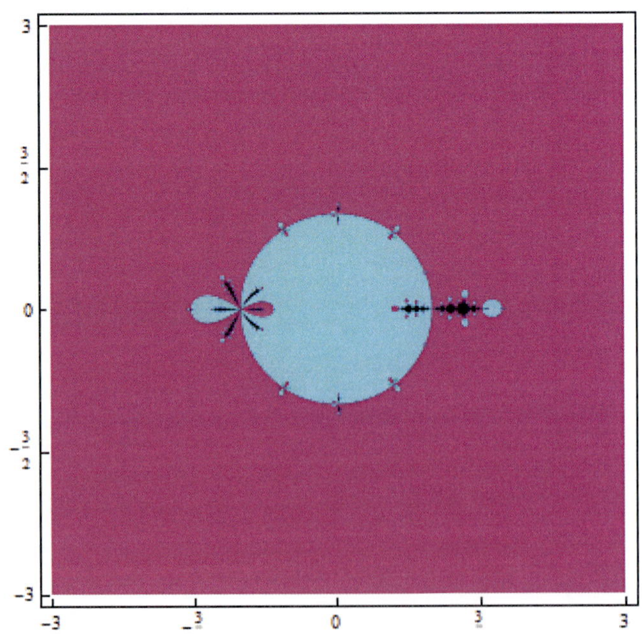

Figure 4.11: Basins of attraction associated to the method with $a = -2.18$.

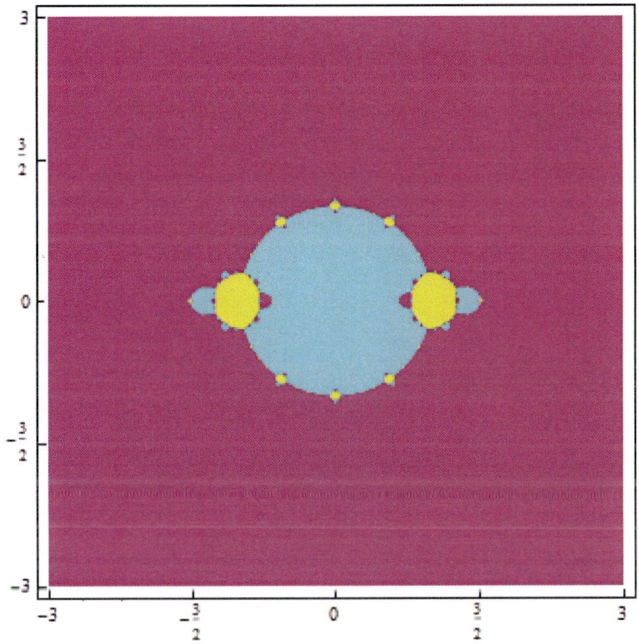

Figure 4.12: Basins of attraction associated to the method with $a = -2$.

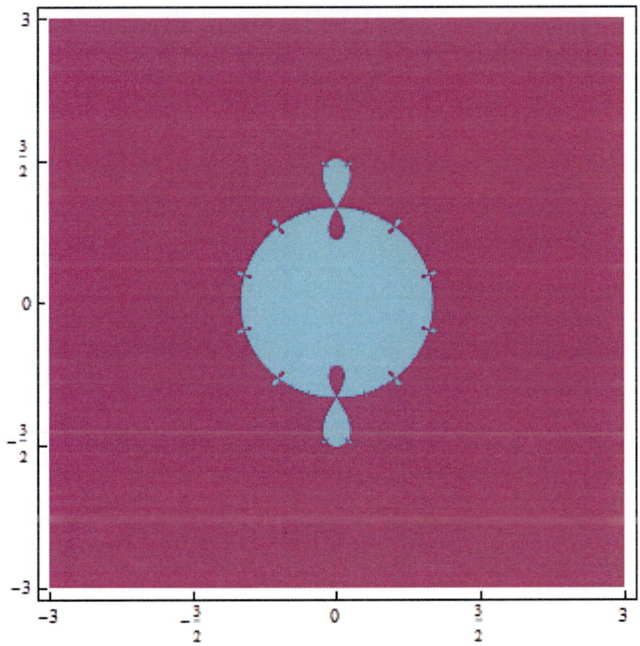

Figure 4.13: Basins of attraction associated to the method with $a = 0$.

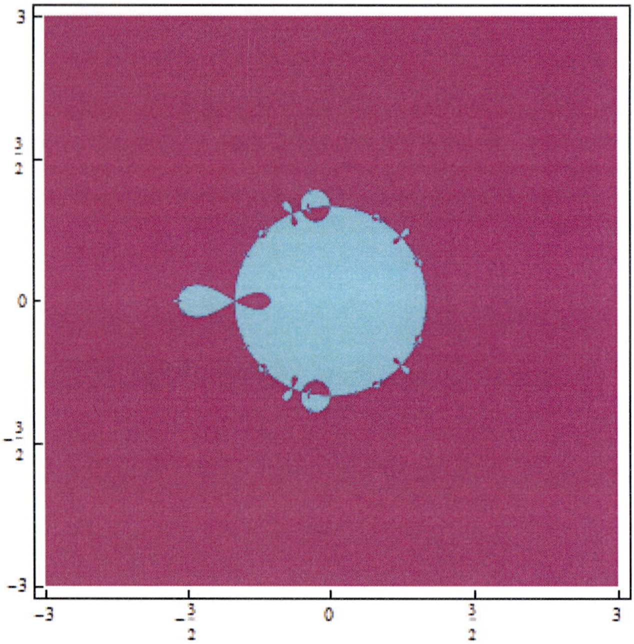

Figure 4.14: Basins of attraction associated to the method with $a = -5$.

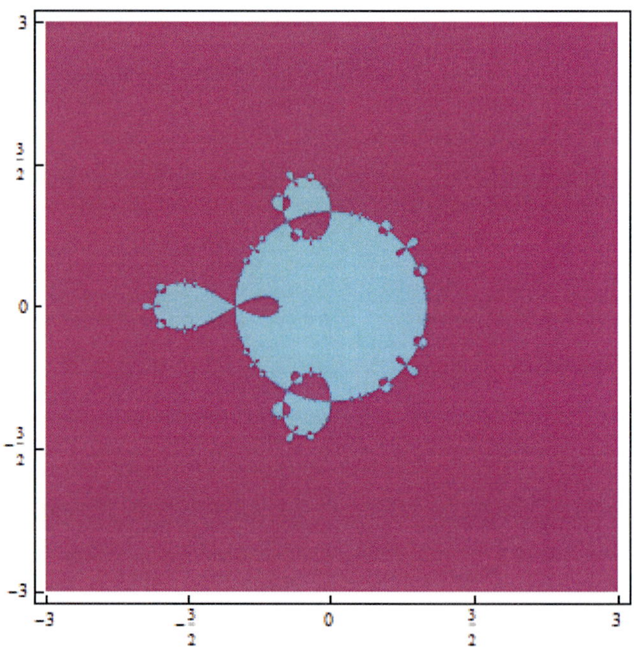

Figure 4.15: Basins of attraction associated to the method with $a = 5$.

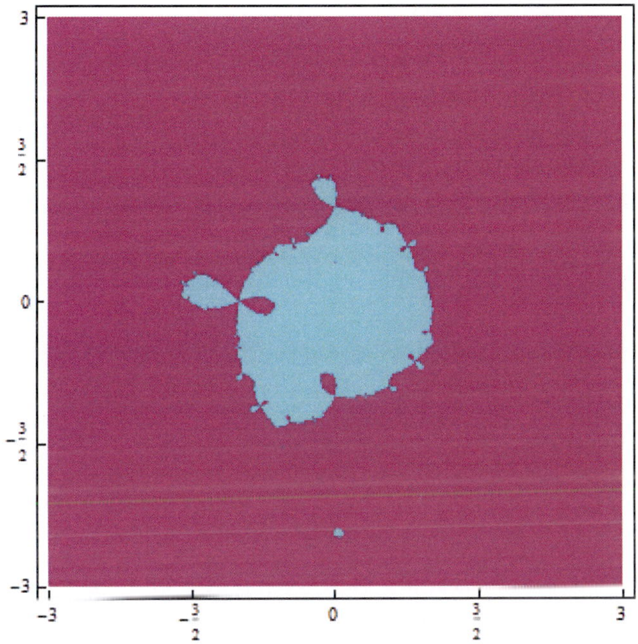

Figure 4.16: Basins of attraction associated to the method with $a = 2i$.

As a consequence, we can state that this family is very stable and that the existing anomalies are very rare in practice.

References

[1] ARPREC. C++/Fortran-90 arbitrary precision package. Available at `http://crd.lbl.gov/~dhbailey/mpdist/`.

[2] S. Amat, S. Busquier and S. Plaza, Dynamics of the King and Jarratt iterations, *Aequationes Math.*, 69, 212–223, 2005.

[3] G. Ardelean, A comparison between iterative methods by using the basins of attraction, *Appl. Math. Comput.*, 218 (1), 88–95, 2011.

[4] I. K. Argyros, Computational theory of iterative methods, Series: Studies in Computational Mathematics, Editors: C. K. Chui and L. Wuytack, Vol. 15, Elsevier Publ. Comp., New York, 2007.

[5] I. K. Argyros and S. Hilout, A unifying theorem for Newton's method on spaces with a convergence structure. *J. Complexity*, 27(1), 39–54, 2011.

[6] I. K. Argyros, D. Chen and Q. Qian, The Jarrat method in Banach space setting, *J. Computing. Appl. Math.*, 51, 103–106, 1994.

[7] I. K. Argyros and Á. A. Magreñán, On the convergence of an optimal fourth-order family of methods and its dynamics, *Appl. Math. Comput.*, 252, 336–346, 2015.

[8] I. K. Argyros and Á. A. Magreñán, Local Convergence and the Dynamics of a Two-Step Newton-Like Method, *Int. J. Bifurcation and Chaos*, 26 (5) 1630012, 2016.

[9] C. Chun and Y. Ham, Some sixth-order variants of Ostrowski root-finding methods, *Appl. Math. Comput.*, 193, 389–394, 2007.

[10] C. Chun and B. Neta, A new sixth-order scheme for nonlinear equations, *Appl. Math. Let.*, 25 (2), 185–189, 2012.

[11] A. Cordero, J. M. Gutiérrez, Á. A. Magreñán and J. R. Torregrosa, Stability analysis of a parametric family of iterative methods for solving nonlinear models, *Appl. Math. Comput.*, 285, 26–40, 2016.

[12] A. Cordero, L. Feng, Á. A. Magreñán and J. R. Torregrosa, A new fourth-order family for solving nonlinear problems and its dynamics, *J. Math. Chemistry*, 53 (3), 893–910, 2015.

[13] P. Fatou, Sur les équations fonctionelles, *Bull. Soc. Math. France*, 47, 161–271, 1919.

[14] A. Fraile, E. Larrodé, Á. A. Magreñán and J. A. Sicilia, Decision model for siting transport and logistic facilities in urban environments: A methodological approach. *J. Comput. Appl. Math.*, 291, 478–487, 2016.

[15] M. García-Olivo, J. M. Gutiérrez and Á. A. Magreñán, A complex dynamical approach of Chebyshev's method,*SeMA Journal*, 71 (1), 57–68, 2015.

[16] Y. H. Geum, Y. I. Kim, and Á. A. Magreñán, A biparametric extension of King's fourth-order methods and their dynamics, *Appl. Math. Comput.*, 282, 254–275, 2016.

[17] R. González-Crespo, R. Ferro, L. Joyanes, S. Velazco and A. G. Castillo, Use of ARIMA mathematical analysis to model the implementation of expert system courses by means of free software OpenSim and Sloodle platforms in virtual university campuses, *Expert Systems with Applications,* 40 (18), 7381–7390, 2013.

[18] M. Grau and J. L. Díaz-Barrero, An improvement to Ostrowski root-finding method, *Appl. Math. Comput.*, 173 (1), 450–456, 2006.

[19] G. Julia, Memoire sur l'iteration des fonctions rationelles, *J. de Math. pures et appliquées*, 8 (1), 47–215, 1918.

[20] S. K. Khattri and I. K. Argyros, Sixth order derivative free family of iterative methods. *Appl. Math. Comput.,* 217 (12), 5500–5507, 2011.

[21] S. K. Khattri and I. K. Argyros, How to develop fourth and seventh order iterative methods? *Novi. Sad. J. Math.*, 40, 61–67, 2010.

[22] S. K. Khattri and T. Log, Constructing third-order derivative-free iterative methods. *Int. J. Comput. Math.*, 88 (7), 1509–1518, 2011.

[23] Y. I. Kim, A new two-step biparametric family of sixth-order iterative methods free from second derivatives for solving nonlinear algebraic equations, *Appl. Math. Comput.* 215, 3418–3424, 2009.

[24] H. T. Kung and J. F. Traub, Optimal order of one-point and multipoint iterations, *J. Assoc. Comput. Mach.*, 21, 643–651, 1974.

[25] W. Lorenzo, R. G. Crespo and A. Castillo, A prototype for linear features generalization. *Int. J. Interac. Multim. Artif. Intell.*, 1 (3), 60–66, 2010.

[26] T. Lotfi, Á. A., Magreñán, K. Mahdiani, K. and J. J. Rainer, A variant of Steffensen-King's type family with accelerated sixth-order convergence and high efficiency index: Dynamic study and approach, *Appl. Math. Comput.*, 252, 347–353, 2015.

[27] T. Lotfi, Á. A. Magreñán, K. Mahdiani and J. J. Rainer, A variant of Steffensen-King's type family with accelerated sixth-order convergence and high efficiency index: Dynamic study and approach, *Appl. Math. Comput.*, 252, 347–353, 2015.

[28] Á. A. Magreñán, Different anomalies in a Jarratt family of iterative root-finding methods, *Appl. Math. Comput.*, 233, 29–38, 2014.

[29] Á. A. Magreñán, A new tool to study real dynamics: The convergence plane, *Appl. Math. Comput.*, 248, 215–224, 2014.

[30] Á. A. Magreñán and I. K. Argyros On the local convergence and the dynamics of Chebyshev-Halley methods with six and eight order of convergence,*J. Comput. Appl. Math.*, 298, 236–251, 2016.

[31] B. Neta, A sixth-order family of methods for nonlinear equations, *Int. J. Comput. Math.*, 7, 157–161 (1979).

[32] A. M. Ostrowski, Solutions of Equations and System Equations, Academic Press, New York, 1960.

[33] H. Ren, Q. Wu and W. Bi, New variants of Jarratt's method with sixth-order convergence, *Numer. Algorithms*, 52, 585–603, 2009.

[34] B. Royo, J. A. Sicilia, M. J. Oliveros and E. Larrodé, Solving a long-distance routing problem using ant colony optimization. *Appl. Math.*, 9 (2L), 415–421, 2015.

[35] J. R. Sharma and R. K. Guha, A family of modified Ostrowski methods with accelerated sixth order convergence, *Appl. Math. Comput.*, 190, 111–115, 2007.

[36] J. A. Sicilia, C. Quemada, B. Royo and D. Escuín, An optimization algorithm for solving the rich vehicle routing problem based on variable neighborhood search and tabu search metaheuristics. *J. Comput. Appl. Math.*, 291, 468–477, 2016.

[37] J. A. Sicilia, D. Escuín, B. Royo, E. Larrodé and J. Medrano, A hybrid algorithm for solving the general vehicle routing problem in the case of the urban freight distribution. In Computer-based Modelling and Optimization in Transportation (pp. 463–475). Springer International Publishing, 2014.

[38] H. Susanto and N. Karjanto, Newton's method basin of attraction revisited, *Appl. Math. Comput.*, 215, 1084–1090, 2009.

[39] J. F. Traub, Iterative Methods for the Solution of Equations, Prentice Hall, 1964. Reissued Chelsea Publishing Company, 1982.

[40] X. Wang, J. Kou and Y. Li, A variant of Jarratt method with sixth-order convergence, *Appl. Math. Comput.*, 204, 14–19, 2008.

Chapter 5

Local convergence and basins of attraction of a two-step Newton-like method for equations with solutions of multiplicity greater than one

Ioannis K. Argyros

Department of Mathematical Sciences, Cameron University, Lawton, OK 73505, USA, Email: iargyros@cameron.edu

Á. Alberto Magreñán

Universidad Internacional de La Rioja, Escuela Superior de Ingeniería y Tecnología, 26002 Logroño, La Rioja, Spain, Email: alberto.magrenan@unir.net

CONTENTS

5.1 Introduction

Let $S=\mathbb{R}$ or $S=\mathbb{C}$, $D\subseteq S$ be convex and let $F:D\rightarrow S$ be a differentiable function. We shall approximate solutions x^* of the equation

$$F(x)=0, \tag{5.1}$$

Many problems from Applied Sciences including engineering can be solved finding the solutions of equations in a form like (5.1) [4,5,20,25,31,39,43–45].

We study the local convergence and the dynamics of the two-step Newton-like method defined for each $n=0,1,2,\ldots$ by

$$\begin{aligned} y_n &= x_n-\alpha F'(x_n)^{-1}F(x_n) \\ x_{n+1} &= x_n-\phi(x_n)F'(x_n)^{-1}F(x_n) \end{aligned} \tag{5.2}$$

where x_0 is an initial point, $\alpha\in S$ is a given parameter and $\phi:S\rightarrow S$ is a given function.

Special choices of parameter α and function ϕ lead to well known methods.

- **Newton's method.**

$$\text{Choose } \alpha=0 \text{ and } \phi(x)=1 \text{ for each } x\in\mathcal{D} \tag{5.3}$$

to obtain

$$x_{n+1}=x_n-F'(x_n)^{-1}F(x_n), \qquad \text{for each } n=0,1,2,\ldots. \tag{5.4}$$

- **Laguerre's method.** Choose $\alpha=1$ and

$$\phi(x)=-1+\frac{q}{1+sign(q-m)\sqrt{\left(\frac{q-m}{m}\right)\left[(q-1)-q\frac{F(x)F''(x)}{F'(x)^2}\right]}},$$

where $q\neq 0$, m is a real parameter. In particular for special choices of q, we obtain: $q=2m$.

- **Euler-Chebyshev method.** Choose $\alpha=1$ and

$$\phi(x)=-1+\frac{m(3-m)}{2}+\frac{m^2}{2}\frac{F(x)F''(x)}{F'(x)^2}.$$

- **Osada's method.** Choose $\alpha = 1$ and

$$\phi(x) = -1 + \frac{1}{g}m(m+1) - \frac{(m-1)^2}{2}\frac{F'(x)^2}{F''(x)F(x)}.$$

Other choices of ϕ can be found in [4, 5, 13, 29, 30, 33, 34, 36–41, 45, 46] and the references therein.

The rest of the chapter is organized as follows: Section 5.2 contains the local convergence analysis of method (5.2). The dynamics of method (5.2) are discussed in Section 5.3. Finally, the numerical examples are presented in the concluding Section 5.4.

5.2 Local convergence

We present the local convergence analysis of method (5.2) in this section. It is convenient for the local convergence analysis that follows to introduce some constants and functions. Let $L_0 > 0$, $L > 0$ $\alpha \in S$ be given parameters and $m > 1$ be a given integer. Set $r_0 = \frac{2}{L_0}\left(\frac{3}{2}\right)^{m-2}$.

Define functions g_0, h_m and H_1 on the interval $[0, r_0)$ by

$$g_0(t) = \frac{L_0}{2}\left(\frac{1}{3}\right)^{m-2} t + 1 - \left(\frac{1}{2}\right)^{m-2}$$

$$H_0(t) = g_0(t) - 1,$$

$$h_m(t) = \begin{cases} \frac{L_0}{3}t + \frac{1}{2} & ,m = 2 \\ \frac{L_0}{2}\left(\frac{1}{3}\right)^{m-1} t + \frac{1}{3}\left(\frac{1}{2}\right)^{m-3} & ,m \geq 3 \end{cases}$$

$$g_1(t) = \frac{h_m(t) + \frac{|1-\alpha|L}{m}\left(\frac{1}{2}\right)^{m-2}}{1 - g_0(t)}$$

and

$$H_1(t) = h_m(t) + \frac{|1-\alpha|L}{m}\left(\frac{1}{2}\right)^{m-2} + g_0(t) - 1.$$

Notice that

$$g_0(r_0) = 1, \quad H_0(r_0) = 0 \text{ and } 0 \leq g_0(t) < 1 \text{ for each } t \in [0, r_0).$$

Hence, functions g_1 and H_1 are well defined on the interval $[0, r_0)$.

Suppose that

$$h_m(0) + \left(\frac{1}{2}\right)^{m-2}\left(\frac{|1-\alpha|L}{m} - 1\right) < 0.$$

Then, we have that

$$H_1(0) = h_m(0) + \left(\frac{1}{2}\right)^{m-2}\left(\frac{|1-\alpha|L}{m} - 1\right) < 0$$

and

$$\begin{aligned} H_1(r_0) &= h_m(r_0) + \left(\frac{1}{2}\right)^{m-2}\frac{|1-\alpha|L}{m} + \frac{L_0}{2}\left(\frac{1}{3}\right)^{m-2} r_0 - \left(\frac{1}{2}\right)^{m-2} \\ &= h_m(r_0) + \left(\frac{1}{2}\right)^{m-2}\frac{|1-\alpha|L}{m} > 0, \end{aligned}$$

since $g_0(r_0) = 1$. It follows from the intermediate value theorem that function H_1 has zeros in the interval $(0, r_0)$. Denote by r_1 the smallest such zero. Then, we have that

$$H_1(r_1) = 0, \quad r_1 < r_0 \text{ and } 0 \leq g_1(t) < 1 \text{ for each } t \in, (0, r_1).$$

Suppose that there exists a continuous function $\psi : [0, r_1) \rightarrow [0, +\infty)$ such that

$$h_m(r_0) + \left(\frac{1}{2}\right)^{m-2}\left[\frac{L}{m}(|1-\alpha| + \psi(0)) - 1\right] < 0. \tag{5.5}$$

Define function on the interval $[0, r_1)$ by

$$g_2(t) = \frac{1}{1 - g_0(t)}\left[h_m(t) + \left(\frac{1}{2}\right)^{m-2}\frac{L}{m}(|1-\alpha| + \psi(t))\right]$$

and function on the interval $[0, r_1]$ by

$$\begin{aligned} H_2(t) &= H_1(t) + \frac{L}{m}\left(\frac{1}{2}\right)^{m-2}\psi(t) \\ &= h_m(t) + \left(\frac{1}{2}\right)^{m-2}\frac{L}{m}(|1-\alpha|L + \psi(t)) \\ &\quad + \frac{L_0}{2}\left(\frac{1}{3}\right)^{m-2} t - \left(\frac{1}{2}\right)^{m-2}. \end{aligned}$$

By (5.5) we have that $H_2(0) < 0$. Moreover, we have that

$$H_2(r_1) = H_1(r_1) + \frac{L}{m}\left(\frac{1}{2}\right)^{m-2}\psi(r_1) \geq 0.$$

Then, function H_2 has zeros in the interval $(0, r_1]$. Denote the smallest such zero by r. Then, we have that

$$r \leq r_1 < r_0, \tag{5.6}$$

$$0 \leq g_0(t) < 1, \tag{5.7}$$

$$0 \leq g_1(t) < 1 \tag{5.8}$$

and

$$0 \leq g_2(t) < 1 \quad \text{for each } t \in [0, r). \tag{5.9}$$

Let $U(v, \rho)$, $\bar{U}(v, \rho)$ stand, respectively for the open and closed balls in S, respectively with center $v \in S$ and of radius $\rho > 0$. Next, using the preceding notation we present the local convergence analysis of method (5.2).

Theorem 5.2.1 *Let $m > 1$ be an integer and let $F : \mathcal{D} \subset S \to S$ be an m-times differentiable function. Suppose that there exist $L_0 > 0$, $L > 0$, $\alpha \in S$, $x^* \in \mathcal{D}$, functions $\varphi : S \to S$, $\psi : [0, r_1) \to [0, +\infty)$ such that Φ is continuous, ψ is continuous and non-decreasing satisfying*

$$h_m(0) + \left(\frac{1}{2}\right)^{m-2} \left[\frac{L}{m}(|1-\alpha| + \psi(0)) - 1\right] < 0,$$

$$F(x^*) = F'(x^*) = \cdots = F^{(m-1)}(x^*) = 0, \quad F^{(m)}(x^*) \neq 0, \tag{5.10}$$

$$|1 - \phi(x)| \leq \psi(\|x - x^*\|), \textit{ for each } x \in \bar{U}(x^*, r_1), \tag{5.11}$$

$$\|F^{(m)}(x^*)^{-1}(F^{(m)}(x) - F^{(m)}(x^*)\| \leq L_0\|x - x^*\|, \textit{ for each } x \in \mathcal{D} \tag{5.12}$$

$$\|F^{(m)}(x^*)^{-1}F^{(m)}(x)\| \leq L, \textit{ for each } x \in \mathcal{D} \tag{5.13}$$

and

$$\bar{U}(x^*, r) \subseteq \mathcal{D}, \tag{5.14}$$

where r_1, r are defined above this Theorem. Then, the sequence $\{x_n\}$ generated for $x_0 \in U(x^*, r) \setminus \{x^*\}$ by method (5.2) is well defined, remains in $\bar{U}(x^*, r)$ for each $n = 0, 1, 2, \ldots$ and converges to x^*. Moreover, the following estimates hold for each $n = 0, 1, 2, \ldots$

$$\|y_n - x^*\| \leq g_1(\|x_n - x^*\|)\|x_n - x^*\| \leq \|x_n - x^*\| < r \tag{5.15}$$

and

$$\|x_{n+1} - x^*\| \leq g_2(\|x_n - x^*\|)\|x_n - x^*\| < \|x_n - x^*\|, \tag{5.16}$$

where the "g" functions are defined previously. Furthermore, for $R \in [r, r_0)$ the limit point x^* is the only solution of equation $F(x) = 0$ in $\bar{U}(x^*, R) \cap D$, where r_0 is defined previously.

Proof: We shall show estimates (5.15) and (5.16) using mathematical induction. By hypothesis $x_0 \in U(x^*, r) \setminus \{x^*\}$. Then, we have in turn the approximation

$$F'(x_0)-F^{(m)}(x^*)(x_0-x^*)^{m-1}$$

$$=F'(x_0)-F'(x^*)-F^{(m)}(x^*)(x_0-x^*)^{(m-1)}$$

$$=\int_0^1 F''(x^*+\theta_1(x_0-x^*))(x_0-x^*)d\theta_1-F^{(m)}(x^*)(x_0-x^*)^{m-1}$$
$$=\int_0^1\left[F''(x^*+\theta_1(x_0-x^*))-F''(x^*)\right](x_0-x^*)d\theta_1-F^{(m)}(x^*)(x_0-x^*)^{m-1}$$

$$=\int_0^1\int_0^1 F'''(x^*+\theta_2(x^*+\theta_1(x_0-x^*)-x^*))\theta_1(x_0-x^*)^2d\theta_1d\theta_2$$
$$-F^{(m)}(x^*)(x_0-x^*)^{m-1}$$

$$=\int_0^1\int_0^1 F'''(x^*+\theta_2\theta_1(x_0-x^*))\theta_1(x_0-x^*)d\theta_1d\theta_2-F^{(m)}(x^*)(x_0-x^*)^{m-1}$$

$$=\cdots$$

$$=\int_0^1\int_0^1\cdots\int_0^1\left[F^{(m)}(x^*+\theta_{m-1}\theta_{m-2}\ldots\theta_1(x_0-x^*))-F^{(m)}(x^*)\right]$$

$$\times\theta_{m-2}\cdots\theta_1(x_0-x^*)^{(m-1)}d\theta_1d\theta_2\ldots d\theta_{m-1}$$

$$-\int_0^1\int_0^1\cdots\int_0^1(1-\theta_{m-2}\ldots\theta_1)F^{(m)}(x^*)(x_0-x^*)^{(m-1)}d\theta_1d\theta_2\ldots d\theta_{m-1}$$

(5.17)

Then, using (5.6), (5.7), (5.12) and (5.17), we get for $x_0\neq x^*$ that

$$\|\left(F^{(m)}(x^*)(x_0-x^*)^{m-1}\right)^{-1}\left[F'(x_0)-F^{(m)}(x^*)(x_0-x^*)^{m-1}\right]\|$$

$$\leq\|x_0-x^*\|^{1-m}\Big\{\|\int_0^1\int_0^1\cdots\int_0^1\left(F^{(m)}(x^*)\right)^{-1}\left[F^{(m)}(x^*+\theta_{m-1}\theta_{m-2}\ldots\theta_1\right.$$

$$\left.(x_0-x^*))-F^{(m)}(x^*)\right]$$

$$\times\theta_{m-2}\cdots\theta_1(x_0-x^*)^{(m-1)}d\theta_1d\theta_2\ldots d\theta_{m-1}\|$$

$$+\int_0^1\int_0^1\cdots\int_0^1(1-\theta_{m-2}\ldots\theta_1)\|x_0-x^*\|^{m-1}d\theta_1d\theta_2\ldots d\theta_{m-1}\Big\}$$

$$\leq\|x_0-x^*\|^{1-m}\Big[L_0\int_0^1\int_0^1\cdots\int_0^1\theta_{m-1}(\theta_{m-2}\cdots\theta_1)^2\|x_0-x^*\|^m d\theta_1\cdots\theta_{m-1}$$

$$+\int_0^1\int_0^1(1-\theta_{m-2}\cdots\theta_1)\|x_0-x^*\|^{m-1}d\theta_1\ldots d\theta_{m-1}\Big]$$

$$=\frac{L_0}{2}\left(\frac{1}{3}\right)^{m-2}\|x_0-x^*\|+1-\left(\frac{1}{3}\right)^{m-2}=g_0(\|x_0-x^*\|)<g_0(r)$$

$$=\frac{L_0}{2}\left(\frac{1}{3}\right)^{m-2}r+1-\left(\frac{1}{3}\right)^{m-2}<1.$$

(5.18)

It follows from (5.18) and the Banach lemma on invertible functions [4,5,39, 45] that $F'(x_0)^{-1} \in Ł(\mathbb{R},\mathbb{R})$ and

$$\|F'(x_0)^{-1}F^{(m)}(x^*)\| \leq \frac{1}{\|x_0-x^*\|^{m-1}\left[1-\left(\frac{L_0}{2}\left(\frac{1}{3}\right)^{m-2}\|x_0-x^*\|+1-\left(\frac{1}{2}\right)^{m-2}\right)\right]} \tag{5.19}$$

Hence, y_0 is well defined by the first substep of method (5.2) for $n=0$. Using method (5.2) for $n=0$ and (5.10) we obtain the approximation

$$\begin{aligned}
y_0-x^* &= x_0-x^*-F'(x_0)^{-1}F(x_0)+(1-\alpha)F'(x_0)^{-1}F(x_0)\\
&= F'(x_0)^{-1}\left[F'(x_0)(x_0-x^*)-F(x_0)\right]+(1-\alpha)F'(x_0)^{-1}F(x_0)\\
&= -\left[F'(x_0)^{-1}F^{(m)}(x^*)\right]\int_0^1 F^{(m)}(x^*)^{-1}\left[F'(x^*+\theta_1(x_0-x^*))-F'(x_0)\right]\\
&\quad (x_0-x^*)d\theta_1+(1-\alpha)\left[F'(x_0)^{-1}F^{(m)}(x^*)\right]F^{(m)}(x^*)^{-1}F(x_0).
\end{aligned} \tag{5.20}$$

We shall find some upper bounds on the norms of the integral expression and $F(x_0)$ in (5.20). We have in turn the following

$$\begin{aligned}
&\int_0^1\left[F'(x^*+\theta_1(x_0-x^*))-F'(x_0)\right](x_0-x^*)d\theta_1\\
&= -\int_0^1\int_0^1 F''(x_0+\theta_2(x^*+\theta_1(x_0-x^*)-x_0))(1-\theta_1)(x_0-x^*)^2d\theta_1d\theta_2\\
&= -\int_0^1\int_0^1\left[F''(x_0+\theta_2(\theta_1-1)(x_0-x^*))-F''(x^*)\right](1-\theta_1)(x_0-x^*)^2d\theta_1d\theta_2\\
&= \int_0^1\int_0^1\int_0^1 F'''(x^*+\theta_3(x_0+\theta_2(\theta_1-1)(x_0-x^*)-x^*))(1-(1-\theta_1)\theta_2)(1-\theta_1)\\
&(x_0-x^*)^3d\theta_1d\theta_3\\
&= \int_0^1\int_0^1\int_0^1\left[F'''(x^*+\theta_3(x_0+\theta_2(\theta_1-1)(x_0-x^*)-x^*))-F'''(x^*)\right]\\
&\times(1-(1-\theta_1)\theta_2)(1-\theta_1)(x_0-x^*)^3d\theta_1d\theta_2\theta_3\\
&= \cdots\\
&= \int_0^1\int_0^1\cdots\int_0^1\left[F^{(m)}(x^*+\theta_m\theta_{m-2}\ldots\theta_3(1-(1-\theta_1)\theta_2)(x_0-x^*))-F^{(m)}(x^*)\right]\\
&\times\theta_{m-1}\ldots\theta_3(1-(1-\theta_1)\theta_2)(1-\theta_1)(x_0-x^*)^m d\theta_1d\theta_2\ldots\theta_m\\
&+\int_0^1\int_0^1\cdots\int_0^1 F^{(m)}(x^*+\theta_{m-1}\ldots\theta_3(1-(1-\theta_1)\theta_2)(1-\theta_1)(x_0-x^*)^m d\theta_1d\theta_2\ldots\theta_m
\end{aligned} \tag{5.21}$$

Then, using (5.21) the definition of function h_m and (5.12) we get that

$$\| \int_0^1 F^{(m)}(x^*)^{-1} \left[F'(x^*+\theta_1(x_0-x^*)) - F'(x_0)\right](x_0-x^*)d\theta_1 \| \le h_m(\|x_0-x^*\|)\|x_0-x^*\|^m. \tag{5.22}$$

We also have that

$$\begin{aligned} F'(x_0) &= F'(x_0) - F'(x^*) = \textstyle\int_0^1 F''(x^*+\theta_1(x_0-x^*))(x_0-x^*)d\theta_1 \\ &= \cdots = \textstyle\int_0^1 \int_0^1 \cdots \int_0^1 F^{(m)}(x^*+\theta_{m-1}\theta_{m-2}\ldots\theta_1(x_0-x^*)) \\ &\times \theta_{m-2}\ldots\theta_1(x_0-x^*)^{m-1} d\theta_1 d\theta_2 \ldots d\theta_{m-1} \end{aligned} \tag{5.23}$$

Using (5.13) and (5.23), we get that

$$\|F^{(m)}(x^*)^{-1}F'(x_0)\| \le L\left(\frac{1}{2}\right)^{m-2} \|x_0-x^*\|^{m-1}. \tag{5.24}$$

Therefore, from (5.24), we get that since

$$F(x_0) = F(x_0) - F(x^*) = \int_0^1 F'(x^*+\theta(x_0-x^*))(x_0-x^*)d\theta \tag{5.25}$$

$$\begin{aligned} \|F^{(m)}(x^*)^{-1}F(x_0)\| &= \| \textstyle\int_0^1 F^{(m)}(x^*)^{-1}F'(x^*+\theta(x_0-x^*))(x_0-x^*)d\theta\| \\ &\le L \textstyle\int_0^1 \theta^{m-1}\|x_0-x^*\|^m \left(\frac{1}{2}\right)^{m-2} = \frac{L}{m}\left(\frac{1}{2}\right)^{m-2} \|x_0-x^*\|^m. \end{aligned} \tag{5.26}$$

Then, using the definition of function g_1, (5.6), (5.8), (5.19), (5.20), (5.22) and (5.26) we get in turn that

$$\begin{aligned} \|y_0-x^*\| &\le \|F'(x_0)^{-1}F^{(m)}(x^*)\| \| \textstyle\int_0^1 F^{(m)}(x^*)^{-1}\left[F'(x^*+\theta(x_0-x^*)) - F'(x_0)\right] \\ &(x_0-x^*)d\theta\| \\ &+|1-\alpha|\|F'(x_0)^{-1}F^{(m)}(x^*)\|\|F^{(m)}(x^*)^{-1}F(x_0)\| \\ &\le \frac{\left[h_m(\|x_0-x^*\|) + \frac{|1-\alpha|L}{m}\left(\frac{1}{2}\right)^{m-2}\right]\|x_0-x^*\|^m}{\|x_0-x^*\|^{m-1}(1-g_0(\|x_0-x^*\|))} \\ &= g_1(\|x_0-x^*\|)\|x_0-x^*\| < \|x_0-x^*\| < r, \end{aligned} \tag{5.27}$$

which shows (5.15) for $n=0$ and $y_0 \in U(x^*,r)$. We also have that x_1 is well defined by the second substep of method (5.2) for $n=0$. Using method (5.2) for

$n = 0$, we have that

$$\begin{aligned} x_1 - x^* &= x_0 - x^* - \phi(x_0)F'(x_0)^{-1}F(x_0) \\ &= x_0 - x^* - F'(x_0)^{-1}F(x_0) + (1-\phi(x_0))F'(x_0)^{-1}F(x_0) \\ &= y_0 - x^* + (1-\phi(x_0))\left[F'(x_0)^{-1}F^{(m)}(x^*)\right]\left[F^{(m)}(x^*)^{-1}F(x_0)\right] \end{aligned} \tag{5.28}$$

In view of the definition of function g_2, (5.6), (5.9), (5.11), (5.19), (5.26) and (5.28) we get that

$$\begin{aligned} &\|x_1 - x^*\| \leq \|y_0 - x^*\| + |1-\phi(x_0)|\|F'(x_0)^{-1}F^{(m)}(x^*)\|\|F^{(m)}(x^*)^{-1}F(x_0)\| \\ &\leq \left[g_1(\|x_0 - x^*\|) + \frac{\psi(\|x_0 - x^*\|)\frac{L}{m}\left(\frac{1}{2}\right)^{m-1}\|x_0 - x^*\|^{m-1}}{\|x_0 - x^*\|^{m-1}(1-g_0(\|x_0 - x^*\|))}\right]\|x_0 - x^*\| \\ &= g_2(\|x_0 - x^*\|)\|x_0 - x^*\| < \|x_0 - x^*\| < r, \end{aligned} \tag{5.29}$$

which shows (5.16) for $n = 0$ and $x_1 \in U(x^*, r)$. By simply replacing x_0, y_0, x_1 by x_k, y_k, x_{k+1} in the preceding estimates we arrive at estimates (5.15) and (5.16). Then, from the estimates $\|x_{k+1} - x^*\| \leq c\|x_k - x^*\| < r, x = g_2(\|x_0 - x^*\|) \in [0, 1)$, we get that $\lim_{k\to\infty} x_k = x^*$ and $x_{k+1} \in U(x^*, r)$. Finally to show the uniqueness part, let $y^* \in \bar{U}(x^*, R)$ be such that $F(y^*) = F'(y^*) = \cdots = F^{(m-1)}(y^*) = 0, F^{(m)}(y^*) \neq 0$. Set $T = \int_0^1 F'(y^* + \theta(x^* - y^*))d\theta$. Then, in view of (5.18) for x_0 replaced by $z = y^* + \theta(x^* - y^*)$ we get in turn that

$$\begin{aligned} &\|F^{(m)}(x^*)^{-1}(z - x^*)^{-1}(T - F^{(m)}(x^*)(z - x^*)^{m-1})\| \\ &\leq \frac{L_0}{2}\left(\frac{1}{3}\right)^{m-2}\int_0^1 (1-\theta)\|y^* - x^*\|d\theta + 1 - \left(\frac{1}{2}\right)^{m-2} \\ &\leq \frac{L_0}{2}\left(\frac{1}{3}\right)^{m-2} R + 1 - \left(\frac{1}{2}\right)^{m-2} < 1. \end{aligned}$$

It follows that T^{-1} exists. Then, from the identity $0 = F(y^*) - F(x^*) = T(y^* - x^*)$, we conclude that $x^* = y^*$. ⊠

5.3 Basins of attraction

The dynamical properties related to an iterative method applied to polynomials give important information about its stability and reliability. Many authors such as Amat et al. [1–3], Gutiérrez et al. [21], Magreñán [33, 34] and many others [6–13, 15–19, 21–24, 26–29, 32, 35, 38–41, 45] have found interesting dynamical planes, including periodical behavior and others anomalies. One of our main interests in this chapter is the study of the parameter spaces associated to a family of iterative methods, which allow us to distinguish between the good and bad methods in terms of its numerical properties.

Firstly, some dynamical concepts of complex dynamics that are used in this work are shown. Given a rational function $R : \hat{\mathbb{C}} \to \hat{\mathbb{C}}$, where $\hat{\mathbb{C}}$ is the Riemann sphere, the *orbit of a point* $z_0 \in \hat{\mathbb{C}}$ is defined as

$$\{z_0, R(z_0), R^2(z_0), ..., R^n(z_0), ...\}.$$

A point $z_0 \in \hat{\mathbb{C}}$, is called a *fixed point* of $R(z)$ if it verifies that $R(z) = z$. Moreover, z_0 is called a *periodic point* of period $p > 1$ if it is a point such that $R^p(z_0) = z_0$ but $R^k(z_0) \neq z_0$, for each $k < p$. Moreover, a point z_0 is called *pre-periodic* if it is not periodic but there exists a $k > 0$ such that $R^k(z_0)$ is periodic.

There exist different types of fixed points depending on its associated multiplier $|R'(z_0)|$. Taking the associated multiplier into account, a fixed point z_0 is called:

- *superattractor* if $|R'(z_0)| = 0$,
- *attractor* if $|R'(z_0)| < 1$,
- *repulsor* if $|R'(z_0)| > 1$,
- and *parabolic* if $|R'(z_0)| = 1$.

The fixed points that do not correspond to the roots of the polynomial $p(z)$ are called *strange fixed points*. On the other hand, a *critical point* z_0 is a point which satisfies that $R'(z_0) = 0$.

The basin of attraction of an attractor α is defined as

$$\mathcal{A}(\alpha) = \{z_0 \in \hat{\mathbb{C}} \ : \ R^n(z_0) \to \alpha, \ n \to \infty\}.$$

The *Fatou set* of the rational function R, $\mathcal{F}(R)$, is the set of points $z \in \hat{\mathbb{C}}$ whose orbits tend to an attractor (fixed point, periodic orbit or infinity). Its complement in $\hat{\mathbb{C}}$ is the *Julia set*, $\mathcal{J}(R)$. That means that the basin of attraction of any fixed point belongs to the Fatou set and the boundaries of these basins of attraction belong to the Julia set.

Now, we are going to see some of the basin of attractions related to different choices of α and $\phi(x)$. In concrete we have chosen 3 values of α

- $\alpha = \frac{1}{2}$
- $\alpha = 1$
- $\alpha = -1$

and three choices of functions ϕ

- $\phi(x) = x^2$

- $\phi(x) = \dfrac{x}{2}$
- $\phi(x) = x^4 + 1$

when method (5.2) is applied to three different polynomials

- $F(x) = (x-1)^2(x+1)$
- $F(x) = (x-1)^3(x+1)$
- $F(x) = (x-1)^4(x+1)$.

A point is painted in cyan if the iteration of the method starting in x_0 converges to the fixed point 1 (multiplicity greater than one), in magenta if it converges to -1 (simple one) and in yellow if the iteration diverges to ∞. If after 100 iterations the sequence does not converge to any of the roots with a tolerance of 10^{-3} or diverges to infinity that point is painted in black.

5.3.1 Basins of $F(x) = (x-1)^2(x+1)$

In this subsection we are going to see how basins of attraction change when α and function $\phi(x)$ change when $F(x) = (x-1)^2(x+1)$.

In Figure 5.1 we see how the basin change when α change with $\phi(x) = x^2$.

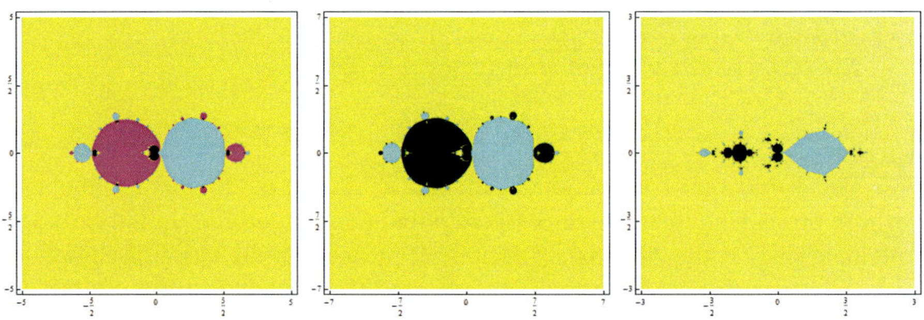

$\alpha = 0.5$ $\alpha = 1$ $\alpha = -1$

Figure 5.1: Basins of attraction of the method (5.2) associated to the roots of $F(x) = (x-1)^2(x+1)$ with $\phi(x) = x^2$.

In Figure 5.2 we see how the basin change when α change with $\phi(x) = \frac{x}{2}$.
In Figure 5.3 we see how the basin change when α change with $\phi(x) = x^4 + 1$.

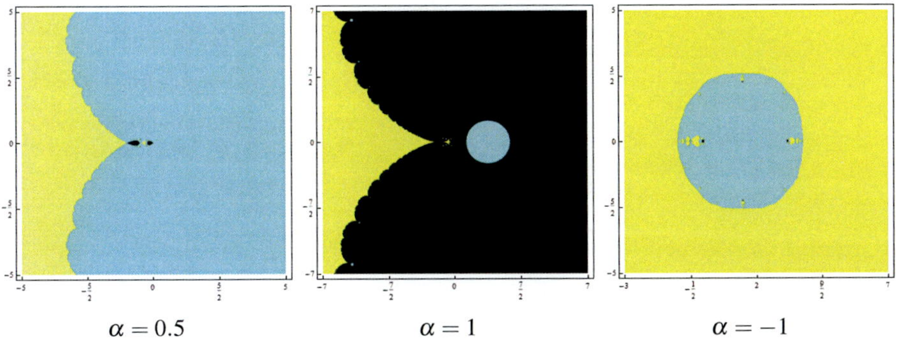

Figure 5.2: Basins of attraction of the method (5.2) associated to the roots of $F(x) = (x-1)^2(x+1)$ with $\phi(x) = \frac{x}{2}$.

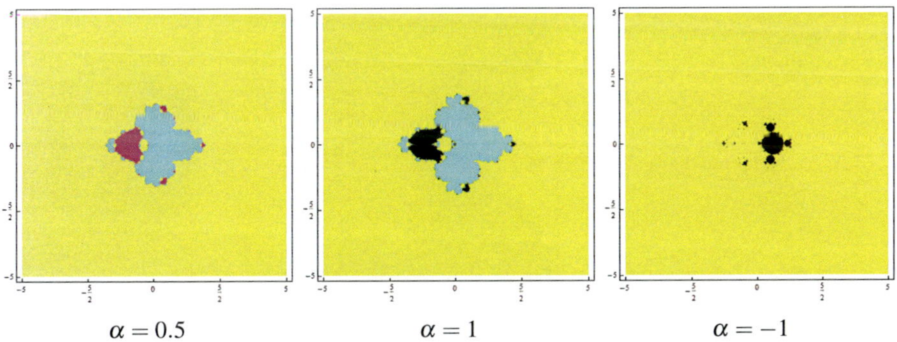

Figure 5.3: Basins of attraction of the method (5.2) associated to the roots of $F(x) = (x-1)^2(x+1)$ with $\phi(x) = x^4 + 1$.

5.3.2 Basins of $F(x) = (x-1)^3(x+1)$

In this subsection we are going to see how basins of attraction change when α and function $\phi(x)$ change when $F(x) = (x-1)^3(x+1)$.

In Figure 5.4 we see how the basin change when α change with $\phi(x) = x^2$.

In Figure 5.5 we see how the basin change when α change with $\phi(x) = \frac{x}{2}$.

In Figure 5.6 we see how the basin change when α change with $\phi(x) = x^4 + 1$.

5.3.3 Basins of $F(x) = (x-1)^4(x+1)$

In this subsection we are going to see how basins of attraction change when α and function $\phi(x)$ change when $F(x) = (x-1)^4(x+1)$.

In Figure 5.7 we see how the basin change when α change with $\phi(x) = x^2$.

In Figure 5.8 we see how the basin change when α change with $\phi(x) = \frac{x}{2}$.

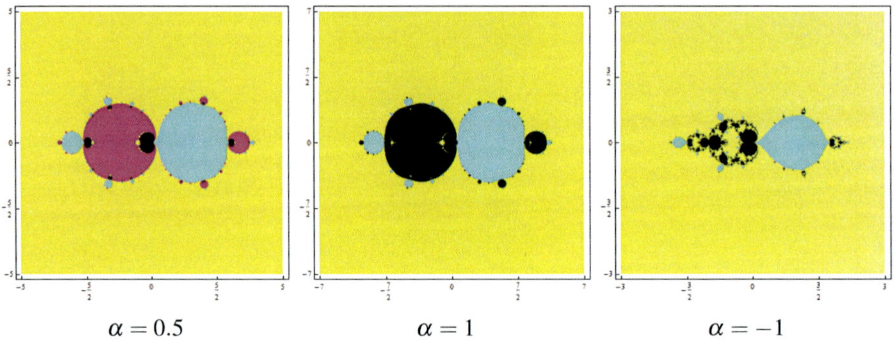

Figure 5.4: Basins of attraction of the method (5.2) associated to the roots of $F(x) = (x-1)^3(x+1)$ with $\phi(x) = x^2$.

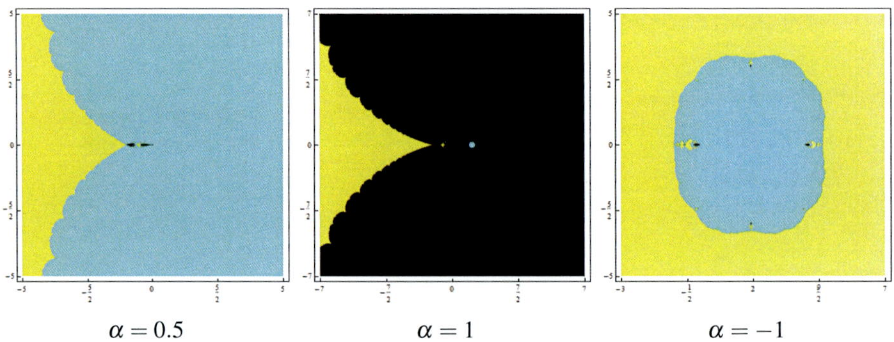

Figure 5.5: Basins of attraction of the method (5.2) associated to the roots of $F(x) = (x-1)^3(x+1)$ with $\phi(x) = \frac{x}{2}$.

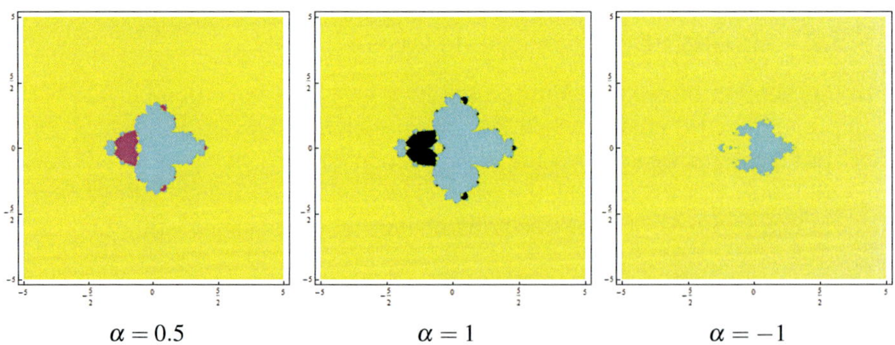

Figure 5.6: Basins of attraction of the method (5.2) associated to the roots of $F(x) = (x-1)^3(x+1)$ with $\phi(x) = x^4 + 1$.

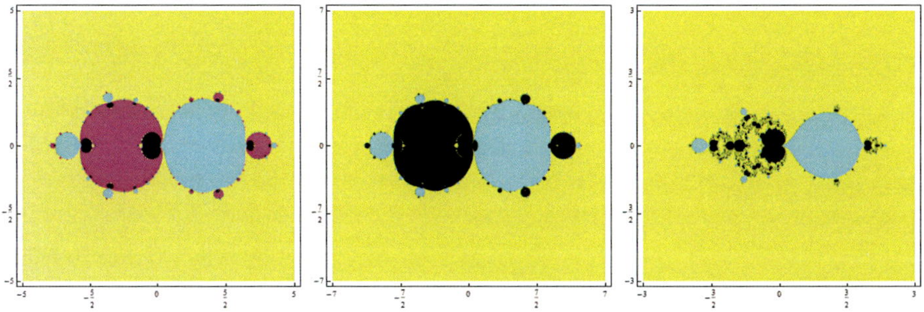

$\alpha = 0.5$ $\alpha = 1$ $\alpha = -1$

Figure 5.7: Basins of attraction of the method (5.2) associated to the roots of $F(x) = (x-1)^4(x+1)$ with $\phi(x) = x^2$.

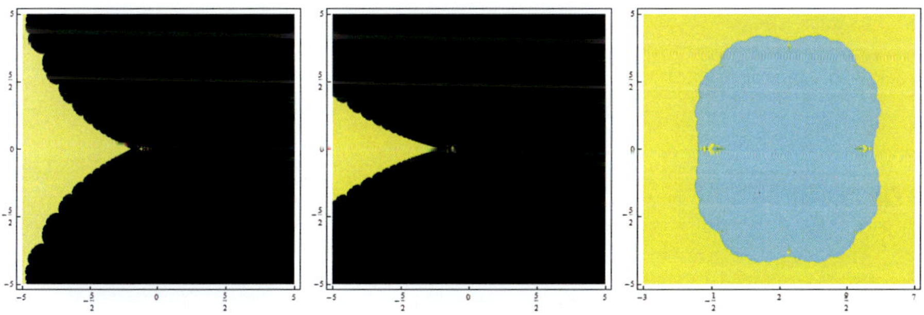

$\alpha = 0.5$ $\alpha = 1$ $\alpha = -1$

Figure 5.8: Basins of attraction of the method (5.2) associated to the roots of $F(x) = (x-1)^4(x+1)$ with $\phi(x) = \frac{x}{2}$.

In Figure 5.9 we see how the basin change when α change with $\phi(x) = x^4 + 1$.

5.4 Numerical examples

We present numerical examples in this section.

Example 5.4.1 Let $S = \mathbb{R}, D = [-1, 1]$ and define function F on D by

$$F(x) = x(e^x - 1)$$

Then, we have that $x^* = 0$, $m = 2$, $F''(x^*) = 2$

$$L = \frac{3}{2}e$$

and

$$L_0 = \frac{1}{2}(3e - 4) = \frac{3e}{2} - 2,$$

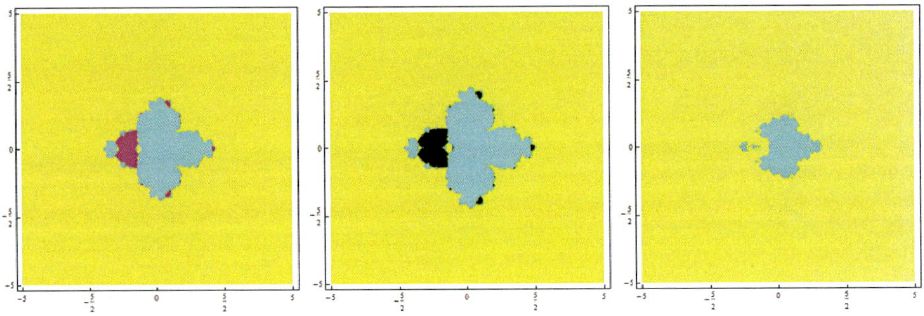

$\alpha = 0.5$ $\alpha = 1$ $\alpha = -1$

Figure 5.9: Basins of attraction of the method (5.2) associated to the roots of $F(x) = (x-1)^4(x+1)$ with $\phi(x) = x^4 + 1$.

since $L = \frac{1}{2} \max_{-1 \leq x \leq 1} |(x+2)e^x| = \frac{3}{2}e$ and

$$L_0 = \frac{1}{2} \max_{-1 \leq x \leq 1} |2(1+\frac{x}{2!}+\cdots+\frac{x^{n-1}}{x!}+\cdots)+e^x| = \frac{1}{2}\left[2(1+\frac{1}{2}+\cdots+\frac{1}{n!}+\cdots)+e\right]$$
$$= \frac{1}{2}(3e-4).$$

Now choosing

$$\alpha = 0.85,$$

$$\phi(x) = \phi = 1-x$$

and

$$\psi(x) = |x|$$

we obtain that conditions of Theorem hold and method (5.2) with $\alpha = 0.85$ and $\phi(x) = 1-x$ converges to $x^* = 0$. Moreover,

$$r_0 = 0.962731\ldots,$$

$$r_1 = 0.112174\ldots$$

and

$$r = 0.051511\ldots.$$

Example 5.4.2 Let $S = \mathbb{R}, D = (1,3)$ and define function F on D by

$$F(x) = (x-2)^6(x-3)^4$$

Then, we have that $x^* = 2$, $m = 6$, $F^{(6)}(x^*) = 720$ and

$$L = 743$$

and

$$L_0 = 742.$$

Now choosing

$$\alpha = 1,$$

$$\phi(x) = 3x$$

and

$$\psi(x) = t - 3$$

we obtain that conditions of Theorem hold and method (5.2) with $\alpha = 1$ and $\phi(x) = 3x$ converges to $x^* = 2$. Moreover,

$$r_0 = 0.0136456\ldots,$$

$$r_1 = 0.00341139\ldots$$

and

$$r = 0.000710411\ldots.$$

References

[1] S. Amat, S. Busquier and S. Plaza, Dynamics of the King and Jarratt iterations, *Aequationes Math.*, 69 (3), 212–223, 2005.

[2] S. Amat, S. Busquier and S. Plaza, Chaotic dynamics of a third-order Newton-type method, *J. Math. Anal. Appl.*, 366 (1), 24–32, 2010.

[3] S. Amat, M. A. Hernández and N. Romero, A modified Chebyshev's iterative method with at least sixth order of convergence, *Appl. Math. Comput.*, 206 (1), 164–174, 2008.

[4] I. K. Argyros, Convergence and Application of Newton-type Iterations, Springer, 2008.

[5] I. K. Argyros and S. Hilout, Numerical methods in Nonlinear Analysis, World Scientific Publ. Comp. New Jersey, 2013.

[6] I. K. Argyros and Á. A. Magreñán, On the convergence of an optimal fourth-order family of methods and its dynamics, *Appl. Math. Comput.* 252, 336–346, 2015.

[7] I. K. Argyros and Á. A. Magreñán, Local Convergence and the Dynamics of a Two-Step Newton-Like Method, *Int. J. Bifurcation and Chaos*, 26 (5), 1630012 2016.

[8] R. Behl, Development and analysis of some new iterative methods for numerical solutions of nonlinear equations (PhD Thesis), Punjab University, 2013.

[9] D. D. Bruns and J. E. Bailey, Nonlinear feedback control for operating a nonisothermal CSTR near an unstable steady state, *Chem. Eng. Sci.*, 32, 257–264, 1977.

[10] V. Candela and A. Marquina, Recurrence relations for rational cubic methods I: The Halley method, *Computing*, 44, 169–184, 1990.

[11] V. Candela and A. Marquina, Recurrence relations for rational cubic methods II: The Chebyshev method, *Computing*, 45 (4), 355–367, 1990.

[12] F. Chicharro, A. Cordero and J. R. Torregrosa, Drawing dynamical and parameters planes of iterative families and methods, *The Scientific World Journal*, Volume 2013, Article ID 780153.

[13] C. Chun, Some improvements of Jarratt's method with sixth-order convergence, *Appl. Math. Comput.*, 190 (2), 1432–1437, 1990.

[14] A. Cordero, J. García-Maimó, J. R. Torregrosa, M. P. Vassileva and P. Vindel, Chaos in King's iterative family, *Appl. Math. Let.*, 26, 842–848, 2013.

[15] A. Cordero, J. M. Gutiérrez, Á. A., Magreñán and J. R. Torregrosa, Stability analysis of a parametric family of iterative methods for solving nonlinear models, *Appl. Math. Comput.*, 285, 26–40, 2016.

[16] A. Cordero, L. Feng, L., Á. A., Magreñán and J. R. Torregrosa, A new fourth-order family for solving nonlinear problems and its dynamics, *J. Math. Chemistry*, 53 (3), 893–910, 2015.

[17] J. A. Ezquerro and M. A. Hernández, Recurrence relations for Chebyshev-type methods, *Appl. Math. Optim.*, 41(2), 227–236, 2000.

[18] J. A. Ezquerro and M. A. Hernández, New iterations of R-order four with reduced computational cost, *BIT Numer. Math.*, 49, 325–342, 2009.

[19] J. A. Ezquerro and M. A. Hernández, On the R-order of the Halley method, *J. Math. Anal. Appl.*, 303, 591–601, 2005.

[20] A. Fraile, E. Larrodé, Á. A. Magreñán and J. A. Sicilia, Decision model for siting transport and logistic facilities in urban environments: A methodological approach. *J. Comput. Appl. Math.*, 291, 478–487, 2016.

[21] J. M. Gutiérrez and M. A. Hernández, Recurrence relations for the super-Halley method, *Comput. Math. Appl.*, 36 (7), 1–8, 1998.

[22] M. Ganesh and M. C. Joshi, Numerical solvability of Hammerstein integral equations of mixed type, *IMA J. Numer. Anal.*, 11, 21–31, 1991.

[23] M. García-Olivo, J. M. Gutiérrez and Á. A. Magreñán, A complex dynamical approach of Chebyshev's method,*SeMA Journal*, 71 (1), 57–68, 2015.

[24] Y. H. Geum, Y. I., Kim and Á. A., Magreñán, A biparametric extension of King's fourth-order methods and their dynamics, *Appl. Math. Comput.*, 282, 254–275, 2016.

[25] R. González-Crespo, R. Ferro, L. Joyanes, S. Velazco and A. G. Castillo, Use of ARIMA mathematical analysis to model the implementation of expert system courses by means of free software OpenSim and Sloodle platforms in virtual University campuses, *Expert Systems with Applications*, 40 (18), 7381–7390, 2013.

[26] M. A. Hernández, Chebyshev's approximation algorithms and applications, *Computers Math. Applic.*, 41 (3-4), 433–455, 2001.

[27] M. A. Hernández and M. A. Salanova, Sufficient conditions for semilocal convergence of a fourth order multipoint iterative method for solving equations in Banach spaces, *Southwest J. Pure Appl. Math.*, 1, 29–40, 1999.

[28] P. Jarratt, Some fourth order multipoint methods for solving equations, *Math. Comput.*, 20 (95), 434–437, 1966.

[29] J. Kou and Y. Li, An improvement of the Jarratt method, *Appl. Math. Comput.*, 189, 1816–1821, 2007.

[30] J. Kou and X. Wang, Semilocal convergence of a modified multi-point Jarratt method in Banach spaces under general continuity conditions, *Numer. Algorithms*, 60, 369–390, 2012.

[31] W. Lorenzo, R. G. Crespo and A. Castillo, A prototype for linear features generalization, *Int. J. Interac. Multim. Artif. Intell.*, 1(3), 60–66, 2010.

[32] T. Lotfi, Á. A., Magreñán, K. Mahdiani, K. and J. J. Rainer, A variant of Steffensen-King's type family with accelerated sixth-order convergence and high efficiency index: Dynamic study and approach, *Appl. Math. Comput.*, 252, 347–353, 2015.

[33] Á. A. Magreñán, Different anomalies in a Jarratt family of iterative root-finding methods, *Appl. Math. Comput.*, 233, 29–38, 2014.

[34] Á. A. Magreñán, A new tool to study real dynamics: The convergence plane, *Appl. Math. Comput.*, 248, 215–224, 2014.

[35] Á. A. Magreñán and I. K. Argyros, On the local convergence and the dynamics of Chebyshev-Halley methods with six and eight order of convergence,*J. Comput. Appl. Math.*, 298, 236–251, 2016.

[36] B. Neta and C. Chun, A family of Laguerre methods to find multiple roots of nonlinear equations, *Appl. Math. Comput.*, 219, 10987–11004, 2013.

[37] B. Neta and C. Chun, Erratum on a family of Laguerre methods to find multiple roots of nonlinear equations, *Appl. Math. Comput.*, 248, 693–696, 2014.

[38] M. Petković, B. Neta, L. Petković and J. Džunić, Multipoint Methods for Solving Nonlinear Equations, Elsevier, 2013.

[39] L. B. Rall, Computational solution of nonlinear operator equations, Robert E. Krieger, New York, 1979.

[40] H. Ren, Q. Wu and W. Bi, New variants of Jarratt method with sixth-order convergence, *Numer. Algorithms*, 52(4), 585–603, 2009.

[41] W. C. Rheinboldt, An adaptive continuation process for solving systems of nonlinear equations, Polish Academy of Science, *Banach Ctr. Publ.*, 3, 129–142, 1978.

[42] B. Royo, J. A. Sicilia, M. J. Oliveros and E. Larrodé, Solving a long-distance routing problem using ant colony optimization. *Appl. Math.*, 9 (2L), 415–421, 2015.

[43] J. A. Sicilia, C. Quemada, B. Royo and D. Escuín, An optimization algorithm for solving the rich vehicle routing problem based on variable neighborhood search and tabu search metaheuristics. *J. Comput. Appl. Math.*, 291, 468–477, 2016.

[44] J. A. Sicilia, D. Escuín, B. Royo, E. Larrodé and J. Medrano, A hybrid algorithm for solving the general vehicle routing problem in the case of the urban freight distribution. In Computer-based Modelling and Optimization in Transportation (pp. 463–475). Springer International Publishing, 2014.

[45] J. F. Traub, Iterative methods for the solution of equations, Prentice-Hall Series in Automatic Computation, Englewood Cliffs, N. J., 1964.

[46] X. Wang, J. Kou and C. Gu, Semilocal convergence of a sixth-order Jarratt method in Banach spaces, *Numer. Algorithms*, 57, 441–456, 2011.

Chapter 6

Extending the Kantorovich theory for solving equations

Ioannis K. Argyros

Department of Mathematical Sciences, Cameron University, Lawton, OK 73505, USA, Email: iargyros@cameron.edu

Á. Alberto Magreñán

Universidad Internacional de La Rioja, Escuela Superior de Ingeniería y Tecnología, 26002 Logroño, La Rioja, Spain, Email: alberto.magrenan@unir.net

CONTENTS

6.1 Introduction

Let X, Y be Banach spaces, $D \subset X$ be convex, $F : D \subset X \to Y$ be a Fréchet differentiable operator. We shall determine a solution x^* of the equation

$$F(x) = 0, \tag{6.1}$$

Many problems from Applied Sciences can be solved finding the solutions of equations in a form like (6.1) [5, 9, 12, 13, 27, 31–33].

The most popular iterative method is undoubtedly Newton's method defined by

$$x_{n+1} = x_n - F'(x_n)^{-1}F(x_n), \quad \text{for each } n = 0,1,2,\cdots, \tag{6.2}$$

where x_0 is an initial point. Newton's method has order of convergence 2 if the initial point x_0 is chosen sufficiently close to the solution $x^\star$.

The computation of the inverse may be expensive in general. That is why the modified Newton's method defined as

$$y_{n+1} = y_n - F'(y_0)^{-1}F(y_n), \quad \text{for each } n = 0,1,2,\cdots, \tag{6.3}$$

where $y_0 = x_0$. The modified Newton's method is only linearly convergent.

The famous Kantorovich theorem [26] is an important result in numerical analysis and in Nonlinear Functional Analysis, where it is also used as a semilocal result for establishing the existence of a solution of a nonlinear equation in an abstract space. The well-known Kantorovich hypothesis

$$h = L\eta \leq \frac{1}{2}. \tag{6.4}$$

is the sufficient criterion for the semilocal convergence of Newton's method. Where, $L > 0$ and $\eta > 0$ are defined by

$$\|F'(x_0)^{-1}(F'(x) - F'(y))\| \leq L\|x - y\| \text{ for each } x, y \in \mathcal{D}.$$

and

$$\|F'(x_0)^{-1}F(x_0)\| \leq \eta$$

respectively.

In a series of papers [1–11, 14–17], we provided weaker sufficient convergence criteria. One of the most recent and weakest criterion is given by

$$h_1 = L_1\eta \leq \frac{1}{2}, \tag{6.5}$$

where

$$L_1 = \frac{1}{8}\left(4L_0 + \sqrt{L_0L} + \sqrt{L^2 + 8L_0L}\right),$$

where L_0 is the center-Lispchitz constant given in

$$\|F'(x_0)^{-1}(F'(x) - F'(x_0))\| \leq L_0\|x - x_0\| \text{ for each } x \in \mathcal{D}.$$

Notice that $L_0 \leq L$,

$$h \leq \frac{1}{2} \Rightarrow h_1 \leq \frac{1}{2} \tag{6.6}$$

and

$$\frac{h_1}{h} \to 0 \text{ as } \frac{L_0}{L} \to 0. \tag{6.7}$$

It is well known that in practice the computation of constant L requires that of constant L_0. Throughout the chapter we denote by $U(x,\rho)$ and $\bar{U}(x,\rho)$ the open and closed balls with center $x \in \mathcal{D}$ and radius $\rho > 0$, respectively.

In this chapter we present two different improvements of criterion (6.5) based on modified Newton's method and Newton's method in Sections 6.2 and 6.3 respectively.

6.2 First convergence improvement

In this section, using the modified Newton's method previously defined, we provide an improvement of the convergence criteria for Newton's method.

Remark 6.2.1 *As it is shown in [8], if*

$$\eta \leq \frac{1}{L_0}$$

then, equation $F(x) = 0$ has a unique solution in $\bar{U}(x_0, r_0)$ where

$$r_0 = \frac{1-\sqrt{1-2L_0\eta}}{L_0}.$$

On the other hand, modified Newton's method

$$y_{n+1} = y_n - F'(y_0)^{-1}F(y_n), y_0 = x_0, \text{ for each } n = 0,1,2,\cdots$$

remains in $\bar{U}(x_0, r_0)$ for each $n = 0,1,2,\cdots$ and converges to $x^\star$. Moreover, the following estimates hold for each $n = 0,1,2,\cdots$

$$\|y_{n+1} - y_{n+1}\| \leq q\|y_{n+1} - y_n\|$$

and

$$\|y_n - x^*\| \leq \frac{q^n\eta}{1-q}$$

where

$$q = 1 - \sqrt{1 - L_0\eta}.$$

The iterates x_n may belong in $\bar{U}(x_0, r)$ for some radius r. This radius r perhaps such that $r > r_0$. However the convergence will be quadratic. Below we present such a case.

Let us now define b, r, η_1, η_2 and η_0 by

$$b = \frac{2-(L+2L_0)r_0}{2-Lr_0}, \; r = \frac{\eta_0}{1-b}, \; \eta_1 = \frac{1}{2L_0}\left(1-\left(\frac{L}{L+2L_0}\right)^2\right), \; \eta_2 = \frac{1-\delta^2}{2L_0},$$

where

$$\delta = 1 - \frac{L + 3L_0 - \sqrt{L^2 + 9L_0^2 + 2L_0 L}}{L}$$

and

$$\eta_0 = \min\{\eta_1, \eta_2\}.$$

If

$$\eta < \eta_1$$

then

$$b \in (0,1) \text{ and } r > r_0.$$

Moreover, if

$$\eta < \eta_2$$

then

$$\frac{L_0 r_0}{2} \leq \frac{2 - (L + 2L_0) r_0}{2 - L r_0}.$$

Due to the choice of b, r and r_0 we get

$$\frac{L b^{k-1} r_0}{2(1 - L_0(1 + b^{k-1}) r_0)} \leq \frac{L r_0}{2(1 - L_0 r)} \leq b \text{ for each } k = 2, 3, \cdots.$$

Now, we present the following result.

Theorem 6.2.1 *Let $F : \mathcal{D} \subseteq X \to Y$ be Fréchet differentiable. Suppose that there exists $x_0 \in \mathcal{D}$ and parameters $L_0 > 0$, $L \geq L_0$, $\eta > 0$ such that*

$$F'(x_0)^{-1} \in Ł(Y, X), \tag{6.8}$$

$$\|F'(x_0)^{-1} F(x_0)\| \leq \eta, \tag{6.9}$$

$$\|F'(x_0)^{-1}[F'(x) - F'(x_0)]\| \leq L_0 \|x - x_0\| \text{ for each } x \in \mathcal{D}, \tag{6.10}$$

$$\|F'(x_0)^{-1}[F'(x) - F'(y)]\| \leq L \|x - y\| \text{ for each } x, y \in \mathcal{D}. \tag{6.11}$$

$$\eta < \eta_0 \tag{6.12}$$

$$U(x_0, r) \subseteq \mathcal{D}, \tag{6.13}$$

Then, the sequence $\{x_n\}$ generated by Newton's method (6.2) is well defined, remains in the open ball $U(x_0, r)$ for each $n = 0, 1, 2, \cdots$ and converges to a *unique solution* x^* *located in the closed ball* $\bar{U}(x_0,r)$ *of equation* $F(x)=0$. *Moreover the following estimates hold for each* $n=0,1,2,\cdots$

$$\|x_{n+1}-x^*\| \leq \frac{L}{2(1-L_0\|x_n-x_0\|)}\|x_n-x^*\|^2. \tag{6.14}$$

Proof: It follows from condition (6.12) that

$$h_0 = L_0\eta \leq \frac{1}{2} \tag{6.15}$$

which implies r_0 is well defined and equation $F(x)=0$ has a unique solution $x^\star \in \bar{U}(x_0,r_0)$.

By means of considering Newton's method we get

$$\begin{aligned} x_{n+1}-x^* &= x_n-x^*-F'(x_n)^{-1}F(x_n) \\ &= -F'(x_n)^{-1}\int_0^1 [F'(x^*+\theta(x_n-x^*))-F'(x_n)](x_n-x^*)d\theta. \end{aligned} \tag{6.16}$$

It is clear that $x_1=y_1\in\bar{U}(x_0,r_0)\subseteq\bar{U}(x_0,r)$. Now considering (6.11), (6.15) and (6.16) for $n=0$ that

$$\|x_1-x^*\| \leq \frac{L_0}{2}\|x_0-x^*\|^2 \leq \frac{L_0r_0}{2}\|x_0-x^*\| < b\|x_0-x^*\|,$$

$$\|x_2-x^*\| \leq \frac{L\|x_1-x^*\|^2}{2(1-L_0\|x_1-x_0\|)} \leq \frac{L\|x_1-x^*\|}{2(1-L_0\eta)}\|x_1-x^*\| \leq \frac{Lbr_0}{2(1-L_0\eta)} < b\|x_1-x^*\|,$$

$$\|x_2-x_0\| \leq \|x_2-x^*\|+\|x^*-x_0\| \leq b^2r_0+r_0 = (b^2+1)r_0 \leq (b+1)r_0 = \frac{1-b^2}{1-b}r_0 < r$$

and

$$\|F'(x_0)^{-1}(F'(x_1)-F'(x_0))\| \leq L_0\|x_1-x_0\| < 1.$$

Now, from the Banach Lemma on invertible operators [26] that the inverse exists $F'(x_1)^{-1}\in Ł(\mathcal{Y},\mathcal{X})$ and the following holds

$$\|F'(x_1)^{-1}F'(x_0)\| \leq \frac{1}{1-L_0\|x_1-x_0\|} \leq \frac{1}{1-L_0\eta}.$$

In a similar way, we get that $F'(x_k)^{-1}\in Ł(\mathcal{Y},\mathcal{X})$ and

$$\|F'(x_k)^{-1}F'(x_0)\| \leq \frac{1}{1-L_0\|x_k-x_0\|}.$$

Finally, for $k\geq 2$ we have

$$\begin{aligned}
\|x_{k+1}-x^*\| &\leq \frac{L\|x_k-x^*\|}{2(1-L_0\|x_k-x_0\|)}\|x_k-x^*\| \\
&\leq \frac{Lb^{k-1}r_0\|x_k-x^*\|}{2(1-L_0(\|x_k-x^*\|+\|x^*-x_0\|))} \\
&\leq \frac{Lb^{k-1}r_0\|x_k-x^*\|}{2(1-L_0(b^{k-1}+1)r_0)} \\
&\leq \frac{Lb^{k-1}r_0\|x_k-x^*\|}{2(1-L_0r_0)} \\
&< b\|x_k-x^*\|
\end{aligned}$$

and

$$\begin{aligned}
\|x_{k+1}-x_0\| &\leq \|x_{k+1}-x^*\|+\|x^*-x_0\| \\
&\leq b^k\|x^*-x_0\|+\|x^*-x_0\| \\
&\leq (b^k+1)\|x^*-x_0\| \\
&\leq (b+1)\|x^*-x_0\| \\
&\leq \frac{1-b^2}{1-b}r_0 \leq r_0.
\end{aligned}$$

As a consequence, $x_{k+1} \in U(x_0,r)$ and $\lim_{k\to\infty} x_{k+1} = x^*$. ⊠

Remark 6.2.2 *It follows from* (6.16) *that Newton's method converges quadratically to* $x^\star$, *r is given in closed form and* (6.12) *can be weaker than the Kantorovich hypothesis [26]*

$$\eta \leq \frac{1}{2L}$$

for sufficiently small L_0.

6.3 Second convergence improvement

In this section we provide another improvement of the convergence criteria for Newton's method (6.2) using (6.5) and a tight majorizing sequence for Newton's method.

Remark 6.3.1 *We have from the proof of semilocal convergence of Theorem 3.2 in [4] that sequence* $\{r_n\}$ *defined by*

$$r_0 = 0,\ r_1 = \eta,$$

$$r_2 = r_1 + \frac{b_1L_0(r_2-r_1)^2}{2(1-a_1L_0r_1)}$$

$$r_{n+2} = r_{n+1}\frac{b_{n+1}L(r_{n+1}-r_n)^2}{2(1-a_{n+1}L_0r_{n+1})} \text{ for each } n = 0,1,2,\cdots$$

where

$$a_{n+1} = \frac{\|x_{n+1} - x_0\|}{r_{n+1}}$$

and

$$b_{n+1} = \left(\frac{\|x_{n+1} - x_n\|}{r_{n+1} - r_n} \right)^2 \text{ for each } n = 0, 1, 2, \cdots$$

is a tighter majorizing sequence than $\{t_n\}$. *Therefore, it may converge under weaker convergence criteria than* $\{t_n\}$. *As an example, let us consider there exists* $a \in [0, 1]$ *and* $b \in [0, 1]$ *such that*

$$a_{n+1} \leq a \text{ and } b_{n+1} \leq b$$

for each $n = 0, 1, 2, \cdots$. *It is clear that one can always choose* $a = b = 1$.

We consider the sequence $\{s_n\}$ defined by

$$s_0 = 0, \; s_1 = \eta,$$

$$s_2 = s_1 + \frac{\bar{L}_0 (s_1 - s_0)^2}{2(1 - \bar{L}_0 s_1)}$$

$$s_{n+2} = s_{n+1} \frac{\bar{L}(s_{n+1} - s_n)^2}{2(1 - \bar{L}_0 s_{n+1})} \text{ for each } n = 0, 1, 2, \cdots$$

where

$$\bar{L} = bL \text{ and } \bar{L}_0 = aL_0.$$

Then, it is clear that $\{s_n\}$ is less tight than $\{r_n\}$ but tighter than $\{t_n\}$ and also still a majorizing sequence for sequence $\{x_n\}$ generated by Newton's method (6.2). Taking this fact into account we will find a weaker convergence criteria using $\bar{L}_0$ and $\bar{L}$ instead of L_0 and L. The new convergence criteria are weaker if $a \neq 1$ or $b \neq 1$. If $a = b = 1$ then, they are the same as the earlier ones. Now, setting

$$K = \max\{aL_0, bL\}$$

and define function

$$f(t) = \frac{K}{2} t^2 - t + \eta$$

and sequence $\{u_n\}$ by

$$u_0 = 0,$$

$$u_{n+1} = u_n - \frac{f(u_n)}{f'(u_n)} \text{ for each } n = 0, 1, 2, \cdots$$

we obtain the Kantorovich-type convergence criterion [26]

$$h = K\eta \leq \frac{1}{2}$$

which is at least as weak as the classical Kantorovich criterion given by (6.4), since

$$K \leq L.$$

Furthermore, it is clear that the classical error bounds in the literature for this setting [3, 5, 6] are also improved if $K < L$.

References

[1] S. Amat, S. Busquier and M. Negra, Adaptive approximation of nonlinear operators, *Numer. Funct. Anal. Optim.*, 25, 397–405, 2004.

[2] I. K. Argyros, Computational theory of iterative methods. Series: Studies in Computational Mathematics, 15, Editors: C. K. Chui and L. Wuytack, Elsevier Publ. Co. New York, U.S.A., 2007.

[3] I. K. Argyros, Y. J. Cho and S. Hilout, Numerical method for equations and its applications. CRC Press/Taylor and Francis, New York, 2012.

[4] I. K. Argyros and S. Hilout, Weaker conditions for the convergence of Newton's method, *J. Complexity*, 28, 364–387, 2012.

[5] I. K. Argyros, A unifying local-semilocal convergence analysis and applications for two-point Newton-like methods in Banach space, *J. Math. Anal. Appl.*, 298, 374–397, 2004.

[6] I. K. Argyros, On the Newton-Kantorovich hypothesis for solving equations, *J. Comput. Appl. Math.*, 169, 315–332, 2004.

[7] I. K. Argyros, Concerning the "terra incognita" between convergence regions of two Newton methods, *Nonlinear Anal.*, 62, 179–194, 2005.

[8] I. K. Argyros, Approximating solutions of equations using Newton's method with a modified Newton's method iterate as a starting point, *Rev. Anal. Numér. Théor. Approx.*, 36 (2), 123–138, 2007.

[9] I. K. Argyros, S. George and Á. A. Magreñán, Local convergence for multi-point-parametric Chebyshev-Halley-type methods of high convergence order, *J. Comp. App. Math.*, 282, 215–224, 2015.

[10] I. K. Argyros, On a class of Newton-like methods for solving nonlinear equations, *J. Comput. Appl. Math.*, 228, 115–122, 2009.

[11] I. K. Argyros, A semilocal convergence analysis for directional Newton methods, *Math. Comput.*, 80, 327–343, 2011.

[12] I. K. Argyros, D. González and Á. A. Magreñán, A semilocal convergence for a uniparametric family of efficient Secant-like methods. *J. Func. Spaces*, 2014.

[13] I. K. Argyros and Á. A. Magreñán, A unified convergence analysis for Secant-type methods. *J. Korean Math. Soc.*, 51 (6), 1155–1175, 2014.

[14] I. K. Argyros, Y. J. Cho and S. Hilout, Numerical method for equations and its applications. CRC Press/Taylor and Francis, New York, 2012.

[15] I. K. Argyros and S. Hilout, Enclosing roots of polynomial equations and their applications to iterative processes, *Surveys Math. Appl.*, 4, 119–132, 2009.

[16] I. K. Argyros and S. Hilout, Extending the Newton-Kantorovich hypothesis for solving equations, *J. Comput. Appl. Math.*, 234, 2993–3006, 2010.

[17] I. K. Argyros and S. Hilout, Computational methods in nonlinear analysis, World Scientific Pub. Comp., New Jersey, 2013.

[18] W. Bi, Q. Wu and H. Ren, Convergence ball and error analysis of Ostrowski-Traub's method, *Appl. Math. J. Chinese Univ. Ser. B*, 25, 374–378, 2010.

[19] E. Cătinaş, The inexact, inexact perturbed, and quasi-Newton methods are equivalent models, *Math. Comp.*, 74 (249), 291–301, 2005.

[20] P. Deuflhard, Newton methods for nonlinear problems. Affine invariance and adaptive algorithms, Springer Series in Computational Mathematics, 35, Springer-Verlag, Berlin, 2004.

[21] J. A. Ezquerro, J. M. Gutiérrez, M. A. Hernández, N. Romero and M. J. Rubio, The Newton method: from Newton to Kantorovich (Spanish), *Gac. R. Soc. Mat. Esp.*, 13 (1), 53–76, 2010.

[22] J. A. Ezquerro and M. A. Hernández, On the *R*-order of convergence of Newton's method under mild differentiability conditions, *J. Comput. Appl. Math.*, 197(1), 53–61, 2006.

[23] J. A. Ezquerro and M. A. Hernández, An improvement of the region of accessibility of Chebyshev's method from Newton's method, *Math. Comp.*, 78(267), 1613–1627, 2009.

[24] J. A. Ezquerro, M. A. Hernández and N. Romero, Newton-type methods of high order and domains of semilocal and global convergence, *Appl. Math. Comput.*, 214 (1), 142–154, 2009.

[25] M. A. Hernández, A modification of the classical Kantorovich conditions for Newton's method, *J. Comp. Appl. Math.*, 137, 201–205, 2001.

[26] L. V. Kantorovich and G. P. Akilov, Functional Analysis, Pergamon Press, Oxford, 1982.

[27] W. Lorenzo, R. G. Crespo and A. Castillo, A prototype for linear features generalization, *Int. J. Interac. Multim. Artif. Intell.*, 1(3), 60–66, 2010.

[28] P. D. Proinov, General local convergence theory for a class of iterative processes and its applications to Newton's method, *J. Complexity*, 25, 38–62, 2009.

[29] P. D. Proinov, New general convergence theory for iterative processes and its applications to Newton-Kantorovich type theorems, *J. Complexity*, 26, 3–42,2010.

[30] H. Ren and Q. Wu, Convergence ball of a modified secant method with convergence order $1.839\ldots$, *Appl. Math. Comput.*, 188, 281–285, 2007.

[31] B. Royo, J. A. Sicilia, M. J. Oliveros and E. Larrodé, Solving a long-distance routing problem using ant colony optimization. *Appl. Math.*, 9 (2L), 415–421, 2015.

[32] J. A. Sicilia, C. Quemada, B. Royo and D. Escuín, An optimization algorithm for solving the rich vehicle routing problem based on variable neighborhood search and tabu search metaheuristics. *J. Comput. Appl. Math.*, 291, 468–477, 2016.

[33] J. A. Sicilia, D. Escuín, B. Royo, E. Larrodé and J. Medrano, A hybrid algorithm for solving the general vehicle routing problem in the case of the urban freight distribution. In Computer-based Modelling and Optimization in Transportation (pp. 463–475). Springer International Publishing, 2014.

[34] J. F. Traub, Iterative methods for the solution of equations, Prentice-Hall Series in Automatic Computation, Englewood Cliffs, N. J., 1964.

[35] Q. Wu and H. Ren, A note on some new iterative methods with third-order convergence, *Appl. Math. Comput.*, 188, 1790–1793, 2007.

Chapter 7

Robust convergence for inexact Newton method

Ioannis K. Argyros

Department of Mathematical Sciences, Cameron University, Lawton, OK 73505, USA, Email: iargyros@cameron.edu

Á. Alberto Magreñán

Universidad Internacional de La Rioja, Escuela Superior de Ingeniería y Tecnología, 26002 Logroño, La Rioja, Spain, Email: alberto.magrenan@unir.net

CONTENTS

7.1 Introduction

The main task this chapter is to use the iterative methods to find solutions x^* of the equation

$$F(x) = 0, \tag{7.1}$$

where $D : D \subset X \to Y$ is a Fréchet-differentiable operator X, Y are Banach spaces and $D \subset X$. Many problems from Applied Sciences can be solved finding the solutions of equations in a form like (7.1) [9, 12, 14, 17, 18, 23, 25, 28–34].

The most well-known and used iterative method is Newton's method due to its simplicity and goodness

$$x_{n+1} = x_n - F'(x_n)^{-1}F(x_n) \qquad \text{for each} \quad n = 0,1,2,\cdots. \tag{7.2}$$

where, $F'(x)$ denotes the Fréchet-derivative of F at $x \in \mathcal{D}$.

Concerning the semilocal convergence of Newton's method, one of the most important results is the famous Kantorovich theorem [24] for solving nonlinear equations. This theorem provides a simple and transparent convergence criterion for operators with bounded second derivatives F'' or the Lipschitz continuous first derivatives. Each Newton iteration requires the solution of a linear system which may be very expensive to compute. The linear systems are solved inexactly in floating point computations. Therefore, it is very important to work with an error tolerance on the residual. The convergence domain of Newton-like methods used to be very small in general. So, it is one of the main purpose of researcher in area is to enlarge the domain without additional hypotheses. In particular, using the elegant work by Ferreira and Svaiter [22] and optimization considerations, we will expand this domain. In this chapter, the semilocal convergence analysis depends on a scalar majorant function. Relevant work in this field can be found in [10–22].

The main idea is that we introduce a center-majorant function (see (7.3)) that can be used to provide more precise information on the upper bounds of $\| F'(x)^{-1}F'(x_0) \|$ than in [22], where the less precise majorant function is used (see (7.4)). This modification provides a flexibility that leads to more precise error estimates on the distances $\| x_{n+1} - x_k \|$ and $\| x_n - x^\star \|$ for each $n = 1,2,\cdots$ under the same hypotheses and computational cost as in [22]. Furthermore, we show that these advantages can even be obtained under weaker convergence criteria in some interesting special cases.

In order to make the paper as self contained as possible we present results taken from other studies such as [12,22,23] for continuity without the proofs. We only provide the proofs of the results related to our new center-majorizing function.

The chapter is organized as follows: In Section 7.2 we present some results on convex functions. The main result is presented in Section 7.3. Finally, some special cases and applications are provided in Section 7.4.

7.2 Standard results on convex functions

Let us denote $U(x,R)$ and $\overline{U}(x,R)$, respectively, for the open and closed ball in the Banach $\mathcal{X}$ with center $x \in X$ and radius $R > 0$. Let also denote $\mathcal{L}(\mathcal{X},\mathcal{Y})$ for the space of bounded linear operators from the Banach space $\mathcal{X}$ into the Banach space $\mathcal{Y}$.

We need the following auxiliary results of elementary convex analysis.

Proposition 7.2.1 *(cf. [23]) Let $I \subset \mathbb{R}$ be an interval and $\varphi : I \to \mathbb{R}$ be convex.*

1. *For each $u_0 \in int(I)$, the function*

$$u \to \frac{\varphi(u_0) - \varphi(u)}{u_0 - u}, u \in I, u \neq u_0,$$

is increasing and there exist (in $\mathbb{R}$)

$$D^- \varphi(u_0) = \lim_{u \to u^-} \frac{\varphi(u_0) - \varphi(u)}{u_0 - u} = \sup_{u < u_0} \frac{\varphi(u_0) - \varphi(u)}{u_0 - u}.$$

2. *If $u, v, w \in I, u < w$ and $u \leq v \leq w$ then*

$$\varphi(v) - \varphi(u) \leq [\varphi(w) - \varphi(u)] \frac{v-u}{w-u}.$$

Proposition 7.2.2 *([22]) If $h : [a,b) \to \mathbb{R}$ is convex, differentiable at a, $h'(a) < 0$ and*

$$\lim_{t \to b-} h(t) = 0,$$

then

$$a - \frac{h(a)}{h'(a)} \leq b,$$

with equality if and only if h is affine in $[a,b)$.

7.3 Semilocal convergence

We present the main semilocal convergence result as follows.

Theorem 7.3.1 *Let $\mathcal{X}$ and $\mathcal{Y}$ be Banach spaces, $R \in \mathbb{R}$, $D \subseteq X$ and $F : D \to \mathcal{Y}$ be continuous and continuously differentiable on $int(D)$. Let $x_0 \in int(D)$ with $F'(x_0)^{-1} \in \mathcal{L}(\mathcal{Y}, \mathcal{X})$. Suppose that there exist $f_0 : [0,R) \to \mathbb{R}$ and $f : [0,R) \to \mathbb{R}$ continuously differentiable, such that $U(x_0,R) \subseteq D$,*

$$\|F'(x_0)^{-1} [F'(x) - F'(x_0)] \| \leq f_0'(\|x - x_0\|) - f_0'(0), \tag{7.3}$$

for each $x \in U(x_0,R)$,

$$\|F'(x_0)^{-1} [F'(y) - F'(x)] \| \leq f'(\|y - x\| + \|x - x_0\|) - f'(\|x - x_0\|), \tag{7.4}$$

for each $x, y \in U(x_0,R)$, $\|x - x_0\| + \|y - x\| < R$ and

$$\|F'(x_0)^{-1} F(x_0)\| \leq f(0), \tag{7.5}$$

(C_1) *$f_0'(0) = -1$, $f(0) > 0$, $f'(0) = -1$, $f_0(t) \leq f(t)$, $f_0'(t) \leq f'(t)$ for each $t \in [0,R)$;*

(C_2) f_0', f' *are strictly increasing and convex;*

(C_3) $f(t) < 0$ *for some* $t \in (0,R)$.

Let

$$\beta := \sup_{t\in[0,R)} -f(t), \quad t_* := \min f^{-1}(\{0\}), \quad \bar{\tau} := \sup\{t \in [0,R) : f(t) < 0\}.$$

Choose $0 \leq \rho < \beta/2$. *Define*

$$\kappa_\rho := \sup_{\rho<t<R} \frac{-(f(t)+2\rho)}{|f_0'(\rho)|(t-\rho)}, \; \lambda_\rho := \sup\{t \in [\rho,R) : \kappa\rho + f'(t) < 0\}, \; \Theta_\rho := \frac{\kappa\rho}{2-\kappa\rho}. \tag{7.6}$$

Then for any $R \in [0,\Theta_\rho]$ *and* $z_0 \in U(x_0,\rho)$, *the sequence* $\{z_n\}$ *generated by the inexact Newton method with initial point* z_0 *and residual relative error tolerance* θ *for* $k = 0,1,\ldots$, $z_{n+1} = z_n + S_n$,

$$\|F'(z_0)^{-1}[F(z_n)+F'(z_n)S_n]\| \leq \theta\|F'(z_0)^{-1}F(z_n)\|, \tag{7.7}$$

is well defined (for any particular choice of each S_n*),*

$$\|F'(z_0)^{-1}F(z_n)\| \leq \left(\frac{1+\theta^2}{2}\right)^n [f(0)+2\rho], k = 0,1,\ldots, \tag{7.8}$$

the sequence z_n *is contained in* $U(z_0,\lambda_\rho)$ *and converges to a point* $x_* \in \bar{U}(x_0,t_*)$, *which is the unique zero of* F *on* $U(x_0,\bar{\tau})$. *Moreover, if*

(C_4) $\lambda_\rho < R-\rho$,

then the sequence $\{z_n\}$ *satisfies, for* $k = 0,1,\ldots$,

$$\|x_* - z_{n+1}\| \leq \left[\frac{1+\theta}{2}\frac{D^- f_0'(\lambda_\rho)}{|f_0'(\lambda_\rho)|}\|x_* - z_n\| + \theta\frac{f_0'(\lambda_\rho+\rho)+2|f_0'(\rho)|}{|f_0'(\lambda_\rho+\rho)|}\right]\|x_* - z_n\|.$$

If, additionally, $0 \leq \theta < \kappa\rho/(4+\kappa\rho)$, *then* $\{z_n\}$ *converges Q-linearly as follows*

$$\|x_* - z_{n+1}\| \leq \kappa\left[\frac{1+\theta}{2}+\frac{2\theta}{\kappa_\rho}\right]\|x_* - z_n\|, \quad k = 0,1,\ldots.$$

Remark 7.3.2 **(a)** If $f_0' = f'$, then, Theorem 7.3.1 reduces to the corresponding one [22, Theorem 2.1]. On the other hand, if $f_0' < f'$ it constitutes a clear improvement, since $\frac{1}{|f_0'(t)|} < \frac{1}{|f'(t)|}$. So our theorem is at least as good as the previous one.

(b) If one choose $\theta = 0$ in Theorem 7.3.1 we have the exact Newton method and its convergence properties. If $\theta = \theta_n$ in each iteration and letting θ_n goes to zero as n goes to infinity, the penultimate inequality of the Theorem 7.3.1 implies that the generated sequence converges to the solution with superlinear rate.

Here after we assume that the hypotheses of Theorem 7.3.1 hold. The scalar function f in the above theorem is called a majorant function for F at point x_0, where as the function f_0 is the center-majorant function for F at point x_0.

Next, we analyze some basic properties of functions f_0 and f. Condition (C_2) implies strict convexity of f and f_0. It is worth noticing that t_* is the smallest root of $f(t) = 0$ and, since f is convex, if the equation has another one, this second root is $\bar{\tau}$.

Define

$$\bar{t} := \sup\{t \in [0,R) : f'(t) < 0\}. \tag{7.9}$$

Proposition 7.3.3 *(cf. [22]) The following statements on the majorant function hold*

i) $f'(t) < 0$ *for any* $t \in [0,\bar{t})$, *(and* $f'(t) \geq 0$ *for any* $t \in [0,R)\ [0,\bar{t})$*);*

ii) $0 < t_* < \bar{t} \leq \bar{\tau} \leq R$;

iii) $\beta = -\lim_{t \to \bar{t}^-} f(t), 0 < \beta < \bar{t}.$

Now, we will prove a special case of Theorem 7.3.1. This case is $\rho = 0$ and $z_0 = x_0$. As in [22], in order to simplify the notation in the case $\rho = 0$, we will use κ, λ and θ instead of κ_0, λ_0 and θ_0 respectively:

$$\kappa := \sup_{0<t<R} \frac{-f(t)}{t}, \quad \lambda := \sup\{t \in [0,R) : \kappa + f'(t) < 0\}, \quad \Theta := \frac{\kappa}{2-\kappa}. \tag{7.10}$$

Proposition 7.3.4 *(cf. [22]) For κ, λ, θ as in (7.9) it holds that*

$$0 < \kappa < 1, \quad 0 < \Theta < 1, \quad t_* < \lambda \leq \bar{t} \leq \bar{\tau}, \tag{7.11}$$

and

$$f'(t) + \kappa < 0, \forall t \in [0,\lambda), \quad \inf_{0 \leq t < R} f(t) + \kappa t = \lim_{t \to \lambda^-} f(t) + \kappa t = 0, \tag{7.12}$$

Firstly, we need a Banach-type result for invertible operators. We will use the following one that appears in [24].

Proposition 7.3.5 *If* $\|x - x_0\| \leq t < \bar{t}$ *then* $F'(x)^{-1} \in \mathcal{L}(\mathcal{Y},\mathcal{X})$ *and*

$$\|F'(x)^{-1}F'(x_0)\| \leq \frac{1}{-f_0'(t)}.$$

Proof: We have by (7.9) that $f_0'(t) < 0$, since $f_0'(t) \leq f'(t)$. Using (7.3) and (C_1) we get

$$\begin{aligned}\|F'(x_0)^{-1}F'(x) - I\| = \|F'(x_0)^{-1}[F'(x) - F'(x_0)]\| &\leq f_0'(\|x - x_0\|) - f_0'(0) \\ &= f_0'(t) + 1 < 1.\end{aligned}$$

Using Banach Lemma [24] and the preceding inequality above we conclude that $F'(x)^{-1} \in \mathcal{L}(\mathcal{Y}, \mathcal{X})$ and

$$\|F'(x)^{-1}F'(x_0)\| = \|(F'(x_0)^{-1}F'(x))^{-1}\| \leq \frac{1}{1 - (f_0'(t) + 1)} = -\frac{1}{f_0'(t)}.$$

□

Notice that the corresponding result in [22] uses f instead of f_0 to obtain

$$\|F'(x)^{-1}F'(x_0)\| \leq -\frac{1}{f'(t)}.$$

But

$$-\frac{1}{f_0'(t)} \leq -\frac{1}{f'(t)}.$$

Therefore, if $f_0'(t) < f'(t)$, then our result is more precise and less expensive to arrive at. Furthermore, this modification influences error bounds and the sufficient convergence criteria.

The linearization errors on F and f are given respectively by

$$E_F(y, x) := F(y) - [F(x) + F'(x)(y - x)], \quad x \in U(x_0, R), \quad y \in D \tag{7.13}$$

$$e_f(v, t) := f(v) - [f(t) + f'(t)(v - t)], \quad t, v \in [0, R). \tag{7.14}$$

The linearization error of the majorant function bounded the linearization error of F.

Lemma 7.3.6 *(cf. [22]) If $x, y \in \mathcal{X}$ and $\|x - x_0\| + \|y - x\| < R$ then*

$$\|F'(x_0)^{-1}E_F(y, x)\| \leq e_f(\|x - x_0\| + \|y - x\|, \|x - x_0\|).$$

Convexity of f and f' guarantee that $e_f(t + s, t)$ is increasing in s and t.

Lemma 7.3.7 *(cf. [22]) If $0 \leq b \leq t, 0 \leq a \leq s$ and $t + s < R$ then*

$$e_f(a + b, b) \leq e_f(t + s, t),$$

$$e_f(a + b, b) \leq \frac{1}{2}\frac{f'(t + s) - f'(t)}{s}a^2, s \neq 0.$$

Next, we bound the linearization error E_F using the linearization error on the majorant function.

Corollary 7.3.8 *(cf. [22]) If $x, y \in \mathcal{X}$, $\|x - x_0\| \le t$, $\|y - x\| \le s$ and $s + t < R$ then*

$$\|F'(x_0)^{-1} E_F(y,x)\| \le e_f(t+s,t),$$

$$\|F'(x_0)^{-1} E_F(y,x)\| \le \frac{1}{2}\frac{f'(s+t) - f'(t)}{s}\|y - x\|^2, s \neq 0.$$

In the following result, it is worth noticing to say that the first inequality is very useful to obtain the asymptotic bounds by the inexact Newton method and the second one will be used to show the robustness of the method with respect to the initial iterate.

Corollary 7.3.9 *([22]) For each $y \in U(x_0, R)$,*

$$-f(\|y - x_0\|) \le \|F'(x_0)^{-1} F(y)\| \le f(\|y - x_0\|) + 2\|y - x_0\|.$$

From the first inequality showed above this corollary proves that F has no zeroes in the region $t_* < \|x - x_0\| < \bar{\tau}$.

Lemma 7.3.10 *If $x \in \mathcal{X}$, $\|x - x_0\| \le t < R$ then*

$$\|F'(x_0)^{-1} F'(x)\| \le 2 + f_0'(t).$$

Proof: From (7.3) and (C_1) we obtain

$$\begin{aligned} \|F'(x_0)^{-1} F'(x)\| &\le I + F'(x_0)^{-1}[F'(x) - F'(x_0)] \\ &\le 1 + f_0'(\|x - x_0\|) - f_0'(0) = 2 + f_0'(t). \end{aligned} \tag{7.15}$$

□

Lemma 7.3.11 *Take $\theta = 0$, $0 \le t \le \lambda$, $x_*, x, y \in \mathcal{X}$. If $\lambda < R$, $\|x - x_0\| \le t$, $\|x_* - x\| \le \lambda - t$, $F(x_*) = 0$ and*

$$\|F'(x_0)^{-1}[F(x) + F'(x)(y - x)]\| \le \theta \|F'(x_0)^{-1} F(x)\| \tag{7.16}$$

then

$$\|x_* - y\| \le \left[\frac{1+\theta}{2} + \frac{2\theta}{\kappa}\right]\|x_* - x\|, \tag{7.17}$$

$$\|x_* - y\| \le \left[\frac{1+\theta}{2}\frac{D^- f'(\lambda)}{|f_0'(\lambda)|}\|x_* - x\| + \theta\frac{2 + f_0'(\lambda)}{|f_0'(\lambda)|}\right]\|x_* - x\|. \tag{7.18}$$

Proof: In view of (7.13) and $F(x^*) = 0$, we get in turn that

$$y - x_* = F'(x)^{-1}\left[E_F(x_*, x) + [F(x) + F'(x)(y - x)]\right].$$

and

$$-F'(x_0)^{-1} F(x) = F'(x_0)^{-1}\left[E_F(x_*, x) + F'(x)(x_* - x)\right].$$

It follows from (7.16) in the preceding identities that,

$$\|x_* - y\| \leq \|F'(x)^{-1}F'(x_0)\| \left[\|F'(x_0)^{-1}E_F(x_*,x)\| + \theta\|F(x_0)^{-1}F(x)\|\right]$$

and

$$\|F'(x_0)^{-1}F(x)\| \leq \|F'(x_0)^{-1}E_F(x_*,x)\| + \|F'(x_0)^{-1}F'(x)\|\|x_* - x\|.$$

Combining two above inequalities with Proposition 7.3.5, Corollary 7.3.8 with $y = x_*$ and $s = \lambda - t$ and Lemma 7.3.10 we have that

$$\|x_* - y\| \leq \frac{1}{|f_0'(t)|}\left[\frac{1+\theta}{2}\frac{f'(\lambda)-f'(t)}{\lambda - t}\|x_* - x\| + \theta[2 + f_0'(t)]\right]\|x_* - x\|.$$

Since $\|x_* - x\| \leq \lambda - t$, $f' < -\kappa < 0$ in $[0,\lambda)$ and f_0', f' are increasing the first inequality follows from last inequality. Using Proposition 7.2.1 and taking in account that $f_0' \leq f' < 0$ in $[0,\lambda)$ and increasing we obtain the second inequality from above inequality. □

In the following result we present a single inexact Newton iteration with relative error θ.

Lemma 7.3.12 *Take t, ε, $\theta = 0$ and $x \in D$ such that*

$$\|x - x_0\| \leq t < \bar{t}, \|F'(x_0)^{-1}F(x)\| \leq f(t) + \varepsilon, t - (1+\theta)\frac{f(t)+\varepsilon}{f_0'(t)} < R. \quad (7.19)$$

If $y \in \mathcal{X}$ and

$$\|F'(x_0)^{-1}[F(x) + F'(x)(y - x)]\| \leq \theta\|F'(x_0)^{-1}F(x)\|, \quad (7.20)$$

then the following estimates hold

1. $\|y - x\| \leq -(1+\theta)\dfrac{f(t)+\varepsilon}{f_0'(t)}$;
2. $\|y - x_0\| \leq t - (1+\theta)\dfrac{f(t)+\varepsilon}{f_0'(t)} < R$;
3. $\|F'(x_0)^{-1}F(y)\| \leq f\left(t - (1+\theta)\dfrac{f(t)+\varepsilon}{f_0'(t)}\right) + \varepsilon + 2\theta(f(t)+\varepsilon)$.

Proof: From 7.3.5 and the first inequality in (7.19) we conclude that $F'(x)^{-1} \in \mathcal{L}(\mathcal{Y},\mathcal{X})$ and $\|F'(x)^{-1}F'(x_0)\| \leq -1/f_0'(t)$. Therefore, using also triangular inequality, (7.20) and the identity

$$y - x = F'(x)^{-1}F'(x_0)\left[F'(x_0)^{-1}[F(x) + F'(x)(y-x)] - F'(x_0)^{-1}F(x)\right],$$

we conclude that

$$\|y-x\| \leq \frac{-1}{f_0'(t)}(1+\theta)\|F'(x_0)^{-1}F(x)\|.$$

To end the proof of the first item, use the second inequality on (7.19) and the above inequality. The second item follows from the first one, the first and the third inequalities in (7.19) and triangular inequality. Using the definition of the error (7.13) we have that

$$F(y) = E_F(y,x) + F'(x_0)\left[F'(x_0)^{-1}[F(x)+F'(x)(y-x)]\right].$$

Hence, using (7.20), the second inequality on (7.19) and the triangle inequality we have that

$$\begin{aligned}\|F'(x_0)^{-1}F(y)\| &\leq \|F'(x_0)^{-1}E_F(y,x)\| + \theta\|F'(x_0)^{-1}F(x)\| \\ &\leq \|F'(x_0)^{-1}E_F(y,x)\| + \theta(f(t)+\varepsilon).\end{aligned}$$

Using the first item, (7.19) and Lemma 7.3.6 with $s = -(1+\theta)(f(t)+\varepsilon)/f_0'(t)$ we get

$$\begin{aligned}\|F'(x_0)^{-1}E_F(y,x)\| &\leq e_f\left(t-(1+\theta)\frac{f(t)+\varepsilon}{f_0'(t)},t\right) \\ &= f\left(t-(1+\theta)\frac{f(t)+\varepsilon}{f_0'(t)}\right)+\varepsilon+\theta(f(t)+\varepsilon).\end{aligned}$$

Using the two above equation we obtain the latter inequality in the third item. □

In view of Lemma 7.3.12, define, for $\theta = 0$, the auxiliary map $n_\theta : [0,\bar{t}) \times [0,8) \to \mathbb{R}\times\mathbb{R}$,

$$n_\theta(t,\varepsilon) := \left(t-(1+\theta)\frac{f(t)+\varepsilon}{f_0'(t)}, \varepsilon+2\theta(f(t)+\varepsilon)\right). \tag{7.21}$$

Let

$$\Omega := \{(t,\varepsilon)\in\mathbb{R}\times\mathbb{R} : 0\leq t<\lambda, 0\leq\varepsilon\leq\kappa t, 0<f(t)+\varepsilon\}. \tag{7.22}$$

Lemma 7.3.13 *If* $0\leq\theta\leq\Theta$, $(t,\varepsilon)\in\Omega$ *and* $(t_+,\varepsilon_+) = n_\theta(t,\varepsilon)$, *that is,*

$$t_+ = t-(1+\theta)\frac{f(t)+\varepsilon}{f_0'(t)}, \qquad \varepsilon_+ = \varepsilon+2\theta(f(t)+\varepsilon),$$

then $n_\theta(t,\varepsilon)\in\Omega$, $t<t_+$, $\varepsilon\leq\varepsilon_+$ *and*

$$f(t+)+\varepsilon+ < \left(\frac{1+\theta^2}{2}\right)(f(t)+\varepsilon).$$

Proof: Since $0 \leq t < \lambda$, according to (7.10) we have $f'(t) < -\kappa < 0$. Therefore $t < t_+$ and $\varepsilon \leq \varepsilon_+$.

As $\varepsilon \leq \kappa t, f(t)+\varepsilon > 0$ and $-1 \leq f'(t) < f'(t)+\kappa < 0$,

$$\begin{aligned} -\frac{f(t)+\varepsilon}{f_0'(t)} &\leq -\frac{f(t)+\kappa t}{f_0'(t)} \\ &= -\frac{f(t)+\kappa t}{f_0'(t)+\kappa}\left[1+\frac{\kappa}{f_0'(t)}\right] \leq -\frac{f(t)+\kappa t}{f_0'(t)+\kappa}(1-\kappa) \end{aligned} \tag{7.23}$$

The function $h(s) := f(s)+\kappa s$ is differentiable at the point t, $h'(t) < 0$, is strictly convex and

$$\lim_{s\to\lambda^-} h(t) = 0.$$

Hence, using Proposition 7.2.2, we have $t - h(t)/h'(t) < \lambda$, which is equivalent to

$$-\frac{f(t)+\kappa t}{f_0'(t)+\kappa} < \lambda - t. \tag{7.24}$$

Using the above inequality with (7.23) and the definition of t_+ we conclude that

$$t+ < t+(1+\theta)(1-\kappa)(\lambda-t).$$

Using (7.10) and (7.12), we have $(1+\theta)(1-\kappa) \leq 1-\theta < 1$, which combined with the above inequality yields $t_+ < \lambda$. Using (7.10), the definition of ε_+ and inequality $\varepsilon \leq \kappa t$ we obtain

$$\begin{aligned} \varepsilon_+ &\leq 2\theta(f(t)+\varepsilon)+\kappa t \\ &= \kappa(t+(1+\theta)(f(t)+\varepsilon)). \end{aligned}$$

Taking into account the inequalities $f(t)+\varepsilon > 0$ and $-1 \leq f'(t) < 0$, we have that

$$f(t)+\varepsilon \leq -\frac{f(t)+\varepsilon}{f_0'(t)}.$$

Using the two above inequalities with the definition of t_+ we obtain $\varepsilon_+ \leq \kappa t_+$.

In order to prove the two last inequalities, note that using the definition (7.14), we get

$$\begin{aligned} f(t_+)+\varepsilon_+ &= f(t)+f'(t)(t+-t)+ef(t+,t)+2\theta(f(t)+\varepsilon)+\varepsilon \\ &= \theta(f(t)+\varepsilon)+e_f(t_+,t) \\ &= \theta(f(t)+\varepsilon)+\int_t^{t_+}(f'(u)-f'(t))\,du. \end{aligned}$$

Taking into account that $f'' > 0$ for every point we conclude that the integral is positive. So, last equality implies that $f(t_+)+\varepsilon_+ = \theta(f(t)+\varepsilon) > 0$. Taking $s \in [t_+,\lambda)$ and using the convexity of f', we have that

$$\begin{aligned}\int_t^{t_+}(f'(u)-f'(t))\,du &\leq \int_t^{t_+}(f'(s)-f'(u))\frac{u-t}{s-t}du\\ &= \frac{1}{2}\frac{(t_+-t)^2}{s-t}(f'(s)-f'(t)).\end{aligned}$$

If we now substitute last inequality into above equation, we have that

$$\begin{aligned}f(t_+)+\varepsilon_+ &\leq \theta(f(t)+\varepsilon)+\frac{1}{2}\frac{(t_+-t)^2}{s-t}(f'(s)-f'(t))\\ &= \left(\theta+\frac{1}{2}\frac{(1+\theta)^2}{(s-t)}\frac{f(t)+\varepsilon}{-f_0'(t)}\frac{f'(s)-f'(t)}{-f_0'(t)}\right)(f(t)+\varepsilon).\end{aligned}$$

On the other hand, due to $f'(s)+\kappa < 0$ and $-1 \leq f'(t)$ we can conclude that

$$\frac{f'(s)-f'(t)}{-f_0'(t)} = \frac{f'(s)+\kappa-f'(t)-\kappa}{-f_0'(t)} \leq 1 \quad \kappa.$$

Using (7.23), (7.24) and the last two above inequalities and taking in account that $(1+\theta)(1-\kappa) \leq 1-\theta$ we conclude that

$$\begin{aligned}f(t_+)+\varepsilon_+ &\leq \left(\theta+\frac{1}{2}(1+\theta)^2(1-\kappa)^2\frac{\lambda-t}{s-t}\right)(f(t)+\varepsilon)\\ &= \left(\theta+\frac{1}{2}(1-\theta)^2\frac{\lambda-t}{s-t}\right)(f(t)+\varepsilon),\end{aligned}$$

and the result follows taking the limit $s \to \lambda_-$. □

Therefore, we shall deal with a *family* of mappings, describing all possible inexact iterations.

Definition 7.3.14 (cf. [22]) For $0 \leq \theta$, N_θ is the family of maps $N_\theta : U(x_0,\bar{t}) \to \mathcal{X}$ such that

$$\|F'(x_0)^{-1}[F(x)+F'(x)(N_\theta(x)-x)]\| \leq \theta\|F'(x_0)^{-1}F(x)\|, \tag{7.25}$$

for each $x \in U(x_0,\bar{t})$.

If $x \in U(x_0,\bar{t})$, then $F'(x)^{-1} \in \mathcal{L}(\mathcal{Y},\mathcal{X})$. Hence, for $\theta = 0$, N_0 has a single element, namely the exact Newton iteration map given by

$$N_0 : U(x_0,\bar{t}) \to X, x \mapsto N_0(x) = x - F'(x)^{-1}F(x).$$

Easily, if $0 \leq \theta \leq \theta'$ then $N_0 \subset N_\theta \subset N_{\theta'}$. Therefore, N_θ is non-empty for all $\theta = 0$.

Remark 7.3.15 (cf. [22]) For any $\theta \in (0,1)$ and $N_\theta \in \mathcal{N}_\theta$

$$N_\theta(x) = x \Leftrightarrow F(x) = 0, \quad x \in U(x_0, \bar{t}).$$

In other words, the fixed points of the exact Newton iteration and the fixed points of the inexact Newton iteration N_θ and the zeros of F are the same.

Using a combination of both Lemmas 7.3.12 and 7.3.13 we found the main tool for the analysis of the inexact Newton method. Define

$$K(t,\varepsilon) := \{x \in \mathcal{X} : \|x - x_0\| \leq t, \|F'(x_0)^{-1}F(x)\| \leq f(t) + \varepsilon\}, \tag{7.26}$$

and

$$K := \bigcup_{(t,\varepsilon)\in\Omega} K(t,\varepsilon). \tag{7.27}$$

Where n_θ and N_θ were defined in (7.21), (7.22) and Definition 3.14 respectively.

Proposition 7.3.16 *(cf. [22]) Take $0 \leq \theta \leq \Theta$ and $N_\theta \in \mathcal{N}_\theta$. Then for any $(t,\varepsilon) \in \Omega$ and $x \in K(t,\varepsilon)$*

$$N_\theta(K(t,\varepsilon)) \subset K(n_\theta(t,\varepsilon)) \subset K, \quad \|N_\theta(x) - x\| \leq t_+ - t,$$

where t_+ is the first component of $n_\theta(t,\varepsilon)$. Moreover,

$$n_\theta(\Omega) \subset \Omega, \quad N_\theta(K) \subset K. \tag{7.28}$$

Proof: Using definitions (7.21), (7.22), Definition 3.14, (7.26), (7.27) with Lemmas 7.3.12 and 7.3.13. □

Theorem 7.3.17 *Take $0 \leq \theta \leq \Omega$ and $N_\theta \in \mathcal{N}_\theta$. For any $(t_0,\varepsilon_0) \in \Omega$ and $y_0 \in K(t_0,\varepsilon_0)$, the sequences*

$$y_{n+1} = N_\theta(y_n),\ (t_{n+1},\varepsilon_{n+1}) = n_\theta(t_n,\varepsilon_n),\ k = 0,1,\ldots, \tag{7.29}$$

are well defined,

$$y_n \in K(t_n,\varepsilon_n),\ (t_n,\varepsilon_n) \in \Omega,\ k = 0,1,\ldots, \tag{7.30}$$

the sequence $\{t_n\}$ is strictly increasing and converges to some $(0,\lambda]$, the sequence $\{\varepsilon_n\}$ is non-decreasing and converges to some $\varepsilon \in [0,\kappa\lambda]$,

$$\|F'(x_0)^{-1}F(y_n)\| \leq f(t_n) + \varepsilon_n\| \leq \left(\frac{1+\theta^2}{2}\right)^n (f(t_0) + \varepsilon_0), \quad k = 0,1,\ldots, \tag{7.31}$$

the sequence $\{y_n\}$ is contained in $U(x_0,\lambda)$ and converges to a point $x_* \in \bar{U}(x_0,t_*)$ which is the unique zero of F in $U(x_0,\bar{\tau})$ and

$$\|y_{n+1} - y_n\| \leq t_{n+1} - t_n, \|x_* - y_n\| \leq \tilde{t} - t_n, \quad k = 0,1,\ldots. \tag{7.32}$$

Moreover, if the following condition holds

(C_4') $\lambda < R$,

then the sequence $\{y_n\}$ *satisfies, for* $k = 0, 1, \ldots$

$$\|x_* - y_{n+1}\| \leq \left[\frac{1+\theta}{2} + \frac{D^- f'(\lambda)}{|f_0'(\lambda)|}\|x_* - y_n\| + \theta\,\frac{2 + f_0'(\lambda)}{|f_0'(\lambda)|}\right]\|x_* - y_n\|. \quad (7.33)$$

Hence, if $0 \leq \theta < \kappa/(4+\kappa)$, *then* $\{y_n\}$ *converges Q-linearly as follows*

$$\|x_* - y_{n+1}\| \leq \left[\frac{1+\theta}{2} + \frac{2\theta}{\kappa}\right]\|x_* - y_n\|, \quad k = 0, 1, \ldots. \quad (7.34)$$

Proof: From the assumptions on θ, (t_0, ε_0), y_0 and the two last inclusions on Proposition 7.3.16 it is clear that the sequences $\{(t_n, \varepsilon_n)\}$ and $\{y_n\}$ as defined in (7.29) are well defined. Furthermore, since (7.30) holds for $k = 0$, using the first inclusion in Proposition 7.3.5 and an inductive argument on k, we conclude that (7.30) holds for all k. From Proposition 7.3.5, (7.30) and (7.29) while the first inequality in (7.31) follows from (7.30) and the definition of $K(t, \varepsilon)$ in (7.26) follows the first inequality in (7.32). Using the definition of in (7.22) shows that

$$\Omega \subset [0, \lambda) \times [0, \kappa\lambda).$$

Hence, using (7.30) and the definition of $K(t, \varepsilon)$, we have $t_n \in [0, \lambda)$, $\varepsilon_n \in [0, \kappa\lambda)$, $y_n \in U(x_0, \lambda)$, $k = 0, 1, \ldots$.

Using Lemma 7.3.13 and (7.22) we conclude that $\{t_n\}$ is strictly increasing, $\{\varepsilon_n\}$ is non-decreasing and the second equality in (7.31) holds for all k. Hence, in view of the first two above inclusions, $\{t_n\}$ and $\{\varepsilon_n\}$ converge, respectively, to some $\widetilde{t} \in (0, \lambda]$ and $\widetilde{\varepsilon} \in [0, \kappa\lambda]$. Convergence to $\widetilde{t}$, together with the first inequality in (7.32) and the inclusion $y_n \in U(x_0, \lambda)$ implies that y_n converges to some $x_* \in \bar{U}(0, \lambda)$ and that the second inequality on (7.32) holds for all k.

Using the inclusion $y_n \in U(x_0, \lambda)$, the first inequality in Corollary 7.3.9 and (7.31), we have that

$$-f(\|y_n - x_0\|) \leq \left(\frac{1+\theta^2}{2}\right)^n (f(t_0) + \varepsilon_0), \quad k = 0, 1, \ldots.$$

According to (7.12), $f' < -\kappa$ in $[0, \lambda)$. Therefore, since $f(t_*) = 0$ and $t_* < \lambda$,

$$f(t) \leq -\kappa(t - t_*), \quad t_* \leq t < \lambda.$$

Hence, if $\|y_n - x_0\| = t_*$, we can combine the two above inequalities, setting $t = \|y_n - x_0\|$ in the second, to obtain

$$\|y_n - x_0\| - t_* \leq \left(\frac{1+\theta^2}{2}\right)^n \frac{f(t_0) + \varepsilon_0}{\kappa}.$$

Note that the above inequality remain valid even if $\|y_n - x_0\| < t_*$. Thus, taking the limit $k \to \infty$ in the above inequality we conclude that $\|x_* - x_0\| \leq t_*$.

Furthermore, now that we know that x_* is in the interior of the domain of F, we can conclude that $F(x_*) = 0$. The classical version of Kantorovich's theorem on Newtons method for a generic majorant function (cf. [6]) guarantee that under the assumptions of Theorem 7.3.1, F has a unique zero in $U(x_0, \bar{\tau})$. Therefore x_* must be this zero of F. To prove (7.33) and (7.34), note that we have $\|y_n - x_0\| \leq t_n$, for all $k = 0, 1, \ldots$. Now, we obtain from second inequality in (7.32) that $\|x_* - y_n\| \leq \lambda - t_n$, for all $k = 0, 1, \ldots$. Hence, using (C_4'), $F(x_*) = 0$ and the first equality in (7.29), applying the Lemma 7.3.11 we arrive at the desire inequalities. In order to conclude the proof, note that for $0 \leq \theta < \kappa/(4+\kappa)$ the quantity in the bracket in (7.34) is less than one, which implies that the sequence $\{y_n\}$ converges Q-linearly. □

Proposition 7.3.18 *If* $0 \leq \rho < \beta/2$ *then*

$$\rho < \bar{t}/2 < \bar{t}, f'(\rho) < 0.$$

Proof: Assumption $\rho < \beta/2$ and Proposition 7.3.3-item (*iii*) proves the first two inequalities of the proposition. The last inequality follows from the first inequality and Proposition 7.3.3-item (*i*). □

Now, using all the preceding results we can prove Theorem 7.3.1.

Proof: (Proof of Theorem 7.3.1) We will begin with the case $\rho = 0$ and $z_0 = x_0$. Note that, from the definition in (7.10), we have that

$$\kappa_0 = \kappa, \lambda_0 = \lambda, \Theta_0 = \Theta.$$

Since

$$(0,0) \in \Omega, x_0 \in K(0,0),$$

using Theorem 7.3.17 we conclude that Theorem 7.3.1 holds for $\rho = 0$. For proving the general case, take

$$0 \leq \rho < \beta/2, z_0 \in \bar{U}(x_0, \rho). \tag{7.35}$$

Using Proposition 7.3.18 and (7.9) we conclude that $\rho < \bar{t}/2$ and $f'(\rho) < 0$. Define

$$g : [0, R-\rho) \to \mathbb{R}, \quad g(t) = \frac{-1}{f_0'(\rho)}[f(t+\rho) + 2\rho]. \tag{7.36}$$

We claim that g is a majorant function for F at point z_0. Trivially, $U(z_0, R-\rho) \subset D$, $g'(0) = -1$, $g(0) > 0$. Moreover g' is also convex and strictly increasing. To end the proof that g satisfies C_1, C_2 and C_3, using Proposition 7.3.3-item (*iii*) and second inequality in (7.35), we have that

$$\lim_{t \to \bar{t} - \rho} g(t) = \frac{-1}{f_0'(\rho)}(2\rho - \beta) < 0.$$

Using Proposition 7.2.1, we obtain that

$$\|F'(z_0)^{-1}F'(x_0)\| \le \frac{-1}{f_0'(\rho)}. \tag{7.37}$$

Hence, using also the second inequality of Corollary 7.3.9, we get

$$\begin{aligned}\|F'(z_0)^{-1}F(z_0)\| &\le \|F'(z_0)^{-1}F'(x_0)\|\|F'(x_0)^{-1}F(z_0)\| \\ &\le \frac{-1}{f_0'(\rho)}\left[f(\|z_0-x_0\|)+2\|z_0-x_0\|\right].\end{aligned}$$

As $f' \ge -1$, the function $t \mapsto f(t)+2t$ is (strictly) increasing. Combining this fact with the above inequality and (7.36) we conclude that

$$\|F'(z_0)^{-1}F'(z_0)\| \le g(0).$$

To end the proof that g is a majorant function for F at z_0, take $x, y \in \mathcal{X}$ such that

$$x, y \in U(z_0, R-\rho), \quad \|x-z_0\|+\|y-x\| < R-\rho.$$

Therefore $x, y \in U(x_0, R)$, $\|x-x_0\|+\|y-x\| < R$ and using (7.37) together with (7.3), we obtain that

$$\begin{aligned}\|F'(z_0)^{-1}\left[F'(y)-F'(x)\right]\| &\le \|F'(z_0)^{-1}F'(x_0)\|\,\|F'(x_0)^{-1}\left[F'(y)-F'(x)\right]\| \\ &\le \frac{-1}{f_0'(\rho)}\left[f'(\|y-x\|+\|x-x_0\|)-f'(\|x-x_0\|)\right].\end{aligned}$$

Since f' is convex, the function $t \mapsto f'(s+t)-f'(s)$ is increasing for $s=0$ and $\|x-x_0\| \le \|x-z_0\|+\|z_0-x_0\| \le \|x-z_0\|+\rho$,

$$f'(\|y-x\|+\|x-x_0\|)-f'(\|x-x_0\|) \le f'(\|y-x\|+\|x-z_0\|+\rho)-f'(\|x-z_0\|+\rho).$$

Combining the two above inequalities with the definition of g, we obtain

$$\|F'(z_0)^{-1}\left[F'(y)-F'(x)\right]\| \le g'(\|y-x\|+\|x-z_0\|)-g'(\|x-z_0\|).$$

Note that for κ_ρ, λ_ρ and Θ_ρ as defined in (7.5), we have that

$$\kappa_\rho = \sup_{0<t<R-\rho} \frac{-g(t)}{t}, \quad \lambda_\rho = \sup\{t \in [0, R-\rho) : \kappa\rho + g'(t) < 0\}, \quad \Theta_\rho = \frac{\kappa\rho}{2-\kappa\rho},$$

which are the same as (7.5) taking g instead of f. Hence, applying Theorem 7.3.1 for F and the majorant function g at point z_0 and $\rho = 0$, we conclude that the sequence z_n is well defined, remains in $U(z_0, \lambda_\rho)$, satisfies (7.8) and converges to some $z_* \in \bar{U}(z_0, t_*, \rho)$ which is a zero of F, where $t_{*,\rho}$ is the smallest solution of $g(t)=0$. Using (7.36) we conclude that $t_{*,\rho}$ is the smallest solution of

$$f(\rho+t)+2\rho = 0.$$

Hence, in view of item *ii* of Proposition 7.3.3, we have $\rho + t_{*,\rho} < \bar{t} \leq \bar{\tau}$ and $\bar{U}(z_0, t_*, \rho) \subset U(x_0, \bar{\tau})$. Hence, z_* is the unique zero of F in $U(x_0, \bar{\tau})$, which we already called x_*. Since

$$g'(t) = \frac{f'(t+\rho)}{|f_0'(\rho)|}, \; D^- g'(t) = \frac{D^- f'(t+\rho)}{|f_0'(\rho)|}, \; t \in [0, R-\rho),$$

applying again Theorem 7.3.1 for F and the majorant function g at point z_0 and $\rho = 0$, we conclude that item (C_4) also holds. □

7.4 Special cases and applications

We present some specializations of Theorem 7.3.1 to some interesting cases such as Kantorovich's theory (cf. [2,6,9,11,24,29,37]) and even Smale's α-theory or Wang's γ-theory (cf. [9,12,13,15,19–21,33,35]).

Theorem 7.4.1 *Let $\mathcal{X}$ and $\mathcal{Y}$ be Banach spaces, $D \subseteq \mathcal{X}$ and $F : D \to \mathcal{Y}$ Fréchet-differentiable on $int(D)$. Take $x_0 \in int(D)$ with $F'(x_0)^{-1} \in \mathcal{L}(\mathcal{Y}, \mathcal{X})$*

Suppose that $U(x_0, 1/\gamma) \subset D$, $b > 0$, $\gamma_0 > 0$, $\gamma > 0$ with $\gamma_0 \leq \gamma$ and that

$$\|F'(x_0)^{-1}F(x_0)\| \leq \beta, \quad \alpha = \beta\gamma < 3 - 2\sqrt{2}, \quad 0 \leq \theta \leq \frac{1 - 2\sqrt{\alpha} - \alpha}{1 + 2\sqrt{\alpha} + \alpha}.$$

$$\|F'(x_0)^{-1}[F(x) - F(y)]\| \leq \frac{2\gamma}{(1-\gamma r)^3} \|x - y\|,$$
$$\text{for each} \quad x, y \in U(x_0, r), \; 0 < r < \frac{1}{\gamma}.$$

$$\|F'(x_0)^{-1}[F(x) - F(x_0)]\| \leq \frac{\gamma_0(2 - \gamma_0 r)}{(1-\gamma_0 r)^2} \|x - x_0\|,$$
$$\text{for each} \quad x \in U(x_0, r), \; 0 < r < \frac{1}{\gamma}.$$

Then, the sequence generated by the inexact Newton method for solving $F(x) = 0$ with starting point x_0 and residual relative error tolerance θ: For $k = 0, 1, \ldots,$

$$x_{n+1} = x_n + S_n, \quad \|F'(x_0)^{-1}\left[F(x_n) + F'(x_n)S_n\right]\| \leq \theta \|F'(x_0)^{-1}F(x_n)\|,$$

is well defined, the generated sequence $\{x_n\}$ converges to a point x_* which is a zero of F,

$$\|F'(x_0)^{-1}F(x_n)\| \leq \left(\frac{1+\theta^2}{2}\right)^n \beta, \quad k = 0, 1, \ldots,$$

the sequence $\{x_n\}$ is contained in $U(x_0, \lambda)$, $x_* \in \bar{U}(x_0, t_*)$ and x_* is the unique zero of F in $U(x_0, \bar{\tau})$, where

$$\lambda := \frac{\beta}{\sqrt{\alpha} + \alpha},$$

$$t_* = \frac{1+\alpha-\sqrt{1-6\alpha+\alpha^2}}{4}, \quad \bar{\tau} = \frac{1+\alpha+\sqrt{1-6\alpha+\alpha^2}}{4}.$$

On the other hand, the sequence $\{x_n\}$ *satisfies, for* $k = 0, 1, \ldots,$

$$\|x_* - x_{k+1}\| \leq \left[\frac{1+\theta}{2} \frac{D^- f'(\lambda)}{|f_0'(\lambda)|} \|x_* - x_n\| + \theta \frac{f_0'(\lambda)+2}{|f_0'(\lambda)|} \right] \|x_* - x_n\|.$$

If, additionally, $0 \leq \theta < (1-2\sqrt{\alpha}-\alpha)/(5-2\sqrt{\alpha}-\alpha)$, *then* $\{x_n\}$ *converges Q-linearly as follows*

$$\|x_* - x_{n+1}\| \leq \left[\frac{1+\theta}{2} + \frac{2\theta}{1-2\sqrt{\alpha}-\alpha} \right] \|x_* - x_n\|, \quad k = 0, 1, \ldots.$$

Proof: The function $f : [0, 1/\gamma) \to \mathbb{R}$

$$f(t) = \frac{t}{1-\gamma t} - 2t + \beta,$$

is a majorant function for F in x_0 (cf. [20]). Therefore, all results follow from Theorem 7.3.1, applied to this particular context by choosing $f_0 : [0, 1/\gamma) \to \mathbb{R}$ such that

$$f_0'(t) = -2 + \frac{1}{(1-\gamma_0 t)^2}.$$

□

Notice that if F is analytic on $U(x_0, 1/\gamma)$ we can choose

$$\gamma_0 = \sup_{k>1} \left\| \frac{F'(x_0)^{-1} F^{(k)}(x_0)}{k!} \right\|^{\frac{1}{k-1}},$$

(cf. [1, 11, 12, 16, 37]). A semilocal convergence result for Newton method is instrumental in the complexity analysis of linear and quadratic minimization problems by means of self-concordant functions (cf. [23, 28]). Also in this setting, Theorem 7.3.1 provides a semilocal convergence result for Newton method with a relative error tolerance.

Theorem 7.4.2 *Let* $\mathcal{X}$ *and* $\mathcal{Y}$ *be a Banach spaces,* $D \subseteq \mathcal{X}$ *and* $F : D \to \mathcal{Y}$ *Fréchet-differentiable on* $int(D)$. *Take* $x_0 \in int(D)$ *with* $F'(x_0)^{-1} \in \mathcal{L}(\mathcal{Y}, \mathcal{X})$. *Suppose that exist constants* $L_0 > 0$, $L > 0$ *with* $L_0 \leq L$ *and* $\beta > 0$ *such that* $\beta L < 1/2$, $U(x_0, 1/L) \subset D$ *and*

$$\|F'(x_0)^{-1} [F'(x) - F'(x_0)]\| \leq L_0 \|x - x_0\|, \quad x \in U(x_0, 1/L),$$

$$\|F'(x_0)^{-1} [F'(y) - F'(x)]\| \leq L \|x - y\|, \quad x, y \in U(x_0, 1/L),$$

$$\|F'(x_0)^{-1} F(x_0)\| \leq \beta, \quad 0 \leq \theta \leq \frac{1-\sqrt{2\beta L}}{1+\sqrt{2\beta L}}.$$

Then, the sequence generated by the inexact Newton method for solving $F(x)=0$ with starting point x_0 and residual relative error tolerance θ: For $k=0,1,\ldots,$

$$x_{n+1}=x_n+S_n, \quad \|F'(x_0)^{-1}[F(x_n)+F'(x_n)S_n]\| \leq \theta\|F'(x_0)^{-1}F(x_n)\|,$$

is well defined,

$$\|F'(x_0)^{-1}F(x_n)\| \leq \left(\frac{1+\theta^2}{2}\right)^n \beta, k=0,1,\ldots.$$

the sequence $\{x_n\}$ is contained in $U(x_0,\lambda)$, converges to a point $x_* \in \bar{U}(x_0,t_*)$ which is the unique zero of F in $U(x_0,1/L)$ where

$$\lambda := \frac{\sqrt{2\beta L}}{L}, \quad t_* = \frac{1-\sqrt{1-2L\beta}}{L}.$$

Moreover, the sequence $\{x_n\}$ satisfies, for $k=0,1,\ldots$, $\delta=\frac{L_0}{L}$

$$\|x_*-z_{n+1}\| \leq \left[\frac{1+\theta}{2}\frac{L}{1-\delta\sqrt{2\beta L}}\|x_*-x_n\|+\theta\frac{1+\delta\sqrt{2\beta L}}{1-\delta\sqrt{2\beta L}}\right]\|x_*-x_n\|.$$

If, additionally, $0\leq\theta<(1-\sqrt{2\beta L})/(5-\sqrt{2\beta L})$, then the sequence $\{x_n\}$ converges Q-linearly as follows

$$\|x_*-x_{n+1}\| \leq \left[\frac{1+\theta}{2}+\frac{2\theta}{1-\sqrt{2\beta L}}\right]\|x_*-x_n\|, \quad k=0,1,\ldots.$$

Proof: The function $f:[0,1/L)\to\mathbb{R}$,

$$f(t) := \frac{L}{2}t^2-t+\beta,$$

is a majorant function for F at point x_0. Then, all result follow from Theorem 7.3.1, applied to this particular context by choosing $f_0:[0,\frac{1}{L})\to\mathbb{R}$ such that $f_0'(t)=L_0t-1$. □

Remark 7.4.3 **(a)** If $f_0=f$ all preceding results reduce to the corresponding ones in [22]. If $f_0<f$ the earlier results are improved, since the error bounds on the distances are more precise. Moreover, these advantages are obtained under the same convergence criteria and computational cost, since function f_0 is a special case of f.

(b) Let us show that obtaining a even weaken criteria is possible in the special case when $\theta=0$. That is when, $\theta_n=0$ for each $n=0,1,2\ldots$. Then, we have Newton's method defined by (7.2).

Using Proposition 7.2.1 and Corollary 7.3.9 for $f(t) = \frac{L}{2}t^2 - t + \beta$ and $f_0'(t) = L_0 t - 1$ we deduce that

$$\|x_{n+1} - x_n\| \leq \frac{\bar{L}\|x_{n+1} - x_n\|^2}{2(1 - L_0\|x_{n+1} - x_0\|)} \text{ for each } n = 0, 1, 2, \ldots,$$

where

$$\bar{L} = \begin{cases} L_0 & \text{if} \quad n = 1 \\ L & \text{if} \quad k > 1 \end{cases}$$

Define scalar sequence $\{r_n\}$ by

$$\begin{aligned} r_0 = 0, r_1 = \beta, r_2 &= r_1 + \frac{L_0(r_1 - r_0)^2}{2(1 - L_0 r_1)}, \\ r_{n+1} &= r_n + \frac{L(r_n - r_{n-1})^2}{2(1 - L_0 r_n)} \quad \text{for each } n = 2, 3, \ldots. \end{aligned}$$

Then, we have that (cf. [11])

$$\|x_{n+1} - x_n\| \leq r_{n+1} - r_n \text{ for each } n = 0, 1, 2, \ldots.$$

It was shown in [11] that $\{r_n\}$ is a majorizing sequence for $\{x_n\}$ and converges provided that

$$H_1 = L_1\beta \leq \frac{1}{2},$$

where

$$L_1 = \frac{1}{8}\left(4L_0 + \sqrt{L_0 L} + \sqrt{L_0 L + 8L_0^2}\right).$$

The corresponding results deduced from Theorem 7.3.1 (with $f_0 = f$) are given by

$$\|x_{n+1} - x_n\| \leq \frac{L\|x_{n+1} - x_n\|^2}{2(1 - L_0\|x_{n+1} - x_0\|)} \text{ for each } n = 0, 1, 2, \ldots$$

and the corresponding majorizing sequence $\{v_n\}$ is defined by

$$v_0 = 0,\ v_{n+1} = v_n + \frac{L(v_n - v_{n-1})^2}{2(1 - L_0 v_n)} = v_n - \frac{f(v_n)}{f'(v_n)} \text{ for each } n = 1, 2, \ldots.$$

The sufficient convergence criterion for sequence $\{v_n\}$ is given by the famous for its simplicity and clarity Kantorovich hypothesis (cf. [24])

$$h = L\beta \leq \frac{1}{2}.$$

Notice that

$$h \leq \frac{1}{2} \Rightarrow h_1 \leq \frac{1}{2}$$

but not necessarily vice versa unless if $L_0 = L$.

We also have that $L_0 \leq L$ holds in general and L/L_0 can be arbitrarily large (cf. [11–13]). Moreover, an inductive argument shows

$$r_n \leq v_n$$

$$r_{n+1} - r_n \leq v_{n+1} - v_n$$

and

$$r_* = \lim_{k \to \infty} r_n \leq v_* = \lim_{k \to \infty} v_n.$$

The inequality in the first two preceding inequalities is strict if $L_0 < L$ for each $n = 2, 3, \ldots$. Finally, notice that

$$\frac{h_1}{h} \to 0 \text{ as } \frac{L_0}{L} \to 0.$$

The preceding estimate shows by how many times at most the applicability of Newton's method is expanded with our approach. Examples where $L_0 < L$ and where the old convergence criteria are not satisfied but the new convergence criteria are applied can be found for example in [11].

Notice that the Lipschitz conditions in the last two theorems imply the center-Lipschitz conditions with $\gamma_0 \leq \gamma$ and $L_0 \leq L$. That is the center-Lipschitz conditions are not additional hypotheses to the Lipschitz conditions.

(c) Another important consideration is that in case of modified inexact Newton method defined by

$$\bar{x}_{n+1} = \bar{x}_n + S_n, \; \|F'(x_0)^{-1}\left[F(\bar{x}_n) + F'(x_0)S_n\right]\| \leq \theta \|F'(x_0)^{-1}F(x_n)\|$$

is used, then f_0 can replace f in all preceding results. The resulting convergence criteria are weaker but the rate of convergence is only linear. However, we can start with the slower method until a certain finite step $N \geq 1$ such that the convergence criteria of Theorem 7.3.1 hold for $x_0 = \bar{x}_N$. Then, we can continue with the faster method. Such an approach has been given by us in [11–13] for Newton's method (7.2). Notice that the approach in [22] shall use f instead of f_0. That means, e.g., in the case of Newton's method the sufficient convergence criterion shall be

$$H = L\beta \leq \frac{1}{2}$$

for the modified Newton's method whereas in our case it will be

$$H_0 = L_0\beta \leq \frac{1}{2}.$$

Notice again that

$$H \leq \frac{1}{2} \Rightarrow H_0 \leq \frac{1}{2} \quad \text{and} \quad \frac{H_0}{H} \to 0 \text{ as } \frac{L_0}{L} \to 0.$$

Finally, the error bounds are also more precise in this case (cf. [5, 11]).

References

[1] F. Alvarez, J. Bolte and J. Munier, A unifying local convergence result for Newton's method in Riemannian manifolds. *Found. Comput. Math.*, 8, 197–226, 2008.

[2] S. Amat, C. Bermúdez, S. Busquier, M. J. Legaz and S. Plaza, On a family of high-order iterative methods under Kantorovich conditions and some applications. *Abstr. Appl. Anal.*, Art. ID 782170, 14 pages, 2012.

[3] S. Amat, S. Busquier and J. M. Gutiérrez, Geometric constructions of iterative functions to solve nonlinear equations. *J. Comput. Appl. Math.*, 157, 197–205, 2003.

[4] I. K. Argyros, Forcing sequences and inexact Newton iterates in Banach space. *Appl. Math. Let.*, 13, 69–75, 2000.

[5] I. K. Argyros, Approximating solutions of equations using Newton's method with a modified Newton's method iterate as a starting point. *Rev. Anal. Numer. Theor. Approx.*, 36, 123–138, 2007.

[6] I. K. Argyros, Computational theory of iterative methods. Series: Studies in Computational Mathematics, 15, Editors: C. K. Chui and L. Wuytack, Elsevier Publ. Co. New York, U.S.A., 2007.

[7] I. K. Argyros, On the semilocal convergence of inexact Newton methods in Banach spaces, *J. Comput. Appl. Math.*, 228, 434–443, 2009.

[8] I. K. Argyros, Y. J. Cho and S. Hilout, On the local convergence analysis of inexact Gauss–Newton–like methods. *Panamer. Math. J.*, 21, 11–18, 2011.

[9] I. K. Argyros, Y. J. Cho and S. Hilout, Numerical methods for equations and its applications, CRC Press/Taylor and Francis Group, New York, 2012.

[10] I. K. Argyros and S. Hilout, Improved local convergence of Newton's method under weak majorant condition *J. Comput. Appl. Math.*, 236, 1892–1902, 2012.

[11] I. K. Argyros and S. Hilout, Weaker conditions for the convergence of Newton's method. *J. Complexity*, 28, 364–387, 2012.

[12] I. K. Argyros and S. Hilout, Computational Methods in Nonlinear Analysis: Efficient Algorithms, Fixed Point Theory and Applications, World Scientific, London, 2013.

[13] I. K. Argyros and S. Hilout, Improved local convergence analysis of inexact Gauss-Newton like methods under the majorant condition in Banach spaces. *J. Franklin Inst.*, 350, 1531–1544, 2013.

[14] L. Blum, F. Cucker, M. Shub and S. Smale, Complexity and Real Computation, Springer-Verlag, New York, 1998.

[15] J. Chen and W. Li, Convergence behaviour of inexact Newton methods under weak Lipschitz condition *J. Comput. Appl. Math.*, 191, 143–164, 2006.

[16] F. Cianciaruso, Convergence of Newton–Kantorovich approximations to an approximate zero. Numer. Funct. Anal. Optim., 28, 631–645 (2007).

[17] R. Dembo, S. C. Eisenstat and T. Steihaug, Inexact Newton methods. SIAM J. Numer. Anal., 19, 400–408 (1982).

[18] J. E. Jr. Dennis and R. B. Schnabel, Numerical methods for unconstrained optimization and nonlinear equations (Corrected reprint of the 1983 original), *Classics in Appl. Math. SIAM*, 16, Philadelphia, PA, 1996.

[19] O. P. Ferreira and B. F. Svaiter, Kantorovich's theorem on Newton's method in Riemannian manifolds, *J. Complexity*, 18, 304–329, 2002.

[20] O. P. Ferreira and B. F. Svaiter, Kantorovich's majorant principle for Newton's method, *Comput. Optim. Appl.*, 42, 213–229, 2009.

[21] O. P. Ferreira, M. L. N. Gonçalves and P. R. Oliveira, Local convergence analysis of inexact Gauss-Newton like methods under majorant condition, *J. Comput. Appl. Math.*, 236, 2487–2498, 2012.

[22] O. P. Ferreira and B. F. Svaiter, A robust Kantorovich's theorem on inexact Newton method with relative residual error tolerance. *J. Complexity*, 28, 346–363, 2012.

[23] J. B. Hiriart-Urruty and C. Lemaréchal, Convex analysis and minimization algorithms (two volumes). I. Fundamentals. II. Advanced theory and bundle methods, 305 and 306, Springer–Verlag, Berlin, 1993.

[24] L. V. Kantorovich and G. P. Akilov, Functional Analysis, Pergamon Press, Oxford, 1982.

[25] W. Lorenzo, R. G. Crespo and A. Castillo, A prototype for linear features generalization, *Int. J. Interac. Multim. Artif. Intell.*, 1(3), 60–66, 2010.

[26] I. Moret, A Kantorovich–type theorem for inexact Newton methods, *Numer. Funct. Anal. Optim.*, 10, 351–365, 1989.

[27] B. Morini, Convergence behaviour of inexact Newton methods. Math. Comp., 68, 1605–1613, 1999.

[28] Y. Nesterov and A. Nemirovskii, Interior–point polynomial algorithms in convex programming, *SIAM Studies in Appl. Math.*, 13, Philadelphia, PA, 1994.

[29] F. A. Potra, The Kantorovich theorem and interior point methods, *Math. Program.*, 102, 47–70, 2005.

[30] B. Royo, J. A. Sicilia, M. J. Oliveros and E. Larrodé, Solving a long-distance routing problem using ant colony optimization. *Appl. Math.*, 9 (2L), 415–421, 2015.

[31] J. A. Sicilia, C. Quemada, B. Royo and D. Escuín, An optimization algorithm for solving the rich vehicle routing problem based on variable neighborhood search and tabu search metaheuristics. *J. Comput. Appl. Math.*, 291, 468–477, 2016.

[32] J. A. Sicilia, D. Escuín, B. Royo, E. Larrodé and J. Medrano, A hybrid algorithm for solving the general vehicle routing problem in the case of the urban freight distribution. In Computer-based Modelling and Optimization in Transportation (pp. 463–475). Springer International Publishing, 2014.

[33] S. Smale, Newton method estimates from data at one point. *The merging of disciplines: New Directions in Pure, Applied and Computational Mathematics (R. Ewing, K. Gross, C. Martin, eds.)*, Springer–Verlag, New York, 185–196, 1986.

[34] J. F. Traub, Iterative methods for the solution of equations, Englewood Cliffs, New Jersey: Prentice Hall, 1964.

[35] X. Wang, Convergence of Newton's and uniqueness of the solution of equations in Banach space, *IMA J. Numer. Anal.*, 20, 123–134, 2001.

[36] T. J. Ypma, Local convergence of inexact Newton methods, *SIAM J. Numer. Anal.*, 21, 583–590, 1984.

[37] P. P. Zabrejko and D. F. Nguen, The majorant method in the theory of Newton–Kantorovich approximations and the Ptăk error estimates, *Numer. Funct. Anal. Optim.*, 9, 671–684, 1987.

Chapter 8

Inexact Gauss-Newton-like method for least square problems

Ioannis K. Argyros

Department of Mathematical Sciences, Cameron University, Lawton, OK 73505, USA, Email: iargyros@cameron.edu

Á. Alberto Magreñán

Universidad Internacional de La Rioja, Escuela Superior de Ingeniería y Tecnología, 26002 Logroño, La Rioja, Spain, Email: alberto.magrenan@unir.net

CONTENTS

8.1 Introduction

In this chapter we are interested in locating a solution x^* of the nonlinear least squares problem:

$$\min G(x) := \frac{1}{2}F(x)^T F(x), \tag{8.1}$$

where F is Fréchet-differentiable defined on $\mathbb{R}^n$ with values in $\mathbb{R}^m$, $m \geq n$.

In Computational Sciences the practice of numerical analysis is essentially connected to variants of Newton's method [1]– [15]. In the present chapter we consider the local convergence of inexact Gauss-Newton method (IGNM) for solving problem (8.1).

In particular, we define the method as follows [5, 8, 10, 11]:
Let $U(x,r)$ and $\overline{U}(x,r)$ denote, respectively, for the open and closed ball in $\mathbb{R}^n$ with center x and radius $r > 0$. Denote by $L(\mathbb{R}^n, \mathbb{R}^m)$ the space of bounded linear operators from $\mathbb{R}^n$ into $\mathbb{R}^m$. For $k = 0$ step 1 until convergence do.
Find the step Δ_k which satisfies

$$B_k\Delta_k = -F'(x_k)^T F(x_k) + r_k, \text{ where } \frac{\|r_k\|}{\|F'(x_k)^T F(x_k)\|} \leq \eta_k. \tag{8.2}$$

Set $x_{k+1} = x_k + \Delta_k$.

Here, B_k is an $n \times n$ nonsingular matrix, and $\{\eta_k\}$ is a sequence of forcing terms such that $\eta_k \in [0,1]$ for each k.

The process is called inexact Gauss-Newton method (IGNM) if $B_k = F'(x_k)^T F(x_k)$ and it is called inexact Gauss-Newton-like method (IGNLM) if $B_k = B(x_k)$ approximates $F'(x_k)^T F(x_k)$. There is a plethora of convergence results for (IGNM) and (IGNLM) under various Lipschitz-type conditions [1]–[15]. Recently, studies have been focused on the analysis of stopping residual controls $\frac{\|r_k\|}{\|F(x_k)\|} \leq \eta_k$ and its effect on convergence properties [3, 5–8, 10, 11] by considering iterative form where a scaled residual control is performed at each iteration as follows:

For $k = 0$ step 1 until convergence do.
Find the step Δ_k which satisfies

$$B_k\Delta_k = -F(x_k) + r_k, \text{ where } \frac{\|P_k r_k\|}{\|P_k F(x_k)\|} \leq \theta_k. \tag{8.3}$$

Set $x_{k+1} = x_k + \Delta_k$.

Here, $P_k^{-1} \in L(\mathbb{R}^m, \mathbb{R}^n)$ for each k.

In this chapter, we are motivated by different studies [5–7] and optimization considerations. We show our advantages by considering a combination of weak Lipschitz and weak center Lipschitz conditions.

The chapter is organized as follows. In Section 8.2, we present some auxiliary results on Moore-Penrose inverses. The local convergence of (IGNM) and (IGNLM) is presented in Section 8.3. Numerical examples are presented in Section 8.4.

8.2 Auxiliary results

We denote as $\mathbb{R}^{m\times n}$ the set of all $m \times n$ matrices A. We also denote by $A^\dagger$ the Moore-Penrose inverse of matrix A [12, 13]. If A has full rank, i.e., if $rank(A) = \min\{m,n\} = n$, then $A^\dagger = (A^t A)^{-1}A^T$. Next, we present some standard Lemmas.

Lemma 8.2.1 *[12]. Suppose that* $A, E \in \mathbb{R}^{m\times n}$, $B = A + E$, $\|A^\dagger\|\|E\| < 1$, $rank(A) = rank(B)$, *then*

$$\|B^\dagger\| \leq \frac{\|A^\dagger\|}{1 - \|A^\dagger\|\|E\|}. \tag{8.4}$$

Moreover, if $rank(A) = rank(B) = \min\{m,n\}$, *then*

$$\|B^\dagger - A^\dagger\| \leq \frac{\sqrt{2}\|A^\dagger\|^2\|E\|}{1 - \|A^\dagger\|\|E\|}. \tag{8.5}$$

Lemma 8.2.2 *[12]. Suppose that* $A, E \in \mathbb{R}^{m\times n}$, $B = A + E$, $\|EA^\dagger\|\|E\| < 1$, $rank(A) = n$, *then* $rank(B) = n$.

Now, we present results involving weak Lipschitz condition ([15]) in the case $F'(x)^\dagger = F'(x)^{-1}$.

Lemma 8.2.3 *Suppose that* F *is continuously Fréchet-differentiable in* $U(x^\star, r)$, $F(x^\star) = 0$, $F'(x^\star)$ *has full rank:*

(i) If F' *satisfies the center-Lipschitz condition with* $\mathcal{L}_0$ *average:*

$$\|F'(x^\theta) - F'(x^\star)\| \leq \int_0^{\theta s(x)} \mathcal{L}_0(u)du \text{ for each } x \in U(x^\star, r),\ 0 \leq \theta \leq 1, \tag{8.6}$$

where $x^\theta = x^\star + \theta(x - x^\star)$, $s(x) = \|x - x^\star\|$ *and* $\mathcal{L}_0$ *is a positive integrable function. Then, for* $\beta = \|F'(x^\star)^\dagger\|$, *if*

$$\beta \int_0^{s(x)} \mathcal{L}_0(u)du < 1, \tag{8.7}$$

$$\int_0^1 \|F'(x^\theta) - F'(x^\star)\|s(x)d\theta \leq \int_0^{s(x)} \mathcal{L}_0(u)(s(x) - u)du. \tag{8.8}$$

and

$$\|F'(y)^\dagger F(x)\| \leq \frac{\|x - x^\star\| + \beta \int_0^{s(x)} \mathcal{L}_0(u)(s(x) - u)du}{1 - \beta \int_0^{s(y)} \mathcal{L}_0(u)du} \tag{8.9}$$

for each x *and* $y \in U(x^\star, r)$.

(ii) If F' satisfies the radius Lipschitz condition with the $\mathcal{L}$ average:

$$\|F'(x)-F'(x^\theta)\| \leq \int_{\theta s(x)}^{s(x)} \mathcal{L}(u)du \text{ for each } x \text{ and } y \in U(x^\star,r),\ 0\leq\theta\leq 1, \tag{8.10}$$

where $\mathcal{L}$ is a positive integrable function. Then,

$$\int_0^1 \|F'(x)-F'(x^\theta)\|s(x)d\theta \leq \int_0^{s(x)} \mathcal{L}(u)udu \tag{8.11}$$

and if (8.7) is satisfied,

$$\begin{aligned}\|F'(x)^\dagger F(x)\| &\leq s(x)+\frac{\beta\int_0^{s(x)}\mathcal{L}(u)udu-\beta\int_0^{s(x)}(\mathcal{L}(u)-\mathcal{L}_0(u)s(x)du}{1-\beta\int_0^{s(x)}\mathcal{L}_0(u)du}\\ &\leq s(x)+\frac{\beta\int_0^{s(x)}\mathcal{L}(u)udu}{1-\beta\int_0^{s(x)}\mathcal{L}_0(u)du}.\end{aligned} \tag{8.12}$$

Proof: (i) Simply use $\mathcal{L}_0$ instead of $\mathcal{L}$ in the proof of (ii) in Lemma 2.3 [5, Page 101].
(ii) Using (8.10), we get

$$\begin{aligned}\int_0^1 \|F'(x)-F'(x^\theta)\|s(x)d\theta &\leq \int_0^1\int_{\theta s(x)}^{s(x)}\mathcal{L}(u)dus(x)d\theta\\ &= \int_0^{s(x)}\mathcal{L}(u)udu,\end{aligned}$$

which shows (8.11). Moreover, since $F'(x^\star)$ has full rank, it follows from Lemmas 8.2.1, 8.2.2, 8.2.3 and

$$\|F'(x^\star)^\dagger\|\|F'(x)-F'(x^\star)\| \leq \beta\int_0^{s(x)}\mathcal{L}_0(u)du < 1 \text{ for each } x\in U(x^\star,r) \tag{8.13}$$

that $F'(x)$ has full rank and

$$\|[F'(x)^TF'(x)]^{-1}F'(x)^T\| \leq \frac{\beta}{1-\beta\int_0^{s(x)}\mathcal{L}_0(u)du} \text{ for each } x\in U(x^\star,r). \tag{8.14}$$

Notice that we have the estimates

$$F'(y)^\dagger F'(x^\theta) = I - F'(y)^\dagger(F'(y)-F'(x^\theta)) \tag{8.15}$$

and

$$\begin{aligned}F'(y)^\dagger F(x) &= F'(y)^\dagger(F(x)-F(x^\star))\\ &= \int_0^1 F'(y)^\dagger F'(x^\theta)d\theta(x-x^\star).\end{aligned} \tag{8.16}$$

Then, in view of (8.11), (8.14)-(8.16), we obtain

$$\begin{aligned}\|F'(x)^{\dagger}F(x)\| &= \|\int_0^1 (I-F'(x)^{\dagger}(F'(x)-F'(x^{\theta})))d\theta(x-x^{\star})\| \\ &\leq (1+\|F'(x)^{\dagger}\|\int_0^1 \|F'(x)-F'(x^{\theta})\|d\theta)\|x-x^{\star}\|\end{aligned}$$

which implies (8.12).

Remark 8.2.4 *If F' satisfies the radius condition with $\mathcal{L}$ average, then, F' satisfies the center Lipschitz condition with $\mathcal{L}_0$ average but not necessarily vice versa even if $\mathcal{L}_0 = \mathcal{L}$. Notice also that in general*

$$\mathcal{L}_0(u) \leq \mathcal{L}(u) \quad \text{for each} \quad u \in [0,R] \tag{8.17}$$

holds and $\frac{\mathcal{L}}{\mathcal{L}_0}$ can be arbitrarily large.

If $\mathcal{L}_0 = \mathcal{L}$, then, our results reduce to the corresponding ones in Lemma 8.2.3 [5]. In other case, if strict inequality holds in (8.17), then our estimates (8.7), (8.8), (8.9) and (8.12) are more precise than the corresponding ones obtained from the preceding ones for $\mathcal{L}_0 = \mathcal{L}$, see (3.1)-(3.4) given in [5]. This improvement is obtained under the same computational cost, since the computation of function $\mathcal{L}$ involves the computation of $\mathcal{L}_0$ as a special case.

Next, we complete this auxiliary section with another two results involving functions appearing in the convergence analysis that follows in the next section.

Lemma 8.2.5 *[15] Let*

$$\varphi(t) := \frac{1}{t^{\alpha}}\int_0^t \mathcal{L}(u)u^{\alpha-1}du, \ \alpha \geq 1, \ 0 \leq t \leq r, \tag{8.18}$$

where $\mathcal{L}$ is a positive integrable function and monotonically increasing in $[0,r]$. Then, function φ is nondecreasing for each α.

Lemma 8.2.6 *[15] Let*

$$\psi(t) := \frac{1}{t^2}\int_0^t \mathcal{L}(u)(\alpha t-u)du, \ \alpha \geq 1, \ 0 \leq t \leq r, \tag{8.19}$$

where $\mathcal{L}$ is as in Lemma 8.2.5. Then, function ψ is monotonically increasing.

8.3 Local convergence analysis

In this section, we present the main local convergence results for (IGNM) and (IGNLM).

Proof are omitted since are analogous to the corresponding ones in [5], where we replace old estimates

$$\beta \int_0^{s(x)} \mathcal{L}(u)du < 1, \tag{8.20}$$

$$\int_0^1 \|F'(x^\theta) - F'(x^\star)\| s(x) d\theta \leq \int_0^{s(x)} \mathcal{L}(u)(s(x)-u)du, \tag{8.21}$$

$$\|F'(y)^\dagger F(x)\| \leq \frac{\|x-x^\star\| + \beta \int_0^{s(x)} \mathcal{L}(u)(s(x)-u)du}{1-\beta \int_0^{s(y)} \mathcal{L}(u)du} \tag{8.22}$$

and

$$\|F(x)^\dagger F(x)\| \leq s(x0 + \frac{\beta \int_0^{s(x)} \mathcal{L}(u)du}{1-\beta \int_0^{s(x)} \mathcal{L}(u)du} \tag{8.23}$$

by the new and more precise (8.7), (8.8), (8.9) and (8.12), respectively.

Next, we present the local convergence results. The first one involving (IGNM)

Theorem 8.3.1 *Suppose $x^\star$ satisfies (8.1), F has a continuous derivative in $U(x^\star, r)$,*
$F'(x^\star)$ has full rank and $F'(x)$ satisfies the radius Lipschitz condition with $\mathcal{L}$ average and the center-Lipschitz condition with $\mathcal{L}_0$ average where $\mathcal{L}$ and $\mathcal{L}_0$ are nondecreasing. Assume $B_k = F'(x_k)^T F'(x_k)$, for each k in (8.2), $v_k = \theta_k \|(P_k F'(x_k)^T F'(x_k))^{-1}\|$
$\|P_k F'(x_k)^T F'(x_k)\| = \theta_k Cond(P_k F'(x_k)^T F'(x_k))$ with $v_k \leq v < 1$. Let $r > 0$ satisfy

$$(1+v)\frac{\beta \int_0^r \mathcal{L}(u)udu}{r(1-\beta \int_0^r \mathcal{L}_0(u)du)} + \frac{\sqrt{2}c\beta^2 \int_0^r \mathcal{L}_0(u)du}{r(1-\beta \int_0^r \mathcal{L}_0(u)du)} + v \leq 1. \tag{8.24}$$

Then (IGNM) is convergent for all $x_0 \in U(x^\star, r)$ and

$$\begin{aligned} \|x_{k+1} - x^\star\| \quad \leq \quad & (1+v)\frac{\beta \int_0^{s(x_0)} \mathcal{L}(u)du}{s(x_0)^2(1-\beta \int_0^{s(x_0)} \mathcal{L}_0(u)du)} \|x_k - x^\star\|^2 \\ & + \left(\frac{\sqrt{2}c\beta^2 \int_0^{s(x_0)} \mathcal{L}_0(u)du}{s(x_0)(1-\beta \int_0^{s(x_0)} \mathcal{L}_0(u)du)} + v \right) \|x_k - x^\star\|, \end{aligned} \tag{8.25}$$

where $c = \|F(x^\star)\|$, $\beta = \|(F'(x^\star)^T F'(x^\star))^{-1} F'(x^\star)^T\|$,

$$q = (1+v)\frac{\beta \int_0^{s(x_0)} \mathcal{L}(u)du}{s(x_0)(1-\beta \int_0^{s(x_0)} \mathcal{L}_0(u)du)} + \frac{\sqrt{2}c\beta^2 \int_0^{s(x_0)} \mathcal{L}_0(u)du}{s(x_0)(1-\beta \int_0^{s(x_0)} \mathcal{L}_0(u)du)} + v < 1. \tag{8.26}$$

Proof: Let $x_0 \in U(x^\star, r)$ where r satisfies (8.11), from the monotonicity of $\mathcal{L}_0, \mathcal{L}$, (8.7) and Lemma 8.2.5, we have in turn

$$\begin{aligned} q &= (1+v)\frac{\beta\int_0^{s(x_0)}\mathcal{L}(u)du}{s(x_0)^2(1-\beta\int_0^{s(x_0)}\mathcal{L}_0(u)du)}s(x_0) + \frac{\sqrt{2}c\beta^2\int_0^{s(x_0)}\mathcal{L}_0(u)du}{s(x_0)(1-\beta\int_0^{s(x_0)}\mathcal{L}_0(u)du)} + v \\ &< (1+v)\frac{\beta\int_0^{r}\mathcal{L}(u)du}{s(x_0)(1-\beta\int_0^{s(x_0)}\mathcal{L}_0(u)du)} + \frac{\sqrt{2}c\beta^2\int_0^{r}\mathcal{L}_0(u)du}{s(x_0)(1-\beta\int_0^{s(x_0)}\mathcal{L}_0(u)du)} + v \le 1 \end{aligned}$$

and

$$\begin{aligned} \|[F'(x^\star)^T F'(x^\star)]^{-1}F'(x^\star)^T\|\|F'(x) - F'(x^\star)\| &\le \beta\int_0^{s(x_0)}\mathcal{L}_0(u)du \\ &\le \beta\int_0^{r}\mathcal{L}_0(u)du < 1, \\ &\quad \text{for each } x \in U(x^\star, r). \end{aligned}$$

That is, q given by (8.26) is less than 1.

By Lemma 8.2.1 and 8.2.2, $F'(x)$ has full rank for each $x \in U(x^\star, r)$ and

$$\begin{aligned} \|[F'(x)^T F'(x)]^{-1}F'(x)^T\| &\le \frac{\beta}{1-\beta\int_0^{s(x_0)}\mathcal{L}_0(u)du}, \\ &\text{for each } x \in U(x^\star, r), \end{aligned}$$

$$\begin{aligned} \|[F'(x)^T F'(x)]^{-1}F'(x)^T - [F'(x^\star)^T F'(x^\star)]^{-1}F'(x^\star)^T\| &\le \frac{\sqrt{2}\beta^2\int_0^{s(x)}\mathcal{L}_0(u)du}{1-\beta\int_0^{s(x)}\mathcal{L}_0(u)du}, \\ &\quad \text{for each } x \in U(x^\star, r). \end{aligned}$$

Then, if $x_k \in U(x^\star, r)$, we have by (8.2) in turn that

$$\begin{aligned} x_{k+1} - x^\star &= x_k - x^\star - F'(x_k)^\dagger(F(x_k) - F(x^\star)) + (F'(x_k)^T F'(x_k))^{-1}r_k \\ &= F'(x_k)^\dagger\int_0^1 (F'(x_k) - F'(x^\theta))(x_k - x^\star)d\theta \\ &\quad + (F'(x_k)^T F'(x_k))^{-1}P_k^{-1}P_k r_k + (F'(x^\star)^T F'(x^\star))^{-1}F'(x^\star)^T F(x^\star) \\ &\quad - (F'(x_k)^T F'(x_k))^{-1}F'(x_k)^T F(x^\star). \end{aligned}$$

Hence, by Lemma 8.2.3 and conditions (8.11) and (8.6) we obtain

$$\begin{aligned}
\|x_{k+1}-x^\star\| &\le \|F'(x_k)^\dagger\| \int_0^1 \|F'(x_k)-F'(x^\theta)\| \|x_k-x^\star\| d\theta \\
&\quad +\theta_k \|(F'(x_k)^T F'(x_k))^{-1}\| \|P_k F'(x_k)^T F'(x_k)\| \\
&\quad +\|(F'(x_k)^T F'(x_k))^{-1} F'(x^\star)^T - (F'(x_k)^T F'(x_k))^{-1} \\
&\quad F'(x_k)^T\| \|F(x^\star)\| \\
&\le \frac{\beta}{1-\beta \int_0^{s(x_k)} \mathcal{L}_0(u)du} \\
&\quad \times \int_0^1 \int_{\theta s(x_k)}^{s(x_k)} \mathcal{L}(u) du s(x_k) d\theta + \theta_k \|(P_k F'(x_k)^T F'(x_k))^{-1}\| \\
&\quad \times \|P_k F'(x_k)^T F'(x_k) (F'(x_k)^T F'(x_k))^{-1} F'(x_k)^T F(x_k)\| \\
&\quad + \frac{\sqrt{2} c \beta^2 \int_0^{s(x_k)} \mathcal{L}_0(u)du}{1-\beta \int_0^{s(x_k)} \mathcal{L}_0(u)du} \\
&\le \frac{\beta \int_0^{s(x_k)} \mathcal{L}(u)du}{1-\beta \int_0^{s(x_k)} \mathcal{L}_0(u)du} + \theta_k Cond(P_k F'(x_k)^T F'(x_k)) \\
&\quad \times \left(s(x_k) + \frac{\beta \int_0^{s(x_k)} \mathcal{L}_0(u)du}{1-\beta \int_0^{s(x_k)} \mathcal{L}_0(u)du} \right) \\
&\quad + \frac{\sqrt{2} c \beta^2 \int_0^{s(x_k)} \mathcal{L}_0(u)du}{1-\beta \int_0^{s(x_k)} \mathcal{L}_0(u)du} \\
&\le (1+v_k) \frac{\beta \int_0^{s(x_k)} \mathcal{L}(u)du}{1-\beta \int_0^{s(x_k)} \mathcal{L}_0(u)du} + v_k s(x_k) \\
&\quad + \frac{\sqrt{2} c \beta^2 \int_0^{s(x_k)} \mathcal{L}_0(u)du}{1-\beta \int_0^{s(x_k)} \mathcal{L}_0(u)du}.
\end{aligned}$$

Taking $k=0$ above, we get $\|x_1-x^\star\| \le q\|x_0-x^\star\| < \|x_0-x^\star\|$. That is $x_1 \in U(x^\star, r)$. By mathematical induction, all x_k belong to $U(x^\star, r)$ and $s(x_k) = \|x_k - x^\star\|$ decreases monotonically. Hence, for all $k \ge 0$, we have

$$\begin{aligned}
\|x_{k+1}-x^\star\| &\leq (1+v_k)\frac{\beta\int_0^{s(x_k)}\mathcal{L}(u)du}{s(x_k)^2(1-\beta\int_0^{s(x_k)}\mathcal{L}_0(u)du)}s(x_k)^2 \\
&\quad +v_k s(x_k)+\frac{\sqrt{2}c\beta^2\int_0^{s(x_k)}\mathcal{L}_0(u)du}{1-\beta\int_0^{s(x_k)}\mathcal{L}_0(u)du} \\
&\leq (1+v)\frac{\beta\int_0^{s(x_k)}\mathcal{L}(u)du}{s(x_k)^2(1-\beta\int_0^{s(x_k)}\mathcal{L}_0(u)du)}s(x_k)^2 \\
&\quad +\left(v+\frac{\sqrt{2}c\beta^2\int_0^{s(x_k)}\mathcal{L}_0(u)du}{1-\beta\int_0^{s(x_k)}\mathcal{L}_0(u)du}\right)s(x_k).
\end{aligned}$$

□

Theorem 8.3.2 *Suppose $x^\star$ satisfies (8.1), $F(x^\star)=0$, F has a continuous derivative in $U(x^\star,r)$, $F'(x^\star)$ has full rank and $F'(x)$ satisfies the center Lipschitz condition (8.6) with $\mathcal{L}_0$ average where $\mathcal{L}_0$ is nondecreasing. Assume $B_k = F'(x_0)^T F'(x_0)$, for each k in (8.2), $v_k = \theta_k\|(P_0F'(x_0)^TF'(x_0))^{-1}\|.\|P_0F'(x_0)^T$
$F'(x_0)\| = \theta_k Cond$
$(P_0F'(x_0)^TF'(x_0))$ with $v_k \leq v < 1$. Let $r>0$ satisfy*

$$(1+v)\frac{\beta\int_0^r\mathcal{L}_0(u)(r-u)du}{r(1-\beta\int_0^r\mathcal{L}_0(u)du)}+\frac{v+\beta\int_0^r\mathcal{L}_0(u)du}{1-\beta\int_0^r\mathcal{L}_0(u)du)}\leq 1. \tag{8.27}$$

Then Modified Inexact Gauss-Newton Method (MIGNM) is convergent for all $x_0 \in U(x^\star,r)$ and

$$\begin{aligned}
\|x_{k+1}-x^\star\| &\leq (1+v)\frac{\beta\int_0^{s(x_0)}\mathcal{L}_0(u)(s(x_0)-u)du}{s(x_0)^2(1-\beta\int_0^{s(x_0)}\mathcal{L}_0(u)du)}\|x_k-x^\star\|^2 \\
&\quad +\frac{v+\beta\int_0^{s(x_0)}\mathcal{L}_0(u)du}{1-\beta\int_0^{s(x_0)}\mathcal{L}_0(u)du},
\end{aligned} \tag{8.28}$$

is less than 1.

Proof: Simply replace $\mathcal{L}$ by $\mathcal{L}_0$ in the proof of Theorem 8.3.2 in [5]. □

Next, we present the corresponding results for Inexact Gauss-Newton-Like Method (IGNLM), where $B_k = B(x_k)$ approximates $F'(x_k)^TF'(x_k)$.

Theorem 8.3.3 *Suppose $x^\star$ satisfies (8.1), F has a continuous derivative in $U(x^\star,r)$, $F'(x^\star)$ has full rank and $F'(x)$ satisfies the radius Lipschitz condition (8.11) with $\mathcal{L}$ average and the center Lipschitz condition with $\mathcal{L}_0$ average where $\mathcal{L}$ and $\mathcal{L}_0$ are nondecreasing. Let $B(x)$ be invertible and*

$$\|B(x)^{-1}F'(x)^TF'(x)\|\leq\omega_1,\ \|B(x)^{-1}F'(x)^TF'(x)-I\|\leq\omega_2, \tag{8.29}$$

$v_k = \theta_k \|(P_k F'(x_k)^T F'(x_k))^{-1}\| \|P_k F'(x_k)^T F'(x_k)\| = \theta_k Cond(P_k F'(x_k)^T F'(x_k))$ *with* $v_k \leq v < 1$. *Let* $r > 0$ *satisfy*

$$(1+v)\frac{\beta\omega_1 \int_0^r \mathcal{L}(u)udu}{r(1-\beta\int_0^r \mathcal{L}_0(u)du)} + \omega_2 + \omega_1 v + \frac{\sqrt{2}\beta^2\omega_1 \int_0^r \mathcal{L}_0(u)du}{r(1-\beta\int_0^r \mathcal{L}_0(u)du))} \leq 1. \quad (8.30)$$

Then (IGNLM) is convergent for all $x_0 \in U(x^\star, r)$ *and*

$$\begin{aligned} \|x_{k+1}-x^\star\| \quad \leq \quad & (1+v)\frac{\beta\omega_1 \int_0^{s(x_0)} \mathcal{L}(u)udu}{s(x_0)^2(1-\beta\int_0^{s(x_0)} \mathcal{L}_0(u)du)}\|x_k - x^\star\|^2 \\ & + \left(\omega_2 + \omega_1 v + \frac{\sqrt{2}\beta^2\omega_1 \int_0^{s(x_0)} \mathcal{L}_0(u)du}{s(x_0)(1-\beta\int_0^{s(x_0)} \mathcal{L}_0(u)du)}\right)\|x_k - x^\star\| \end{aligned} \quad (8.31)$$

where $c = \|F(x^\star)\|$, $\beta = \|[F'(x^\star)^T F'(x^\star)]^{-1} F'(x^\star)^T\|$,

$$\begin{aligned} q \quad = \quad & (1+v)\frac{\beta\omega_1 \int_0^{s(x_0)} \mathcal{L}(u)udu}{s(x_0)(1-\beta\int_0^{s(x_0)} \mathcal{L}_0(u)du)} \\ & + \omega_2 + \omega_1 v + \frac{\sqrt{2}\beta^2\omega_1 \int_0^{s(x_0)} \mathcal{L}_0(u)du}{s(x_0)(1-\beta\int_0^{s(x_0)} \mathcal{L}_0(u)du)} \end{aligned} \quad (8.32)$$

is less than 1.

Proof: Let $x_0 \in U(x^\star, r)$, where r satisfies (8.30), by the monotonicity of $\mathcal{L}_0$, $\mathcal{L}$, (8.17) and Lemma (8.2.5), we have

$$\begin{aligned} q \quad = \quad & (1+v)\frac{\beta\omega_1 \int_0^{s(x_0)} \mathcal{L}(u)udu}{s(x_0)^2(1-\beta\int_0^{s(x_0)} \mathcal{L}_0(u)du)} s(x_0) \\ & + \omega_2 + \omega_1 v + \frac{\sqrt{2}\beta^2\omega_1 \int_0^{s(x_0)} \mathcal{L}_0(u)du}{s(x_0)(1-\beta\int_0^{s(x_0)} \mathcal{L}_0(u)du)} \\ < \quad & (1+v)\frac{\beta\omega_1 \int_0^r \mathcal{L}(u)udu}{r^2(1-\beta\int_0^{r)} \mathcal{L}_0(u)du)} r \\ & + \omega_2 + \omega_1 v + \frac{\sqrt{2}\beta^2\omega_1 \int_0^r \mathcal{L}_0(u)du}{r(1-\beta\int_0^r \mathcal{L}_0(u)du)} \leq 1, \end{aligned}$$

and

$$\begin{aligned} \|[F'(x)^T F'(x)]^{-1} F'(x)^T\| \|F'(x) - F'(x^\star)\| \quad \leq \quad & \beta \int_0^{s(x)} \mathcal{L}_0(u)du \\ \leq \quad & \beta \int_0^r \mathcal{L}_0(u)du < 1, \\ & \text{for each } x \in U(x^\star, r). \end{aligned}$$

That is, q given by (8.32) is less than 1.

Since $F'(x)$, $x \in U(x^\star, r)$ has full rank by Lemma 8.2.1 and 8.2.2, we have

$$\|[F'(x)^T F'(x)]^{-1} F'(x)^T\| \leq \frac{\beta}{1-\beta \int_0^{s(x)} \mathcal{L}_0(u)du}, \text{ for each } x \in U(x^\star, r)$$

$$\|[F'(x)^T F'(x)]^{-1} F'(x)^T - [F'(x^\star)^T F'(x^\star)]^{-1} F'(x^\star)^T\| \leq \frac{\sqrt{2}\beta^2 \int_0^{s(x)} \mathcal{L}_0(u)du}{1-\beta \int_0^{s(x)} \mathcal{L}_0(u)du}, \text{ for each } x \in U(x^\star, r).$$

Then, if $x_k \in U(x^\star, r)$, we have by (8.2) in turn that

$$\begin{aligned}
x_{k+1} - x^\star &= x_k - x^\star - B_k^{-1} F'(x_k)^T (F(x_k) - F(x^\star)) + B_k^{-1} r_k \\
&= x_k - x^\star - \int_0^1 B_k^{-1} F'(x_k)^T F'(x^\theta)(x_k - x^\star) d\theta \\
&\quad + B_k^{-1} P_k^{-1} P_k r_k + B_k^{-1} F'(x_k)^T F(x_k) \{[F'(x^\star)^T F'(x^\star)]^{-1} F'(x^\star)^T F(x^\star) \\
&\quad - (F'(x_k)^T F'(x_k))^{-1} F'(x_k)^T F(x^\star)\} \\
&= B_k^{-1} F'(x_k)^T F'(x_k) \int_0^1 F'(x_k)^\dagger (F'(x_k) - F'(x^\theta))(x_k - x^\star) d\theta \\
&\quad - B_k^{-1} (F'(x_k)^T F'(x_k) - B_k)(x_k - x^\star) + B_k^{-1} P_k^{-1} P_k r_k \\
&\quad + B_k^{-1} F'(x_k)^T F(x_k) \{[F'(x^\star)^T F'(x^\star)]^{-1} F'(x^\star)^T F(x^\star) \\
&\quad - (F'(x_k)^T F'(x_k))^{-1} F'(x_k)^T F(x^\star)\}.
\end{aligned}$$

In view of Lemma 8.2.3 and conditions (8.11) and (8.6) we obtain

$$
\begin{aligned}
\|x_{k+1}-x^\star\| \leq\ & \|B_k^{-1}F'(x_k)^TF'(x_k)\|\| \int_0^1 \|F'(x_k)^\dagger\|\|F'(x_k)-F'(x^\theta)\| \\
& \|x_k-x^\star\|d\theta \\
& +\|B_k^{-1}(F'(x_k)^TF'(x_k)-B_k)\|\|x_k-x^\star\| \\
& +\theta_k\|B_k^{-1}P_k^{-1}\|\|P_kF'(x_k)^TF'(x_k)\| \\
& +\|B_k^{-1}F'(x_k)^TF'(x_k)\|\|[F'(x^\star)^TF'(x^\star)]^{-1}F'(x^\star)^T \\
& -[F'(x_k)^TF'(x_k)]^{-1}F'(x_k)^T\|\|F'(x^\star)\| \\
\leq\ & \frac{\beta\omega_1}{1-\beta\int_0^{s(x_k)}\mathcal{L}_0(u)du}\int_0^1\int_{\theta s(x_k)}^{s(x_k)}\mathcal{L}(u)dus(x_k)d\theta \\
& +\omega_2 s(x_k)+\theta_k\|B_k^{-1}F'(x_k)^TF'(x_k)\| \\
& \times\|(P_kF'(x_k)^TF'(x_k))^{-1}\|\|F'(x_k)^\dagger F(x_k)\| \\
& +\frac{\sqrt{2}c\beta^2\omega_1\int_0^{s(x_k)}\mathcal{L}_0(u)du}{1-\beta\int_0^{s(x_k)}\mathcal{L}_0(u)du} \\
\leq\ & \frac{\beta\omega_1\int_0^{s(x_k)}\mathcal{L}(u)du}{1-\beta\int_0^{s(x_k)}\mathcal{L}_0(u)du} \\
& +\omega_1 v_k\left(s(x_k)+\frac{\beta\int_0^{s(x_k)}\mathcal{L}_0(u)du}{1-\beta\int_0^{s(x_k)}\mathcal{L}_0(u)du}\right) \\
& +\omega_2 s(x_k)+\frac{\sqrt{2}c\beta^2\omega_1\int_0^{s(x_k)}\mathcal{L}_0(u)du}{1-\beta\int_0^{s(x_k)}\mathcal{L}_0(u)du} \\
\leq\ & (1+v_k)\frac{\beta\omega_1\int_0^{s(x_k)}\mathcal{L}(u)du}{1-\beta\int_0^{s(x_k)}\mathcal{L}_0(u)du}+(\omega_2+\omega_1 v_k)s(x_k) \\
& +\frac{\sqrt{2}c\beta^2\omega_1\int_0^{s(x_k)}\mathcal{L}_0(u)du}{1-\beta\int_0^{s(x_k)}\mathcal{L}_0(u)du}.
\end{aligned}
$$

If $k=0$ above, we obtain $\|x_1-x^\star\|\leq q\|x_0-x^\star\|<\|x_0-x^\star\|$. That is $x_1\in U(x^\star,r)$, this shows that (8.2) can be continued an infinite number of times. By mathematical induction, all x_k belong to $U(x^\star,r)$ and $s(x_k)=\|x_k-x^\star\|$ decreases monotonically. Hence, for all $k \geq 0$, we have

$$\begin{aligned}
\|x_{k+1}-x^\star\| &\leq (1+v_k)\frac{\beta\omega_1\int_0^{s(x_k)}\mathcal{L}(u)du}{s(x_k)^2(1-\beta\int_0^{s(x_k)}\mathcal{L}_0(u)du)}s(x_k)^2 \\
&\quad +(\omega_2+\omega_1 v_k)s(x_k)+\frac{\sqrt{2}c\beta^2\omega_1\int_0^{s(x_k)}\mathcal{L}_0(u)du}{s(x_k)(1-\beta\int_0^{s(x_k)}\mathcal{L}_0(u)du)}s(x_k) \\
&\leq (1+v)\frac{\beta\omega_1\int_0^{s(x_0)}\mathcal{L}(u)du}{s(x_0)^2(1-\beta\int_0^{s(x_0)}\mathcal{L}_0(u)du)}s(x_k)^2 \\
&\quad +\left(\omega_2+\omega_1 v+\frac{\sqrt{2}c\beta^2\omega_1\int_0^{s(x_0)}\mathcal{L}_0(u)du}{s(x_0)(1-\beta\int_0^{s(x_0)}\mathcal{L}_0(u)du)}\right)s(x_k).
\end{aligned}$$

□

Remark 8.3.4 *If $\mathcal{L}_0=\mathcal{L}$ our results reduce to the corresponding ones in [5] which in turn improved earlier ones [6, 7, 10]. On the other hand, if $\mathcal{L}_0<\mathcal{L}$, our results constitute an improvement with advantages.*

8.4 Applications and examples

First of all, let $\mathcal{L}_0$ and $\mathcal{L}$ to be positive constant functions. Then, the following corollaries are obtained.

Corollary 8.4.1 *Suppose $x^\star$ satisfies (8.1), F has a continuous derivative in $U(x^\star,r)$ $F'(x_k)$ has full rank and $F'(x)$ satisfies the radius Lipschitz condition with $\mathcal{L}$ average*

$$\|F'(x)-F'(x^\theta)\|\leq(1-\theta)\mathcal{L}\|x-x^\star\|,\ 0\leq\theta\leq 1, \tag{8.33}$$

and the center Lipschitz condition with $\mathcal{L}_0$ average

$$\|F'(x)-F'(x^\star)\|\leq\mathcal{L}_0\theta\|x-x^\star\|, \tag{8.34}$$

where $x^\theta=x^\star+\theta(x-x^\star)$, $s(x)=\|x-x^\star\|$. Assume $B_k=F'(x_k)^TF'(x_k)$ for each k in (8.2), $v_k=\theta_k\|[P_kF'(x_k)^TF'(x_k)]^{-1}\|=\theta_k Cond$ $(P_kF'(x_k)^TF'(x_k))$ with $v_k\leq v<1$. Let $r>0$ satisfy

$$r=\frac{2(1-v-\sqrt{2}c\mathcal{L}_0\beta^2)}{\beta[(1+v)\mathcal{L}+2(1-v)\mathcal{L}_0]} \tag{8.35}$$

where $c=\|F(x^\star)\|$, $\beta=\|[F'(x^\star)^TF'(x^\star)]^{-1}F'(x^\star)^T\|$. Then (IGNM) is convergent for all $x_0\in U(x^\star,r)$,

$$q=v+\frac{\mathcal{L}\beta(1+v)\|x_0-x^\star\|+2\sqrt{2}c\mathcal{L}_0\beta^2}{2(1-\mathcal{L}_0\beta\|x_0-x^\star\|)}<1, \tag{8.36}$$

and the inequality (8.25) holds.

Corollary 8.4.2 *Suppose $x^\star$ satisfies (8.1), $F(x^\star) = 0, F$ has a continuous derivative in $U(x^\star, r)$ $F'(x_k)$ has full rank and $F'(x)$ satisfies the center Lipschitz condition with Lipschitz condition with $\mathcal{L}_0$ average. Assume $B_k = F'(x_0)^T F'(x_0)$ for each k in (8.2), $v_k = \theta_k \| [P_0 F'(x_0)^T F'(x_0)]^{-1} \| \| P_0 F'(x_0)^T F'(x_0) \|$ $= \theta_k Cond(P_0 F'(x_0)^T F'(x_0))$ with $v_k \leq v < 1$. Let $r > 0$ satisfy*

$$r = \frac{2(1-v)}{\beta \mathcal{L}_0 (5+v)} \tag{8.37}$$

where $\beta = \| [F'(x^\star)^T F'(x^\star)]^{-1} F'(x^\star)^T \|$. Then (MIGNM) is convergent for all $x_0 \in U(x^\star, r)$,

$$q = \frac{\mathcal{L}_0 \beta \|x_0 - x^\star\| + v}{1 - \mathcal{L}_0 \beta \|x_0 - x^\star\|} + \frac{\mathcal{L}_0 \beta \|x_0 - x^\star\|(1+v)}{2(1 - \mathcal{L}_0 \beta \|x_0 - x^\star\|)} < 1, \tag{8.38}$$

and the inequality (8.23) holds.

Corollary 8.4.3 *Suppose $x^\star$ satisfies (8.1), F has a continuous derivative in $U(x^\star, r)$ $F'(x_k)$ has full rank and $F'(x)$ satisfies the radius Lipschitz condition (8.33) with $\mathcal{L}$ average and the center Lipschitz condition (8.34) with $\mathcal{L}_0$ average. Assume $B(x)$ and $F'(x)^T F'(x)$ satisfy (8.29), $v_k = \theta_k \| [P_k F'(x_k)^T F'(x_k)]^{-1} \| \| P_k F'(x_k)^T F'(x_k) \| = \theta_k Cond(P_k F'(x_k)^T F'(x_k))$ with $v_k \leq v < 1$. Let $r > 0$ satisfy*

$$r = \frac{2(1 - v\omega_1 - \omega_2 - \sqrt{2} c \mathcal{L}_0 \beta^2 \omega_1)}{\beta[(1+v)\omega_1 \mathcal{L} + 2(1 - \omega_2 - \omega_1 v)\mathcal{L}_0]} \tag{8.39}$$

where $c = \|F(x^\star)\|$, $\beta = \| [F'(x^\star)^T F'(x^\star)]^{-1} F'(x^\star)^T \|$. Then (IGNLM) is convergent for all $x_0 \in U(x^\star, r)$,

$$q = \omega_1 v + \omega_2 + \frac{\mathcal{L} \beta \omega_1 (1+v)\|x_0 - x^\star\| + 2\sqrt{2} c \mathcal{L}_0 \beta^2 \omega_1}{2(1 - \mathcal{L}_0 \beta \|x_0 - x^\star\|)} < 1, \tag{8.40}$$

and the inequality (8.31) holds.

Remark 8.4.4 *(a) If, $\mathcal{L}_0 = \mathcal{L}$ the results reduce to the corresponding ones in [5]. On the other hand, if $\mathcal{L}_0 < \mathcal{L}$ the new results constitute an improvement.*

(b) The results of Section 5 in [5–7] using only center-Lipschitz condition can be improved with $\mathcal{L}_0$ replacing $\mathcal{L}$.

Example 8.4.5 *Let $\mathcal{X} = \mathcal{Y} = \mathbb{R}^3$, $\mathcal{D} = \overline{U}(0,1)$ and $x^* = (0,0,0)^T$. Define function F on $\mathcal{D}$ for $w = (x,y,z)^T$ by*

$$F(w) = \left(e^x - 1, \left(\frac{e-1}{2} \right) y^2 + y, z \right)^T. \tag{8.41}$$

Then, the Fréchet derivative of F is given by

$$F'(w) = \begin{pmatrix} e^x & 0 & 0 \\ 0 & (e-1)y+1 & 0 \\ 0 & 0 & 1 \end{pmatrix}$$

Notice that we have $F(x^\star) = 0$, $F'(x^\star) = F'(x^\star)^{-1} = \text{diag}\{1,1,1\}$ *and* $\mathcal{L}_0 = e - 1 < \mathcal{L} = e$.

References

[1] I. K. Argyros, Y. J. Cho and S. Hilout, Numerical methods for equations and its applications, CRC Press, Taylor and Francis, New York, 2012.

[2] I. K. Argyros and S. Hilout, Weak convergence conditions for inexact Newton-type methods, *Appl. Math. Comput.*, 218, 2800–2809, 2011.

[3] I. K. Argyros and S. Hilout, Extending the applicability of the Gauss-Newton method under average Lipschitz-type conditions, *Numer. Algorithms*, 58, 23–52, 2011.

[4] I. K. Argyros and S. Hilout, Weaker conditions for the convergence of Newton's method, *J. Complexity*, 28, 364–387, 2012.

[5] J. Chen, The convergence analysis of inexact Gauss-Newton methods for nonlinear problems, *Comput. Optim. Appl.*, 40, 97–118, 2008.

[6] J. Chen and W. Li, Convergence of Gauss-Newton's method and uniqueness of the solution, *Appl. Math. Comput.*, 170, 686–705, 2005.

[7] J. Chen and W. Li, Convergence behaviour of inexact Newton methods under weaker Lipschitz condition, *J. Comput. Appl. Math.*, 191, 143–164, 2006.

[8] R. S. Dembo, S. C. Eisenstat and T. Steihaus, Inexact Newton methods, *SIAM J. Numer. Anal.*, 19, 400–408, 1982.

[9] J. E. Jr. Dennis and R. B. Schnabel, Numerical methods for unconstrained optimization and nonlinear equations (Corrected reprint of the 1983 original), *Classics in Appl. Math., SIAM* 16, Philadelphia, PA, 1996.

[10] B. Morini, Convergence behaviour of inexact Newton method, *Math. Comp.*, 61, 1605–1613, 1999.

[11] J. M. Ortega and W. C. Rheinholdt, Iterative Solution of Nonlinear Equations in Several Variables, Academic Press, New York, 1970.

[12] G. W. Stewart, On the continuity of the generalized inverse, *SIAM J. Appl. Math.*, 17, 33–45, 1969.

[13] W. C. Rheinholdt, An adaptive continuation process for solving systems of nonlinear equations, Polish Academy of Science, *Banach Ctr. Publ.*, 3, 129–142, 1977.

[14] J. F. Traub, Iterative methods for the solution of equations, Prentice-Hall Series in Automatic Computation, Englewood Cliffs, N. J., 1964.

[15] X. Wang, The Convergence on Newton's method. *Kexue Tongbao (A special issue of Mathematics, Physics & Chemistry)*, 25, 36–37, 1980.

Chapter 9

Lavrentiev Regularization methods for Ill-posed equations

Ioannis K. Argyros

Department of Mathematical Sciences, Cameron University, Lawton, OK 73505, USA, Email: iargyros@cameron.edu

Á. Alberto Magreñán

Universidad Internacional de La Rioja, Escuela Superior de Ingeniería y Tecnología, 26002 Logroño, La Rioja, Spain, Email: alberto.magrenan@unir.net

CONTENTS

9.1 Introduction

In this chapter, we consider the problem of approximately solving the nonlinear ill-posed operator equation of the form

$$F(x) = y, \tag{9.1}$$

where $F : D(F) \subset X \to X$ is a monotone operator and X is a real Hilbert space. We denote the inner product and the corresponding norm on a Hilbert space by $\langle \cdot, \cdot \rangle$ and $\| \cdot \|$, respectively. Let $U(x, r)$ stand for the open ball in X with center $x \in X$ and radius $r > 0$. Recall that F is said to be a monotone operator if it satisfies the relation

$$\langle F(x_1) - F(x_2), x_1 - x_2 \rangle \geq 0 \tag{9.2}$$

for all $x_1, x_2 \in D(F)$.

We assume, throughout this chapter, that $y^\delta \in Y$ are the available noisy data with

$$\|y - y^\delta\| \leq \delta \tag{9.3}$$

and (9.1) has a solution $\hat{x}$. Since (9.1) is ill-posed, the regularization methods are used ([9–11, 13, 14, 19, 21, 22]) for approximately solving (9.1). Lavrentiev regularization is used to obtain a stable approximate solution of (9.1). In the Lavrentiev regularization, the approximate solution is obtained as a solution of the equation

$$F(x) + \alpha(x - x_0) = y^\delta, \tag{9.4}$$

where $\alpha > 0$ is the regularization parameter and x_0 is an initial guess for the solution $\hat{x}$. For deriving the error estimates, we shall make use of the following equivalent form of (9.4),

$$x_\alpha^\delta = x_0 + (A_\alpha^\delta + \alpha I)^{-1}[y^\delta - F(x_\alpha^\delta) + A_\alpha^\delta(x_\alpha^\delta - x_0)], \tag{9.5}$$

where $A_\alpha^\delta = F'(x_\alpha^\delta)$.

In [15], Mahale and Nair, motivated by the chapter of Tautenhan [22], considered Lavrentiev regularization of (9.1) under a general source condition on $\hat{x} - x_0$ and obtained an order optimal error estimate.

In the present chapter, we are motivated by [15]. In particular, we expand the applicability of the Lavrentiev regularization of (9.1) by weakening one of the major hypotheses in [15] (see below Assumption 2.1 (ii) in the next section).

The chapter is organized as follows: In Section 9.2, we present some basic assumptions and preliminaries required. The main order optimal result using the apriori and a posteriori parameter choice is presented in Section 9.3. Finally some numerical examples are presented in the concluding Section 9.4.

9.2 Basic assumptions and some preliminary results

We use the following assumptions to prove the results in this chapter.

Assumption 10 (1) There exists $r > 0$ such that $U(x_0, r) \subseteq D(F)$ and $F : U(x_0, r) \to X$ is Fréchet differentiable.

(2) There exists $K_0 > 0$ such that, for all $u_\theta = u + \theta(x_0 - u) \in U(x_0, r)$, $\theta \in [0, 1]$ and $v \in X$, there exists an element, say $\phi(x_0, u_\theta, v) \in X$, satisfying

$$[F'(x_0) - F'(u_\theta)]v = F'(u_\theta)\phi(x_0, u_\theta, v), \quad \|\phi(x_0, u_\theta, v)\| \leq K_0\|v\|\|x_0 - u_\theta\|$$

for all $u_\theta \in U(x_0, r)$ and $v \in X$.

(3) $\|F'(u) + \alpha I)^{-1}F'(u_\theta)\| \leq 1$ for all $u_\theta \in U(x_0, r)$

(4) $\|(F'(u) + \alpha I)^{-1}\| \leq \frac{1}{\alpha}$.

The condition (2) in Assumption 10 weakens the popular hypotheses given in [15], [17] and [20].

Assumption 11 There exists a constant $K > 0$ such that, for all $x, y \in U(\hat{x}, r)$ and $v \in X$, there exists an element denoted by $P(x, u, v) \in X$ satisfying

$$[F'(x) - F'(u)]v = F'(u)P(x, u, v), \quad \|P(x, u, v)\| \leq K\|v\|\|x - u\|.$$

Clearly, Assumption 11 implies Assumption 10 (2) with $K_0 = K$, but not necessarily vice versa. Note that $K_0 \leq K$ holds in general and $\frac{K}{K_0}$ can be arbitrarily large [1]– [5]. In fact, there are many classes of operators satisfying Assumption 10 (2), but not Assumption 11 (see the numerical examples at the end of this chapter). Furthermore, if K_0 is sufficiently smaller than K which can happen since $\frac{K}{K_0}$ can be arbitrarily large, then the results obtained in this chapter provide a tighter error analysis than the one in [15].

Finally, note that the computation of constant K is more expensive than the computation of K_0.

Assumption 12 There exists a continuous and strictly monotonically increasing function $\varphi : (0, a] \to (0, \infty)$ with $a \geq \|F'(x_0)\|$ satisfying

(1) $\lim_{\lambda \to 0} \varphi(\lambda) = 0$;

(2) $\sup_{\lambda \geq 0} \frac{\alpha\varphi(\lambda)}{\lambda + \alpha} \leq \varphi(\alpha)$ for all $\alpha \in (0, a]$;

(3) there exists $v \in X$ with $\|v\| \leq 1$ such that

$$\hat{x} - x_0 = \varphi(F'(x_0))v. \tag{9.6}$$

Note that condition (9.6) is suitable for both mildly and severely ill-posed problems [16], [17]. Further note that the source condition (9.6) involves the known initial approximation x_0 whereas the source condition considered in [15] requires the knowledge of the unknown $\hat{x}$.

We need some auxiliary results based on Assumption 10.

Proposition 9.2.1 *For each $u \in U(x_0, r)$ and $\alpha > 0$,*

$$\|(F'(u)+\alpha I)^{-1}[F(\hat{x})-F(u)-F'(u)(\hat{x}-u)\| \leq \frac{5K_0}{2}\|\hat{x}-u\|^2+2K_0\|\hat{x}-x_0\|\|\hat{x}-u\|.$$

Proof: From the fundamental theorem of integration, for any $u \in U(x_0, r)$, we get

$$F(\hat{x})-F(u) = \int_0^1 F'(u+t(\hat{x}-u))(\hat{x}-u)dt.$$

Hence, by Assumption 10,

$$\begin{aligned} & F(\hat{x})-F(u)-F'(u)(\hat{x}-u) \\ = & \int_0^1 [F'(u+t(\hat{x}-u))-F'(x_0)+F'(x_0)-F'(u)](\hat{x}-u)dt \\ = & \int_0^1 F'(x_0)[\phi(x_0, u+t(\hat{x}-u), u-\hat{x})-\phi(x_0, u, \hat{x}-u)]dt. \end{aligned}$$

Then, by (2), (3) in Assumptions 10 and the inequality $\|(F'(u)+\alpha I)^{-1}F'(u_\theta)\| \leq 1$, we get

$$\begin{aligned} & \|(F'(u)+\alpha I)^{-1}[F(\hat{x})-F(u)-F'(u)(\hat{x}-u)\| \\ \leq & \int_0^1 \|\phi(x_0, u+t(\hat{x}-u), u-\hat{x})+\phi(x_0, u, \hat{x}-u)\|dt. \\ \leq & [\int_0^1 K_0(\|u-x_0\|+\|\hat{x}-u\|t)dt+K_0\|u-x_0\|]\|\hat{x}-u\| \\ \leq & [\int_0^1 K_0(\|u-\hat{x}\|+\|\hat{x}-x_0\|+\|\hat{x}-u\|t)dt+K_0(\|u-\hat{x}\|+\|\hat{x}-x_0\|)]\|\hat{x}-u\| \\ \leq & \frac{5K_0}{2}\|\hat{x}-u\|^2+2K_0\|\hat{x}-x_0\|\|\hat{x}-u\|. \end{aligned}$$

□

Theorem 9.2.2 *([16, Theorem 2.2], also see [22]) Let x_α be the solution of (9.4) with y in place of y^δ. Then*

(1) $\|x_\alpha^\delta - x_\alpha\| \leq \frac{\delta}{\alpha}$,

(2) $\|x_\alpha - \hat{x}\| \leq \|x_0 - \hat{x}\|$,

(3) $\|F(x_\alpha^\delta) - F(x_\alpha)\| \leq \delta$.

Remark 9.2.3 *From Theorem 9.2.2 and triangle inequality we have*

$$\|x_\alpha^\delta - \hat{x}\| \leq \frac{\delta}{\alpha} + \|x_0 - \hat{x}\|.$$

So (9.5) is meaningful if

$$r > \frac{\delta}{\alpha} + \|x_0 - \hat{x}\|.$$

9.3 Error estimates

We will try to find error bounds for $\|x_\alpha - \hat{x}\|$ in order to find the error estimates.

Theorem 9.3.1 *Let Assumption 10, 12 hold, and let $K_0\|x_0 - \hat{x}\| < \frac{2}{9}$. Then*

$$\|x_\alpha - \hat{x}\| \leq C\varphi(\alpha)$$

where $C = \frac{1+2K_0\|x_0-\hat{x}\|}{1-9K_0\|x_0-\hat{x}\|/2}$.

Proof: Let $A_\alpha = F'(x_\alpha)$ and $A_0 = F'(x_0)$. Then

$$x_\alpha = x_0 + (A_\alpha + \alpha I)^{-1}[y - F(x_\alpha) + A_\alpha(x_\alpha - x_0)].$$

So

$$\begin{aligned}
x_\alpha - \hat{x} &= x_0 - \hat{x} + (A_\alpha + \alpha I)^{-1}[y - F(x_\alpha) + A_\alpha(x_\alpha - x_0)] \\
&= x_0 - \hat{x} + (A_\alpha + \alpha I)^{-1}[F(\hat{x}) - F(x_\alpha) + A_\alpha(x_\alpha - \hat{x}) + A_\alpha(\hat{x} - x_0)] \\
&= (A_\alpha + \alpha I)^{-1}[\alpha(x_0 - \hat{x}) + F(\hat{x}) - F(x_\alpha) + A_\alpha(x_\alpha - \hat{x})] \\
&= (A_\alpha + \alpha I)^{-1}\alpha(x_0 - \hat{x}) + (A_\alpha + \alpha I)^{-1}[F(\hat{x}) - F(x_\alpha) + A_\alpha(x_\alpha - \hat{x})] \\
&= (A_0 + \alpha I)^{-1}\alpha(x_0 - \hat{x}) + [(A_\alpha + \alpha I)^{-1} - (A_0 + \alpha I)^{-1}]\alpha(x_0 - \hat{x}) \\
&\quad + (A_\alpha + \alpha I)^{-1}[F(\hat{x}) - F(x_\alpha) + A_\alpha(x_\alpha - \hat{x})] \\
&= v_0 + (A_\alpha + \alpha I)^{-1}(A_0 - A_\alpha)v_0 + \Delta_1 \qquad (9.7)
\end{aligned}$$

where $v_0 = (A_0 + \alpha I)^{-1}\alpha(x_0 - \hat{x})$ and $\Delta_1 = (A_\alpha + \alpha I)^{-1}[F(\hat{x}) - F(x_\alpha) + A_\alpha(x_\alpha - \hat{x})]$. By Assumption 12, we have

$$\|v_0\| \leq \varphi(\alpha), \qquad (9.8)$$

by Assumption 10, 12 and Theorem 9.2.2, we get

$$\begin{aligned}
(A_\alpha + \alpha I)^{-1}(A_0 - A_\alpha)v_0 &= (A_\alpha + \alpha I)^{-1}A_0\varphi(x_0, x_\alpha, v_0) \\
&\leq K_0\|x_0 - x_\alpha\|\|v_0\| \\
&\leq K_0(\|x_0 - \hat{x}\| + \|\hat{x} - x_\alpha\|)\|v_0\| \\
&\leq 2K_0\|x_0 - \hat{x}\|\varphi(\alpha) \qquad (9.9)
\end{aligned}$$

and by Proposition 9.2.1

$$\begin{aligned}
\Delta_1 &\leq \frac{5K_0}{2}\|\hat{x} - x_\alpha\|^2 + 2K_0\|\hat{x} - x_0\|\|\hat{x} - x_\alpha\| \\
&\leq \frac{9K_0}{2}\|\hat{x} - x_0\|\|\hat{x} - x_\alpha\|. \qquad (9.10)
\end{aligned}$$

The result now follows from (9.7), (9.8), (9.9) and (9.10).

The following Theorem is a consequence of Theorem 9.2.2 and Theorem 9.3.1.

Theorem 9.3.2 *Under the assumptions of Theorem 9.3.1*

$$\|x_\alpha^\delta - \hat{x}\| \leq C(\frac{\delta}{\alpha} + \varphi(\alpha))$$

where C is as in Theorem 9.3.1.

9.3.1 Apriori parameter choice

Let $\psi : (0, \varphi(a)] \to (0, a\varphi(a)]$ be defined as

$$\psi(\lambda) := \lambda \varphi^{-1}(\lambda). \tag{9.11}$$

Then $\frac{\delta}{\alpha} = \varphi(\alpha) \Leftrightarrow \delta = \psi(\varphi(\alpha))$.

Theorem 9.3.3 *Let the assumptions of Theorem 9.3.1 be satisfied. If we choose the regularization parameter as* $\alpha = \varphi^{-1}(\psi^{-1}(\delta))$ *with* ψ *defined as in (9.11), then*

$$\|x_\alpha^\delta - \hat{x}\| \leq 2C\psi^{-1}(\delta). \tag{9.12}$$

9.3.2 Aposteriori parameter choice

It is worth noticing that the choice of α in the above Theorem depends on the unknown source function φ. In applications, it is preferable that parameter α is chosen independent of the source function φ, but may depend on (δ, y^δ), and hence on the regularized solution.

For Lavrentiev regularization (9.5), Tautenhan in [22] considered the following discrepancy principal for chosing α,

$$\|\alpha(A_\alpha^\delta + \alpha I)^{-1}[F(x_\alpha^\delta) - y^\delta]\| = c\delta \tag{9.13}$$

where $c > 0$ is an appropriate constant. The error estimate in [22] is obtained under the assumption that the solution satisfies a Holder-type condition. In [15], Mahale and Nair considered the discrepancy principle (9.13) and extended the analysis of Tautenhan [22] to include both Holder type and logarithmic type conditions. In this chapter we consider discrepancy principle (9.13) and derive an order optimal error estimate under the Assumption 12.

From the following proposition presented in [22], in which the existence of α satisfying (9.13) is showed.

Proposition 9.3.4 *([22, Proposition 4.1]) Let F be monotone and* $\|F(x_0) - y^\delta\| \geq c\delta$ *with* $c > 2$. *Then there exists an* $\alpha \geq \beta_0 := \frac{(c-1)\delta}{\|x_0 - \hat{x}\|}$ *satisfying (9.13).*

Lemma 9.3.5 *Let Assumption 10 and assumptions in Proposition 9.3.4 hold and $\alpha := \alpha(\delta)$ is chosen according to (9.13). Then*

$$\|\alpha(A_0+\alpha I)^{-1}(F(x_\alpha)-y)\| \geq \frac{(c-2)\delta}{1+k_1} \tag{9.14}$$

where $k_1 = \frac{K_0(2c-1)\|x_0-\hat{x}\|}{c-1}$.

Proof: By (9.13), we obtain

$$\begin{aligned}
|c\delta - \alpha\|(A_\alpha^\delta+\alpha I)^{-1}(F(x_\alpha)-y)\|| &= |\alpha(A_\alpha^\delta+\alpha I)^{-1}(F(x_\alpha^\delta)-y^\delta)\| \\
&\quad -\alpha\|(A_\alpha^\delta+\alpha I)^{-1}(F(x_\alpha)-y)\|| \\
&\leq \|\alpha(A_\alpha^\delta+\alpha I)^{-1}[F(x_\alpha^\delta)-F(x_\alpha) \\
&\quad -(y^\delta-y)]\| \\
&\leq \|F(x_\alpha^\delta)-F(x_\alpha)\|+\|y^\delta-y\| \\
&\leq 2\delta.
\end{aligned}$$

The last step follows from Theorem 9.2.2. So

$$(c-2)\delta \leq \alpha\|(A_\alpha^\delta+\alpha I)^{-1}(F(x_\alpha)-y)\| \leq (c+2)\delta.$$

Let

$$a_\alpha := \alpha(A_0+\alpha I)^{-1}(F(x_\alpha)-y).$$

Then

$$\begin{aligned}
\alpha\|(A_\alpha^\delta+\alpha I)^{-1}(F(x_\alpha)-y)\| &\leq \|a_\alpha\|+\|\alpha[(A_\alpha^\delta+\alpha I)^{-1}-(A_0+\alpha I)^{-1}] \\
&\quad (F(x_\alpha)-y)\| \\
&\leq \|a_\alpha\|+\|[(A_\alpha^\delta+\alpha I)^{-1}(A_0-A_\alpha^\delta)a_\alpha\| \\
&\leq \|a_\alpha\|+\|[(A_\alpha^\delta+\alpha I)^{-1}A_\alpha^\delta\varphi(x_0,x_\alpha^\delta,a_\alpha)\| \\
&\leq \|a_\alpha\|(1+K_0\|x_0-x_\alpha^\delta\|) \\
&\leq \|a_\alpha\|(1+K_0(\|x_0-\hat{x}\|+\|\hat{x}-x_\alpha^\delta\|)) \\
&\leq \|a_\alpha\|(1+K_0(\|x_0-\hat{x}\|+\frac{\delta}{\alpha}+\|\hat{x}-x_0\|)) \\
&\leq \|a_\alpha\|(1+K_0(2\|x_0-\hat{x}\|+\frac{\delta}{\alpha})) \\
&\leq \|a_\alpha\|(1+K_0(2\|x_0-\hat{x}\|+\frac{\|x_0-\hat{x}\|}{c-1})) \\
&\leq \|a_\alpha\|(1+k_1),
\end{aligned}$$

which in turn implies $\|\alpha(A_0+\alpha I)^{-1}(F(x_\alpha)-y)\| \geq \frac{(c-2)\delta}{1+k_1}$.

Lemma 9.3.6 *Let the assumptions of Theorem 9.3.1 and Proposition 9.3.4 hold. Then*

(1) $\|\alpha(A_0+\alpha I)^{-1}(F(x_\alpha)-y)\| \leq \mu\alpha\varphi(\alpha)$,

(2) $\alpha \geq \varphi^{-1}(\psi^{-1}(\xi\delta))$ *where* $\mu = C(1+\frac{3K_0}{2}\|x_0-\hat{x}\|)$ *and* $\xi = \frac{c-2}{1+k_1}$.

Proof: By Assumption 10, it is clear that for $x, z \in U(x_0, r)$ and $u \in X$,

$$F'(z)u = F'(x_0)u - F'(z)\varphi(x_0,z,u), \ \|\varphi(x_0,z,u)\| \leq K_0\|x_0-z\|\|u\|.$$

So

$$\begin{aligned}
F(x_\alpha)-F(\hat{x}) &= \int_0^1 F'(\hat{x}+t(x_\alpha-\hat{x}))(x_\alpha-\hat{x})dt \\
&= A_0(x_\alpha-\hat{x}) - \int_0^1 F'(\hat{x}+t(x_\alpha-\hat{x}))\varphi(x_0,\hat{x}+t(x_\alpha-\hat{x}),x_\alpha-\hat{x})dt.
\end{aligned}$$

Hence

$$\begin{aligned}
& \|\alpha(A_0+\alpha I)^{-1}(F(x_\alpha)-F(\hat{x}))\| \\
= & \|\alpha(A_0+\alpha I)^{-1}A_0(x_\alpha-\hat{x}) \\
& -\int_0^1 \alpha(A_0+\alpha I)^{-1}F'(\hat{x}+t(x_\alpha-\hat{x})) \\
& \varphi(x_0,\hat{x}+t(x_\alpha-\hat{x}),x_\alpha-\hat{x})dt\| \\
\leq & \alpha(\|x_\alpha-\hat{x}\| + K_0(\|x_0-\hat{x}\| + \frac{\|x_\alpha-\hat{x}\|}{2})\|x_\alpha-\hat{x}\|) \\
\leq & \alpha(1+\frac{3K_0}{2}\|x_0-\hat{x}\|)\|x_\alpha-\hat{x}\|.
\end{aligned}$$

The last follows from Theorem 9.2.2. Now by using Theorem 9.3.1, we have $\|\alpha(A_0+\alpha I)^{-1}(F(x_\alpha)-F(\hat{x}))\| \leq C(1+\frac{3K_0}{2}\|x_0-\hat{x}\|)\alpha\varphi(\alpha) = \mu\alpha\varphi(\alpha)$. In view of (9.14), we get

$$\frac{(c-2)\delta}{1+k_1} \leq \mu\alpha\varphi(\alpha)$$

which implies, by definition of ψ

$$\psi(\varphi(\alpha)) = \alpha\varphi(\alpha) \geq \frac{(c-2)\delta}{1+k_1} := \xi\delta$$

where $\xi = \frac{(c-2)}{1+k_1}$. Thus $\alpha \geq \varphi^{-1}(\psi^{-1}(\xi\delta))$. This completes the proof.

Theorem 9.3.7 *Let Assumption 10 be satisfied and* $4K_0\|x_0-\hat{x}\| < 1$. *Then for* $0 < \alpha_0 \leq \alpha$,

$$\|x_\alpha - x_{\alpha_0}\| \leq \frac{\|\alpha(A_0+\alpha I)^{-1}(F(x_\alpha)-F(\hat{x}))\|}{(1-4\|x_0-\hat{x}\|)\alpha_0}.$$

Proof: Since

$$F(x_\alpha) - y + \alpha(x_\alpha - x_0) = 0, \tag{9.15}$$

$$F(x_{\alpha_0}) - y + \alpha_0(x_{\alpha_0} - x_0) = 0, \tag{9.16}$$

and

$$\alpha_0(x_\alpha - x_{\alpha_0}) = (\alpha - \alpha_0)(x_0 - x_\alpha) + \alpha_0(x_0 - x_{\alpha_0}) - \alpha(x_0 - x_\alpha),$$

we have by (9.15) and (9.16),

$$\alpha_0(x_\alpha - x_{\alpha_0}) = \frac{\alpha - \alpha_0}{\alpha}(F(x_\alpha) - y) + F(x_{\alpha_0}) - F(x_\alpha),$$

and

$$(A_\alpha + \alpha_0 I)(x_\alpha - x_{\alpha_0}) = \frac{\alpha - \alpha_0}{\alpha}(F(x_\alpha) - y) + [F(x_{\alpha_0}) - F(x_\alpha) + A_\alpha(x_\alpha - x_{\alpha_0})].$$

Hence

$$x_\alpha - x_{\alpha_0} = \frac{\alpha - \alpha_0}{\alpha}(A_\alpha + \alpha_0 I)^{-1}(F(x_\alpha) - y) + (A_\alpha + \alpha_0 I)^{-1}[F(x_{\alpha_0}) - F(x_\alpha) + A_\alpha(x_\alpha - x_{\alpha_0})].$$

Thus by Proposition 9.3.4

$$\|x_\alpha - x_{\alpha_0}\| \le \|\frac{\alpha - \alpha_0}{\alpha}(A_\alpha + \alpha_0 I)^{-1}(F(x_\alpha) - y)\| + \Gamma_1 \tag{9.17}$$

where $\Gamma_1 = \|(A_\alpha + \alpha_0 I)^{-1}[F(x_{\alpha_0}) - F(x_\alpha) + A_\alpha(x_\alpha - x_{\alpha_0})]\|$. By Fundamental Theorem of Integration and Assumption 10, we have

$$\begin{aligned}
F(x_{\alpha_0}) - F(x_\alpha) + A_\alpha(x_\alpha - x_{\alpha_0}) &= \int_0^1 [F'(x_\alpha + t(x_{\alpha_0} - x_\alpha)) - A_\alpha](x_{\alpha_0} - x_\alpha)dt \\
&= \int_0^1 [F'(x_\alpha + t(x_{\alpha_0} - x_\alpha)) - F'(x_0) \\
&\quad + F'(x_0) - A_\alpha](x_{\alpha_0} - x_\alpha)dt \\
&= \int_0^1 [-F'(u_\theta)\varphi(x_0, u_\theta, x_{\alpha_0} - x_\alpha) \\
&\quad + A_\alpha \varphi(x_0, x_\alpha, x_{\alpha_0} - x_\alpha)]dt
\end{aligned}$$

where $u_\theta = x_0 + \theta(x_0 - (x_\alpha + t(x_{\alpha_0} - x_\alpha)))$. Hence, again by Assumption 10, we get in turn that

$$\begin{aligned}\Gamma_1 &\leq \|\int_0^1 \varphi(x_0,u_\theta,x_{\alpha_0}-x_\alpha)dt\| + \|\varphi(x_0,x_\alpha,x_{\alpha_0}-x_\alpha)\| \\ &\leq K_0[\int_0^1 \|x_0-u_\theta\|dt + \|x_0-x_\alpha\|]\|x_{\alpha_0}-x_\alpha\| \\ &\leq K_0[\|x_0-x_\theta\|/2 + \|x_0-x_\alpha\|]\|x_{\alpha_0}-x_\alpha\| \\ &\leq K_0[\|x_0-x_\alpha\|/2 + \|x_0-x_{\alpha_0}\|/2 + \|x_0-x_\alpha\|]\|x_{\alpha_0}-x_\alpha\| \\ &\leq K_0[3(\|x_0-\hat{x}\| + \|\hat{x}-x_\alpha\|)/2 + (\|x_0-\hat{x}\| + \|\hat{x}-x_{\alpha_0}\|)/2]\|x_{\alpha_0}-x_\alpha\| \\ &\leq 4K_0\|x_0-\hat{x}\|\|x_{\alpha_0}-x_\alpha\|. \qquad (9.18)\end{aligned}$$

The last step follows from Theorem 9.2.2. Now since $\frac{\alpha-\alpha_0}{\alpha} < 1$, we have

$$\begin{aligned}\|\frac{\alpha-\alpha_0}{\alpha}(A_\alpha+\alpha_0 I)^{-1}(F(x_\alpha)-y)\| &\leq \|(A_\alpha+\alpha_0 I)^{-1}(F(x_\alpha)-y)\| \\ &\leq \frac{1}{\alpha_0\alpha}\|\alpha_0(A_\alpha+\alpha_0 I)^{-1} \\ &\quad (A_\alpha+\alpha_0 I)\alpha(A_\alpha+\alpha_0 I)^{-1}(F(x_\alpha)-y)\| \\ &\leq \sup_{\lambda\geq 0}|\frac{\alpha_0(\lambda+\alpha)}{\alpha(\lambda+\alpha_0}| \\ &\quad \frac{\|\alpha(A_\alpha+\alpha_0 I)^{-1}(F(x_\alpha)-y)\|}{\alpha_0} \\ &\leq \frac{\|\alpha(A_\alpha+\alpha_0 I)^{-1}(F(x_\alpha)-y)\|}{\alpha_0}. \qquad (9.19)\end{aligned}$$

So by (9.17), (9.18) and (9.19) we have

$$\|x_\alpha - x_{\alpha_0}\| \leq \frac{\|\alpha(A_\alpha+\alpha_0 I)^{-1}(F(x_\alpha)-y)\|}{\alpha_0(1-4K_0\|x-\hat{x}\|)}.$$

□

Lemma 9.3.8 *Let assumptions of Lemma 9.3.6 hold. Then*

$$\|\alpha(A_\alpha+\alpha_0 I)^{-1}(F(x_\alpha)-y)\| \leq \beta\delta,$$

where $\beta = (c+2)(1+\frac{K_0(4c-3)\|x_0-\hat{x}\|}{c-1})$.

Proof: Observe that

$$\begin{aligned}\|\alpha(A_\alpha+\alpha_0 I)^{-1}(F(x_\alpha)-y)\| &\leq \|\alpha(A_\alpha^\delta+\alpha_0 I)^{-1}(F(x_\alpha)-y)\| \\ &\quad +\|\alpha[(A_\alpha+\alpha_0 I)^{-1}-(A_\alpha^\delta+\alpha_0 I)^{-1}] \\ &\quad (F(x_\alpha)-y)\| \\ &\leq \|\alpha(A_\alpha^\delta+\alpha_0 I)^{-1}(F(x_\alpha)-y)\| \qquad (9.20) \\ &\quad +\|\alpha(A_\alpha+\alpha_0 I)^{-1}(A_\alpha^\delta-A_\alpha)(A_\alpha^\delta+\alpha_0 I)^{-1} \\ &\quad (F(x_\alpha)-y)\|.\end{aligned}$$

Let $a_\delta = \alpha(A_\alpha^\delta + \alpha_0 I)^{-1}(F(x_\alpha) - y)$. Then

$$\begin{aligned}
\|\alpha(A_\alpha + \alpha_0 I)^{-1}(F(x_\alpha)-y)\| &\leq \|a_\delta\| + \|\alpha(A_\alpha+\alpha_0 I)^{-1}(A_\alpha^\delta - A_0 + A_0 - A_\alpha)\\
&\quad (A_\alpha^\delta + \alpha_0 I)^{-1}(F(x_\alpha)-y)\|\\
&\leq \|a_\delta\| + \| - (A_\alpha + \alpha_0 I)^{-1} A_\alpha^\delta \varphi(x_0, x_\alpha^\delta, a_\delta)\|\\
&\quad + \|(A_\alpha + \alpha_0 I)^{-1} A_\alpha \varphi(x_0, x_\alpha, a_\delta)\|\\
&\leq \|a_\delta\|(1 + K_0(\|x_0 - x_\alpha^\delta\| + \|x_0 - x_\alpha\|))\\
&\leq \|a_\delta\|(1 + K_0(\|x_\alpha - x_\alpha^\delta\| + 2\|x_0 - x_\alpha\|))\\
&\leq \|a_\delta\|(1 + K_0(\frac{\delta}{\alpha} + 4\|x_0 - \hat{x}\|))\\
&\leq \|a_\delta\|(1 + K_0(\frac{1}{c-1} + 4)\|x_0 - \hat{x}\|)\\
&\leq \|a_\delta\|(1 + K_0\frac{4c-3}{c-1}\|x_0 - \hat{x}\|).
\end{aligned}$$

Now since $\|a_\delta\| \leq (c+2)\delta$, we have

$$\|\alpha(A_\alpha + \alpha_0 I)^{-1}(F(x_\alpha)-y)\| \leq (1 + K_0\frac{4c-3}{c-1}\|x_0 - \hat{x}\|)(c+2)\delta.$$

□

Now, we present the main result of the chapter.

Theorem 9.3.9 *Let assumptions of Lemma 9.3.6 hold. If, in addition,* $9K_0\|x_0 - \hat{x}\| < 2$, *then*

$$\|x_\alpha^\delta - \hat{x}\| \leq \wp \psi^{-1}(\eta\delta)$$

where $\wp = \frac{1}{\xi} + \frac{1}{1-4K_0\|x_0-\hat{x}\|} + C$ *and* $\eta = \max\{\xi, \beta\}$.

Proof: Firstly, we begin with the case $\alpha := \alpha(\delta) \leq \alpha_0$. Then by Theorem 9.3.2 we get

$$\|x_\alpha^\delta - \hat{x}\| \leq C(\frac{\delta}{\alpha} + \varphi(\alpha_0)). \tag{9.21}$$

Now, we follow with the case $\alpha := \alpha(\delta) > \alpha_0$. In this case by Theorem 9.3.7, we obtain

$$\begin{aligned}
\|x_\alpha^\delta - \hat{x}\| &\leq \|x_\alpha^\delta - x_\alpha\| + \|x_\alpha - x_{\alpha_0}\| + \|x_{\alpha_0} - \hat{x}\|\\
&\leq \frac{\delta}{\alpha} + C\varphi(\alpha_0) + \frac{\|\alpha(A_\alpha + \alpha_0 I)^{-1}(F(x_\alpha)-y)\|}{\alpha_0(1 - 4K_0\|x - \hat{x}\|)}.
\end{aligned}\tag{9.22}$$

From (9.21) and (9.22)

$$\|x_\alpha^\delta - \hat{x}\| \leq \frac{\delta}{\alpha} + C\varphi(\alpha_0) + \frac{\|\alpha(A_\alpha + \alpha_0 I)^{-1}(F(x_\alpha)-y)\|}{\alpha_0(1 - 4K_0\|x - \hat{x}\|)}$$

for all $\alpha \in (0,a]$. Let $\alpha_0 := \varphi^{-1}\psi^{-1}(\beta\delta)$ with $\beta = (1+K_0(4c-3))\|x_0-\hat{x}\|(c+2)/(c-1)$. Then by Lemma 9.3.6 and 9.3.8, we get

$$\begin{aligned}\|x_\alpha^\delta - \hat{x}\| &\le \frac{\delta}{\alpha} + C\varphi(\alpha_0) \\ &\quad + \frac{\|\alpha(A_\alpha+\alpha_0 I)^{-1}(F(x_\alpha)-y)\|}{\alpha_0(1-4K_0\|x-\hat{x}\|)} \\ &\le \frac{\delta}{\varphi^{-1}\psi^{-1}(\xi\delta)} + \frac{\beta\delta}{(1-4K_0\|x-\hat{x}\|)\varphi^{-1}\psi^{-1}(\beta\delta)} + C\psi^{-1}(\beta\delta).\end{aligned}$$

Now since $\varphi^{-1}\psi^{-1}(\lambda) = \frac{\lambda}{\psi^{-1}(\lambda)}$ we have

$$\begin{aligned}\|x_\alpha^\delta - \hat{x}\| &\le \frac{\psi^{-1}(\xi\delta)}{\xi} + \frac{\psi^{-1}(\beta\delta)}{1-4K_0\|x-\hat{x}\|} + C\psi^{-1}(\beta\delta) \\ &\le (\frac{1}{\xi} + \frac{1}{1-4K_0\|x_0-\hat{x}\|} + C)\psi^{-1}(\eta\delta).\end{aligned} \tag{9.23}$$

where $\eta = \max\{\xi,\beta\}$. This completes the proof.

9.4 Numerical examples

Example 9.4.1 Let $X = ([0,1])$. We consider the space of continuous functions defined on $[0,1]$ equipped with the max norm and $D(F) = \overline{U(0,1)}$. Define an operator F on $D(F)$ by

$$F(h)(x) = h(x) - 5\int_0^1 x\theta h(\theta)^3 d\theta. \tag{9.24}$$

Then the Fréchet-derivative is given by

$$F'(h[u])(x) = u(x) - 15\int_0^1 x\theta h(\theta)^2 u(\theta) d\theta \tag{9.25}$$

for all $u \in D(F)$. Using (9.24), (9.25), Assumptions 10 (2), 11 for $\hat{x} = 0$, we get $K_0 = 7.5 < K = 15$.

Example 9.4.2 Let $X = D(F) = \mathbb{R}$, $\hat{x} = 0$ and define a function F on $D(F)$ by

$$F(x) = d_0 x - d_1 \sin 1 + d_1 \sin e^{d_2 x}, \tag{9.26}$$

where d_0, d_1 and d_2 are the given parameters. Note that $F(\hat{x}) = F(0) = 0$. Then it can easily be seen that, for d_2 sufficiently large and d_1 sufficiently small, $\frac{K}{K_0}$ can be arbitrarily large.

We now present two examples where Assumption 11 is not satisfied, but Assumption 10 (2) is satisfied.

Example 9.4.3 Let $X = D(F) = \mathbb{R}$, $\hat{x} = 0$ and define a function F on D by

$$F(x) = \frac{x^{1+\frac{1}{i}}}{1+\frac{1}{i}} + c_1 x - c_1 - \frac{i}{i+1}, \tag{9.27}$$

where c_1 is a real parameter and $i > 2$ is an integer. Then $F'(x) = x^{1/i} + c_1$ is not Lipschitz on D. Hence Assumption 11 is not satisfied. However, the central Lipschitz condition in Assumption 11 (2) holds for $K_0 = 1$. We also have that $F(\hat{x}) = 0$. Indeed, we have

$$\begin{aligned}\|F'(x) - F'(\hat{x})\| &= |x^{1/i} - \hat{x}^{1/i}| \\ &= \frac{|x - \hat{x}|}{\hat{x}^{\frac{i-1}{i}} + \cdots + x^{\frac{i-1}{i}}}\end{aligned}$$

and so

$$\|F'(x) - F'(\hat{x})\| \leq K_0 |x - \hat{x}|.$$

References

[1] I. K. Argyros, Convergence and Application of Newton-type Iterations, Springer, 2008.

[2] I. K. Argyros, Approximating solutions of equations using Newton's method with a modified Newton's method iterate as a starting point, *Rev. Anal. Numer. Theor. Approx.*, 36, 123–138, 2007.

[3] I. K. Argyros, A semilocal convergence for directional Newton methods, *Math. Comput. (AMS)*, 80, 327–343, 2011.

[4] I. K. Argyros and S. Hilout, Weaker conditions for the convergence of Newton's method, *J. Complexity*, 28, 364–387, 2012.

[5] I. K. Argyros, Y. J. Cho and S. Hilout, Numerical methods for equations and its applications, CRC Press, Taylor and Francis, New York, 2012.

[6] A. Bakushinskii and A. Seminova, On application of generalized discrepancy principle to iterative methods for nonlinear ill-posed problems, *Numer. Funct. Anal. Optim.*, 26, 35–48, 2005.

[7] A. Bakushinskii and A. Seminova, A posteriori stopping rule for regularized fixed point iterations, *Nonlinear Anal.*, 64, 1255–1261, 2006.

[8] A. Bakushinskii and A. Seminova, Iterative regularization and generalized discrepancy principle for monotone operator equations, *Numer. Funct. Anal. Optim.*, 28, 13–25, 2007.

[9] A. Binder, H. W. Engl, C. W. Groetsch, A. Neubauer and O. Scherzer, Weakly closed nonlinear operators and parameter identification in parabolic equations by Tikhonov regularization, *Appl. Anal.*, 55, 215–235, 1994.

[10] H. W. Engl, M. Hanke and A. Neubauer, Regularization of Inverse Problems, Dordrecht, Kluwer, 1993.

[11] H. W. Engl, K. Kunisch and A. Neubauer, Convergence rates for Tikhonov regularization of nonlinear ill-posed problems, *Inverse Problems*, 5, 523–540, 1989.

[12] Q. Jin, On the iteratively regularized Gauss-Newton method for solving nonlinear ill-posed problems, *Math. Comp.*, 69, 1603–1623, 2000.

[13] Q. Jin and Z. Y. Hou, On the choice of the regularization parameter for ordinary and iterated Tikhonov regularization of nonlinear ill-posed problems, *Inverse Problems*, 13, 815–827, 1997.

[14] Q. Jin and Z. Y. Hou, On an a posteriori parameter choice strategy for Tikhonov regularization of nonlinear ill-posed problems, *Numer. Math.*, 83, 139–159, 1990.

[15] P. Mahale and M. T. Nair, Lavrentiev regularization of nonlinear ill-posed equations under general source condition, *J. Nonlinear Anal. Optim.*, 4 (2), 193–204, 2013.

[16] P. Mahale and M. T. Nair, Iterated Lavrentiev regularization for nonlinear ill-posed problems, *ANZIAM*, 51, 191–217, 2009.

[17] P. Mahale and M.T. Nair, General source conditions for nonlinear ill-posed problems, *Numer. Funct. Anal. Optim.* 28, 111–126, 2007.

[18] M. T. Nair and U. Tautanhahn, Lavrentiev regularization for linear ill-posed problems under general source conditions, *Z. Anal. Anwendungen*, 23, 167–185, 2004.

[19] O. Scherzer, H. W. Engl and K. Kunisch, Optimal a posteriori parameter choice for Tikhonov regularization for solving nonlinear ill-posed problems, *SIAM J. Numer. Anal.*, 30, 1796–1838, 1993.

[20] E. V. Semenova, Lavrentiev regularization and balancing principle for solving ill-posed problems with monotone operators, *Comput. Methods Appl. Math.*, 4, 444–454, 2010.

[21] U. Tautenhahn, Lavrentiev regularization of nonlinear ill-posed problems, *Vietnam J. Math.*, 32, 29–41, 2004.

[22] U. Tautenhahn, On the method of Lavrentiev regularization for nonlinear ill-posed problems, *Inverse Problems*, 18, 191–207, 2002.

[23] U. Tautenhahn and Q. Jin, Tikhonov regularization and a posteriori rule for solving nonlinear ill-posed problems, *Inverse Problems*, 19, 1–21, 2003.

Chapter 10

King-Werner-type methods of order $1+\sqrt{2}$

Ioannis K. Argyros

Department of Mathematical Sciences, Cameron University, Lawton, OK 73505, USA, Email: iargyros@cameron.edu

Á. Alberto Magreñán

Universidad Internacional de La Rioja, Escuela Superior de Ingeniería y Tecnología, 26002 Logroño, La Rioja, Spain, Email: alberto.magrenan@unir.net

CONTENTS

10.1 Introduction

Iterative methods are used to generate a sequence of approximating a solution x^* of the nonlinear equation

$$F(x) = 0, \tag{10.1}$$

where F is Fréchet-differentiable operator defined on a convex subset D of a Banach space X with values in a Banach space Y.

Newton's method converges quadratically to $x^\star$ if the initial guess is close to enough to the solution. Iterative methods of convergence order higher than two such as Chebyshev-Halley-type methods [1]– [21] require the evaluation of the second Fréchet-derivative, which is very expensive in general. However, there are integral equations where the second Fréchet-derivative is diagonal by blocks and inexpensive [1,2,5] or for quadratic equations, the second Fréchet-derivative is constant. Moreover, in some applications involving stiff systems, high order methods are useful. That is why it is important to study high-order methods.

In particular, Werner in [15,16] studied a method originally proposed by King [8] defined by: Given $x_0, y_0 \in D$, let

$$\begin{aligned} x_{n+1} &= x_n - F'\left(\tfrac{x_n+y_n}{2}\right)^{-1}F(x_n), \\ y_{n+1} &= x_{n+1} - F'\left(\tfrac{x_n+y_n}{2}\right)^{-1}F(x_{n+1}) \end{aligned} \tag{10.2}$$

for each $n = 0,1,2,\ldots$ and $X = \mathbb{R}^i$, $Y = \mathbb{R}$ where i is a whole number. The local convergence analysis is based on assumptions of the form:

(H_0) There exists $x^\star \in D$ such that $F(x^\star) = 0$;

(H_1) $F \in C^{2,a}(D)$, $a \in (0,1]$;

(H_2) $F'(x)^{-1} \in L(Y,X)$ and $\|F'(x)^{-1}\| \le \Gamma$;

(H_3) The Lipschitz condition

$$\|F'(x) - F'(y)\| \le L_1\|x-y\|$$

holds for each $x,y \in D$;

(H_4) The Lipschitz condition

$$\|F''(x) - F''(y)\| \le L_{2,a}\|x-y\|^a$$

holds for each $x,y \in D$;

(H_5) $U(x^\star, b) \subseteq D$ for

$$b = \min\{b_0, \rho_0, \rho_1\},$$

where

$$b_0 = \frac{2}{L_1\Gamma}$$

and ρ_0, ρ_1 solve the system [14, p.337]

$$\begin{aligned} &Bv^{1+a} + Aw = 1 \\ &2Av^2 + Avw = w, \\ &A = \tfrac{1}{2}\Gamma L_1, \quad B = \tfrac{\Gamma L_{2,a}}{4(a+1)(a+2)}. \end{aligned}$$

The convergence order was shown to be $1+\sqrt{2}$. However, there are cases where condition (H_4) is not satisfied.

For example, define function $f : [-1,1] \to (-\infty,\infty)$ by

$$f(x) = x^2 \ln x^2 + x^2 + x + 1, \quad f(0) = 1.$$

Then, we have that $\lim_{x\to 0} x^2 \ln x^2 = 0$, $\lim_{x\to 0} x \ln x^2 = 0$, $f'(x) = 2x\ln x^2 + 4x + 1$ and $f''(x) = 2(\ln x^2 + 4)$. Then, function f does not satisfy (H_4).

Recently, McDougall et al. in [9] studied the King-Werner-type method with repeated initial points, namely:

Given $x_0 \in D$,

let

$$\begin{aligned} y_0 &= x_0 \\ x_1 &= x_0 - F'(\tfrac{x_0+y_0}{2})^{-1}F(x_0), \\ y_n &= x_n - F'(\tfrac{x_{n-1}+y_{n-1}}{2})^{-1}F(x_n), \\ x_{n+1} &= x_n - F'(\tfrac{x_n+y_n}{2})^{-1}F(x_n) \end{aligned} \tag{10.3}$$

for each $n = 1, 2, ...$ and $X = Y = \mathbb{R}$. Notice that the initial predictor step is just a Newton step based on the estimated derivative. The re-use of the derivative means that the evaluations of the y_n values in (10.3) essentially come for free, which then enables the more appropriate value of the derivative to be used in the corrector step in (10.3). Method (10.3) was also shown to be of order $1+\sqrt{2}$.

In the present chapter we study the convergence analysis of a more generalization of method (10.3) in a Banach space setting. The King-Werner-type method is defined by: Given $x_0, y_0 \in D$, let

$$\begin{aligned} x_n &= x_{n-1} - F'(\tfrac{x_{n-1}+y_{n-1}}{2})^{-1}F(x_{n-1}) \ \ for\ each\ n = 1, 2, ... \\ y_n &= x_n - F'(\tfrac{x_{n-1}+y_{n-1}}{2})^{-1}F(x_n) \ \ for\ each\ n = 1, 2, ... \end{aligned} \tag{10.4}$$

We will use only hypotheses up to the first Fréchet derivative of operator F in order to prove the convergence of the method. Hence, the applicability is extended. Notice that the efficiency index of method (10.4) is given in [9] to be $(\sqrt{2}+1)^{\frac{1}{2}} \approx 1.5538$ which is larger than $2^{\frac{1}{2}} \approx 1.4142$ of Newton's method and $3^{\frac{1}{3}} \approx 1.4422$ of the cubic convergence methods such as Halley's method [6], Chebyshev's method [2] and Potra-Ptak's method [10].

The chapter is organized as follows: In Section 10.2 results on majorizing sequences for King-Werner-type methods are presented. The semilocal and local convergence results are provided in Section 10.3. Some numerical examples are presented in Section 10.4.

10.2 Majorizing sequences for King-Werner-type methods

We present some auxiliary results on majorizing sequence that will be used for King-Werner-type methods (10.3) and (10.4).

Lemma 10.2.1

Let $L_0 > 0$, $L > 0$, $s \geq 0$, $\eta > 0$ be given parameters. Denote by α the only positive *root of polynomial p defined by*

$$p(t) = \frac{L_0}{2}t^3 + \frac{L_0}{2}t^2 + Lt - L. \tag{10.5}$$

Suppose that

$$0 < \frac{L(s+\eta)}{2(1-\frac{L_0}{2}(2+\frac{L(s+\eta)}{2-L_0 s})\eta)} \leq \alpha \leq 1 - L_0\eta. \tag{10.6}$$

Then, scalar sequence $\{t_n\}$ *generated by*

$$\begin{aligned} t_0 &= 0, t_1 = \eta, s_0 = s \\ s_n &= t_n + \frac{L(s_{n-1}+t_n-2t_{n-1})(t_n-t_{n-1})}{2(1-\frac{L_0}{2}(t_{n-1}+s_{n-1}))} \\ t_{n+1} &= t_n + \frac{L(s_{n-1}+t_n-2t_{n-1})(t_n-t_{n-1})}{2(1-\frac{L_0}{2}(t_n+s_n))} \quad for\ each\ n = 1,2... \end{aligned} \tag{10.7}$$

is well defined, increasing, bounded above by

$$t^{\star\star} = \frac{\eta}{1-\alpha} \tag{10.8}$$

and converges to its unique least upper bound $t^\star$ *which satisfies* $\eta \leq t^\star \leq t^{\star\star}$. *Moreover, the following estimates hold*

$$0 < s_n - t_n \leq \alpha(t_n - t_{n-1}), \quad 0 < t_{n+1} - t_n \leq \alpha(t_n - t_{n-1}) \tag{10.9}$$

and

$$s_n < t_{n+1} \tag{10.10}$$

for each $n = 1,2,....$

Proof: We have that $p(0) = -L < 0$ and $p(1) = L_0 > 0$. Then if follows by the intermediate value theorem that p has roots $\alpha \in (0,1)$. On the other hand, by the Descartes rule of signs α is the only positive root of polynomial p. Now from (10.6) and the definition of sequence $\{t_n\}$ we have that

$$0 < \frac{L(s_k - t_k + t_{k+1} - t_k)}{2(1-\frac{L_0}{2}(t_k+s_k))} \leq \alpha \tag{10.11}$$

and

$$0 < \frac{L(s_k - t_k + t_{k+1} - t_k)}{2(1-\frac{L_0}{2}(t_{k+1}+s_{k+1}))} \leq \alpha \tag{10.12}$$

hold for $k = 0$. Since, (10.11) and (10.12) are satisfied for $k = 0$ we get from (10.7) that

$$0 < s_k - t_k \leq \alpha(t_k - t_{k-1}) \leq \alpha^k \eta \tag{10.13}$$

$$0 < t_{k+1} - t_k \leq \alpha(t_k - t_{k-1}) \leq \alpha^k \eta \tag{10.14}$$

$$s_k \leq \frac{1-\alpha^{k+1}}{1-\alpha}\eta < t^{\star\star} \tag{10.15}$$

$$t_{k+1} \leq \frac{1-\alpha^{k+1}}{1-\alpha}\eta < t^{\star\star} \tag{10.16}$$

and

$$s_k < t_{k+1}$$

also hold for $k = 1$. Notice that if we show (10.12) using induction then (10.11) follows too for all $k = 0, 1, 2, \ldots$. Evidently, (10.12) holds, if

$$\frac{L}{2}(\alpha^k\eta + \alpha^k\eta) + \frac{L_0}{2}\alpha(\frac{1-\alpha^{k+1}}{1-\alpha} + \frac{1-\alpha^{k+2}}{1-\alpha})\eta - \alpha \leq 0. \tag{10.17}$$

Estimate (10.17) motivates us to introduce recurrent polynomials f_k on $[0,1)$ defined by

$$f_k(t) = Lt^{k-1}\eta + \frac{L_0}{2}(1 + t + \ldots + t^k + 1 + t + \ldots + t^{k+1})\eta - 1 \tag{10.18}$$

and show instead of (10.17) that

$$f_k(\alpha) \leq 0 \; for \; each \; k = 1, 2, \ldots. \tag{10.19}$$

We need a relationship between two consecutive polynomials f_k. Using (10.18) and some algebraic manipulation we get that

$$f_{k+1}(t) = f_k(t) + p(t)t^{k-1}\eta, \tag{10.20}$$

where polynomial p is defined by (10.5). In particular, we have by (10.20), the definition of α and p that

$$f_{k+1}(\alpha) = f_k(\alpha), \tag{10.21}$$

since $p(\alpha) = 0$.

Let us define function f_∞ on $[0,1)$ by

$$f_\infty(t) = \lim_{k\to\infty} f_k(t). \tag{10.22}$$

Moreover, using (10.18) and (10.22) we get that

$$f_\infty(\alpha) = \frac{L_0\eta}{1-\alpha} - 1. \tag{10.23}$$

Then, in view of (10.19), (10.21) and (10.23) we must show instead of (10.19) that

$$f_\infty(\alpha) \leq 0, \tag{10.24}$$

which is true by (10.6). Induction is now complete. It follows that sequence $\{t_k\}$ is increasing, bounded above by $t^{\star\star}$ and as such that it converges to some $t^\star$. □

Remark 10.2.2 Estimate (10.6) is the sufficient convergence condition for sequence $\{t_n\}$. It shows how small η and s should be chosen so that (10.6) is satisfied.

Next, we present a result related to another majorizing, tighter than sequence $\{t_n\}$, sequence for King-Werner-type methods (10.3) and (10.4), but for which the sufficient convergence condition is more complicated than (10.6).

Let us define the scalar sequence $\{r_n\}$ by

$$r_0=0, r_1=\eta, q_0=s, q_1=r_1+\frac{L_0(q_0+r_1-2r_0)(r_1-r_0)}{2(1-\frac{L_0}{2}(r_0+q_0))}, r_2=r_1+\frac{L_0(q_0+r_1-2r_0)(r_1-r_0)}{2(1-\frac{L_0}{2}(r_1+q_1))},$$
$$q_n=r_n+\frac{L(q_{n-1}+r_n-2r_{n-1})(r_n-r_{n-1})}{2(1-\frac{L_0}{2}(r_{n-1}+q_{n-1}))}$$
$$r_{n+1}=r_n+\frac{L(q_{n-1}+r_n-2r_{n-1})(r_n-r_{n-1})}{2(1-\frac{L_0}{2}(r_n+q_n))} \tag{10.25}$$

for each n = 2,3,....

Define polynomial p_1 by

$$p_1(t)=L_0t^2+Lt-L.$$

Then,

$$\beta=\frac{-L+\sqrt{L^2+4L_0L}}{2L_0}$$

is the only positive root of polynomial p_1. Moreover, define recurrent polynomials g_k on $[0,1)$ by

$$g_k(t)=L(r_2-r_1)t^{k-1}+L_0(\frac{1-t^{k+1}}{1-t})(r_2-r_1)+\frac{L_0}{2}(2r_1+(r_2-r_1)t)-1.$$

Then, we get

$$g_{k+1}(t)=g_k(t)+p_1(t)t^{k-1}(r_2-r_1)$$

and

$$g_{k+1}(\beta)=g_k(\beta) \quad for\ each\ k=1,2,....$$

On the other hand, define function g_∞ on $[0,1)$ by

$$g_\infty(t)=\lim_{k\to\infty} g_k(t).$$

Then, we obtain

$$g_\infty(\beta)=\frac{1}{2}(2L_0r_1+(r_2-r_1)\beta L_0-2).$$

Finally, if one defines parameter β_0 by

$$\beta_0=\frac{L(q_1+r_2-2r_1)}{2(1-\frac{L_0}{2}(r_2+q_2))}.$$

It is worth noticing that r_1, q_1 and r_2 depend on the initial data. Then, using the proof of Lemma 10.2.1, the above definitions and β instead of α we arrive at the following result.

Lemma 10.2.3 Suppose that

$$0 < \beta_0 \leq \beta \leq \frac{2(1 - L_0 r_1)}{(r_2 - r_1)L_0}. \tag{10.26}$$

Then, scalar sequence $\{r_n\}$ generated by (10.25) is well defined, increasing, bounded above by $r^{\star\star}$ given by

$$r^{\star\star} = \eta + \frac{r_2 - r_1}{1 - \beta}$$

and converges to its unique least upper bound $r^\star$ which satisfies

$$r_2 \leq r^\star \leq r^{\star\star}.$$

Moreover, the following estimates hold

$$\begin{aligned}
&0 < r_{n+2} - r_{n+1} \leq \beta^n (r_2 - r_1)\\
&0 < q_{n+1} - r_{n+1} \leq \beta^n (r_2 - r_1)\\
&r_{n+2} \leq r_1 + \tfrac{1-\beta^{n+1}}{1-\beta}(r_2 - r_1)\\
&q_{n+2} \leq r_1 + \tfrac{1-\beta^{n+1}}{1-\beta}(r_2 - r_1) + \beta(r_2 - r_1)
\end{aligned}$$

and

$$r_{n+1} \leq q_n$$

for each $n = 1, 2, \ldots$.

From a straight simple induction argument and the hypothesis $L_0 \leq L$ we get in turn that:

Lemma 10.2.4 Suppose that hypotheses (10.6), (2.22) are satisfied and $L_0 \leq L$. Then, the following estimates hold

$$\begin{aligned}
&q_n \leq s_n,\\
&r_n \leq t_n,\\
&q_n - r_n \leq s_n - t_n,\\
&r_{n+1} - q_n \leq t_{n+1} - s_n,\\
&r_{n+1} - r_n \leq t_{n+1} - t_n
\end{aligned}$$

and

$$r^\star \leq t^\star.$$

Moreover strict inequality holds in all but the last inequality for each $n = 2, 3, \ldots$, if $L_0 < L$.

10.3 Convergence analysis of King-Werner-type methods

In this section we provide the convergence results.

Theorem 10.3.1 *Let $F : D \subset X \to Y$ be a Fréchet-differentiable operator. Suppose that there exist $x_0, y_0 \in D$, $L_0 > 0$, $L > 0$, $s \geq 0$, $\eta > 0$ with $L_0 \leq L$ such that for each $x, y \in D$*

$$F'(x_0)^{-1} \in L(Y,X) \tag{10.27}$$

$$\|F'(x_0)^{-1}F(x_0)\| \leq \eta \tag{10.28}$$

$$\|x_0 - y_0\| \leq s \tag{10.29}$$

$$\|F'(x_0)^{-1}(F'(x) - F'(y))\| \leq L\|x-y\| \tag{10.30}$$

$$\|F'(x_0)^{-1}(F'(x) - F'(x_0))\| \leq L_0\|x - x_0\| \tag{10.31}$$

$$\overline{U}(x_0, t^\star) \subseteq D \tag{10.32}$$

and hypotheses of Lemma 10.2.1 hold, where $t^\star$ is given in Lemma 10.2.1. Then, sequence $\{x_n\}$ generated by King-Werner-type (10.4) is well defined, remains in $\overline{U}(x_0, t^\star)$ and converges to a unique solution $x^\star \in \overline{U}(x_0, t^\star)$ of equation $F(x) = 0$. Moreover, the following estimates hold for each $n = 0, 1, 2, ...$

$$\|x_n - x^\star\| \leq t^\star - t_n, \tag{10.33}$$

where, $\{t_n\}$ is given in Lemma 10.2.1. Furthermore, if there exists $R > t^\star$ such that

$$U(x_0, R) \subseteq D \tag{10.34}$$

and

$$L_0(t^\star + R) \leq 2, \tag{10.35}$$

then, the point $x^\star$ is the only solution of equation $F(x) = 0$ in $U(x_0, R)$.

Proof: We shall show using induction that the following assertions hold:

(I_k) $\|x_{k+1} - x_k\| \leq t_{k+1} - t_k$

(II_k) $\|y_k - x_k\| \leq s_k - t_k$

and

(III_k) $\|F'(\frac{x_k+y_k}{2})^{-1}F'(x_0)\| \leq \frac{1}{1-\frac{L_0}{2}(\|x_k-x_0\|+\|y_k-x_0\|)}$.

We have that $(I_k), (II_k)$ hold for $k = 0$ by the initial conditions (10.27)-(10.29). Using (10.31), (II_0), (10.6) and (10.16) we get

$$\begin{aligned} \|F'(x_0)^{-1}(F'(\tfrac{x_0+y_0}{2}) - F'(x_0))\| &\leq L_0\|\tfrac{x_0+y_0}{2} - x_0\| = \tfrac{L_0}{2}\|x_0 - y_0\| \\ &\leq \tfrac{L_0 s}{2} \leq \tfrac{L_0\eta}{2} \leq \tfrac{L_0}{2}t^\star < 1. \end{aligned} \tag{10.36}$$

It follows from (10.36) and the Banach lemma on invertible operators [7] that

$F'(\frac{x_0+y_0}{2})^{-1} \in L(Y,X)$ so that (III_0) is satisfied. Suppose that $(I_k),(II_k),(III_k)$ hold for $k<n$, we get that

$$\|x_{n+1}-x_0\| \leq \sum_{i=1}^{n+1} \|x_i - x_{i-1}\| \leq \sum_{i=1}^{n+1}(t_i - t_{i-1}) = t_{n+1}-t_0 = t_{n+1} < t^\star,$$

$$\|y_n - x_0\| \leq \|y_n - x_n\| + \|x_n - x_0\| \leq s_n - t_n + t_n - t_0 = s_n < t^\star$$
$$\|\frac{x_n+y_n}{2} - x_0\| \leq \frac{1}{2}[\|x_n - x_0\| + \|y_n - x_0\|] < \frac{1}{2}(t^\star + t^\star) = t^\star,$$

and

$$\|x_n + \theta(x_{n+1}-x_n) - x_0\| \leq t_n + \theta(t_{n+1}-t_n) \leq t^\star.$$

Therefore, $x_{n+1}, y_n, \frac{x_n+y_n}{2} \in \overline{U}(x_0,t^\star)$. As in (10.36) we get that

$$\begin{aligned} \|F'(x_0)^{-1}(F'(\tfrac{x_n+y_n}{2}) - F'(x_0))\| &\leq L_0\|\tfrac{x_n+y_n}{2} - x_0\| \\ &\leq \tfrac{L_0}{2}(\|x_n - x_0\| + \|y_n - x_0\|) \\ &\leq \tfrac{L_0}{2}(t_n + s_n) \leq L_0 t^\star < 1. \end{aligned} \tag{10.37}$$

Therefore, $F'(\frac{x_n+y_n}{2})^{-1} \in L(Y,X)$ and (III_n) holds. Using (10.4) we get in turn that

$$\begin{aligned} F(x_n) &= F(x_n) - F(x_{n-1}) - F'(\tfrac{x_{n-1}+y_{n-1}}{2})(x_n - x_{n-1}) \\ &= \int_0^1 [F'(x_{n-1}+\theta(x_n - x_{n-1})) - F'(\tfrac{x_{n-1}+y_{n-1}}{2})](x_n - x_{n-1})d\theta. \end{aligned} \tag{10.38}$$

Then, by (10.7), (10.30), (III_n) and (10.38) we get

$$\begin{aligned} &\|x_{n+1}-x_n\| \leq \\ &\|F'(\tfrac{x_n+y_n}{2})^{-1}F'(x_0)\|\|F'(x_0)^{-1}\int_0^1 [F'(x_{n-1}+\theta(x_n - x_{n-1})) \\ &-F'(\tfrac{x_{n-1}+y_{n-1}}{2})]d\theta\|\|x_n - x_{n-1}\| \\ &\leq \frac{L}{2(1-\frac{L_0}{2}(t_n+s_n))}\int_0^1 \|2x_{n-1}+2\theta(x_n - x_{n-1}) - x_{n-1} - y_{n-1}\|\|x_n - x_{n-1}\|d\theta \\ &\leq \frac{L}{2(1-\frac{L_0}{2}(t_n+s_n))}\int_0^1 \|(x_{n-1}-y_{n-1}) + 2\theta(x_n - x_{n-1})\|\|x_n - x_{n-1}\|d\theta \\ &\leq \frac{L}{2(1-\frac{L_0}{2}(t_n+s_n))}(s_{n-1}-t_{n-1}+t_n - t_{n-1})(t_n - t_{n-1}) = t_{n+1}-t_n, \end{aligned} \tag{10.39}$$

which shows (I_n). Moreover, from (10.4) we get that

$$\begin{aligned} \|y_n - x_n\| &\leq \|F'(\tfrac{x_{n-1}+y_{n-1}}{2})^{-1}F'(x_0)\|\|F'(x_0)^{-1}F(x_n)\| \\ &\leq \frac{L(s_{n-1}-t_{n-1}+t_n-t_{n-1})(t_n-t_{n-1})}{2(1-\frac{L_0}{2}(t_{n-1}+s_{n-1}))} = s_n - t_n, \end{aligned} \tag{10.40}$$

which shows that (II_n) holds. The induction is now complete. Thus, $\{x_n\}$ is a complete sequence in a Banach space X and as such it converges to some $x^\star \in \overline{U}(x_0,t^\star)$ (since $\overline{U}(x_0,t^\star)$ is a closed set). In view of (10.39) we obtain the estimate

$$\|F'(x_0)^{-1}F(x_n)\| \leq \frac{L}{2}((s_{n-1}-t_{n-1}) + (t_n - t_{n-1}))(t_n - t_{n-1}). \tag{10.41}$$

By letting $n \to \infty$ in (10.41) we obtain that $F(x^\star) = 0$. Estimate (10.33) is obtained from (I_n) by using standard majorization techniques [2,5,7,12]. The proof of the uniqueness part can be founded in [6]. □

Remark 10.3.2 It follows from the proof of (10.39) and (10.40) for $n = 1$ and $x_0 = y_0$ that (10.30) is not needed in the computation of the upper bound on $\|x_2 - x_1\|$ and $\|y_1 - x_1\|$. Then, under the hypotheses of Lemma 10.2.3 we obtain (using (10.31) instead of (10.30)) that

$$
\begin{aligned}
&\|x_2 - x_1\| \le \|F'(\tfrac{x_1+y_1}{2})^{-1}F'(x_0)\|\|F'(x_0)^{-1}\int_0^1[F'(x_0+\theta(x_1-x_0))\\
&-F'(\tfrac{x_0+y_0}{2})]\|\|x_1-x_0\|d\theta\\
&\le \frac{\|x_1-x_0\|}{1-\frac{L_0}{2}(\|x_1-x_0\|+\|y_1-x_0\|)}\|F'(x_0)^{-1}\int_0^1[F'(x_0+\theta(x_1-x_0))-F'(x_0)]d\theta\|\\
&\le \frac{L_0(q_0-r_0+r_1-r_0)(r_1-r_0)}{2(1-\frac{L_0}{2}(q_1+r_1))} = r_2 - r_1
\end{aligned}
$$

and similarly

$$\|y_1 - x_1\| \le q_1 - r_1,$$

which justify the definition of q_1, r_1, r_2 and consequently the definition of sequence $\{r_n\}$.

Then, we present the following result.

Theorem 10.3.3 Let $F : D \subseteq X \to Y$ be a Fréchet-differentiable operator. Suppose that there exist $x_0, y_0 \in D$, $L_0 > 0$, $L > 0$, $\eta > 0$ with $L_0 \le L$ such that for each $x, y \in D$

$$
\begin{gathered}
F'(x_0)^{-1} \in L(Y,X)\\
\|F'(x_0)^{-1}F(x_0)\| \le \eta\\
\|F'(x_0)^{-1}(F'(x)-F'(y))\| \le L\|x-y\|\\
\|F'(x_0)^{-1}(F'(x)-F'(x_0))\| \le L_0\|x-x_0\|\\
\overline{U}(x_0, r^\star) \subseteq D
\end{gathered}
$$

and hypotheses of Lemma 10.2.3 hold with $s = 0$, where $r^\star$ is given in Lemma 10.2.3. Then, sequence $\{x_n\}$ generated by King-Werner-type method (10.3) with $y_0 = x_0$ is well defined, remains in $\overline{U}(x_0, r^\star)$ and converges to a unique solution $x^\star \in \overline{U}(x_0, r^\star)$ of equation $F(x) = 0$. Moreover, the following estimates hold for each $n = 0, 1, 2, \ldots$

$$\|x_n - x^\star\| \le r^\star - r_n,$$

where, $\{r_n\}$ is given in Lemma 10.2.3. Furthermore, if there exists $R > r^\star$ such that

$$U(x_0, R) \subseteq D$$

and

$$L_0(r^\star + R) \le 2,$$

then, the point $x^\star$ is the only solution of equation $F(x) = 0$ in $U(x_0, R)$.

Remark 10.3.4 (a) It is worth noticing that (10.31) is not an additional to (10.30) hypothesis, since (10.30) always implies (10.31) but not necessarily vice versa. We also have that

$$L_0 \leq L$$

holds in general and $\frac{L}{L_0}$ can be arbitrarily large [2–6]. In the literature (with the exception of our works) (10.30) is only used for the computation of the upper bounds of the inverses of the operators involved.

(b) The limit point $t^\star$ (or $r^\star$) can replaced by $t^{\star\star}$ (or $r^{\star\star}$) (which are given in closed form) in the hypotheses of Theorem 10.3.1 (or Theorem 10.3.3).

Next, we present the local convergence analysis of King-Werner-type method (10.4).

Theorem 10.3.5 Let $F : D \subseteq X \to Y$ be a Fréchet-differentiable operator. Suppose that there exist $x^\star \in D$, $l_0 > 0$, $l > 0$ with $l_0 \leq l$ such that for each $x, y \in D$

$$F(x^\star) = 0,\ F'(x^\star)^{-1} \in L(Y,X) \tag{10.42}$$

$$\|F(x^\star)^{-1}(F'(x) - F'(y))\| \leq l\|x - y\| \tag{10.43}$$

$$\|F(x^\star)^{-1}(F'(x) - F'(x^\star))\| \leq l_0\|x - x^\star\| \tag{10.44}$$

and

$$\overline{U}(x^\star, \rho) \subseteq D, \tag{10.45}$$

where

$$\rho = \frac{2}{3l + 2l_0}. \tag{10.46}$$

Then, sequence $\{x_n\}$ generated by King-Werner-type method (10.4) is well defined, remains in $\overline{U}(x^\star, \rho)$ and converges to $x^\star$, provided that $x_0, y_0 \in U(x^\star, \rho)$.

Proof: Using the hypotheses that $x_0, y_0 \in U(x^\star, \rho)$, (10.44) and (10.46) we have

$$\|F'(x^\star)^{-1}(F'(x^\star) - F'(x_0))\| \leq l_0\|x^\star - x_0\| < l_0\rho < 1. \tag{10.47}$$

It follows from (10.47) and the Banach lemma on invertible operators that $F'(x_0)^{-1} \in L(Y,X)$ and

$$\|F'(x_0)^{-1}F'(x^\star)\| \leq \frac{1}{1 - l_0\|x^\star - x_0\|}. \tag{10.48}$$

Similarly, we have that

$$\|\frac{x_0 + y_0}{2} - x^\star\| \leq \frac{1}{2}(\|x^\star - x_0\| + \|x^\star - y_0\|) < \frac{1}{2}(\rho + \rho) = \rho,$$

so, $\frac{x_0+y_0}{2} \in U(x^\star, \rho)$ and

$$\|F'(x^\star)^{-1}(F'(x^\star) - F'(\frac{x_0 + y_0}{2}))\| \leq l_0\|x^\star - \frac{x_0 + y_0}{2}\| \leq \frac{l_0}{2}(\|x^\star - x_0\| + \|x^\star - y_0\|) < l_0\rho < 1. \tag{10.49}$$

Then, again we have that $F'(\frac{x_0+y_0}{2})^{-1} \in L(Y,X)$ and

$$\|F'(\frac{x_0+y_0}{2})^{-1}F'(x^\star)\| \leq \frac{1}{1-\frac{l_0}{2}(\|x^\star - x_0\| + \|x^\star - y_0\|)}. \tag{10.50}$$

Using the second substep of (10.4) and (10.42), we get in turn that

$$\begin{aligned} x_1 - x^\star &= x_0 - x^\star - F'(\tfrac{x_0+y_0}{2})^{-1}F(x_0) = F'(\tfrac{x_0+y_0}{2})^{-1} \\ &\quad [F'(\tfrac{x_0+y_0}{2})(x_0 - x^\star) - F(x_0)] \\ &= -F'(\tfrac{x_0+y_0}{2})^{-1}F'(x^\star)\int_0^1 F'(x^\star)^{-1}[F'(x^\star+\theta(x_0-x^\star)) \\ &\quad -F'(\tfrac{x_0+y_0}{2})](x_0 - x^\star)d\theta. \end{aligned} \tag{10.51}$$

Then, in view of (10.43), (10.46), (10.50) and (10.51) we obtain that

$$\begin{aligned} \|x_1 - x^\star\| &\leq \|F'(\tfrac{x_0+y_0}{2})^{-1}F'(x^\star)\| \int_0^1 \|F'(x^\star)^{-1}[F'(x^\star+\theta(x_0-x^\star)) \\ &\quad -F'(\tfrac{x_0+y_0}{2})]\|\|x_0 - x^\star\|d\theta \\ &\leq \frac{\int_0^1 l\|\frac{x_0+y_0}{2} - x^\star - \theta(x_0-x^\star)\|\|x_0-x^\star\|d\theta}{1-\frac{l_0}{2}(\|x^\star-x_0\|+\|x^\star-y_0\|)} \\ &< \frac{\int_0^1 l(|\frac{1}{2}-\theta|\|x_0-x^\star\|+\frac{1}{2}\|y_0-x^\star\|)d\theta}{1-l_0\rho}\|x_0 - x^\star\| \\ &< \frac{\frac{3}{4}l\rho}{1-l_0\rho}\|x_0-x^\star\| < \|x_0-x^\star\| < \rho. \end{aligned} \tag{10.52}$$

Therefore, $x_1 \in U(x^\star,\rho)$.

Notice that $\frac{x_0+y_0}{2} \in U(x^\star,\rho)$, since

$$\|\frac{x_0+y_0}{2} - x^\star\| \leq \frac{1}{2}(\|x_0-x^\star\| + \|y_0-x^\star\|) < \rho,$$

using the first substep of (10.4) and (10.42) we get that

$$\begin{aligned} y_1 - x^\star &= x_1 - x^\star - F'(\tfrac{x_0+y_0}{2})^{-1}F(x_1) \\ &= -F'(\tfrac{x_0+y_0}{2})^{-1}F'(x^\star)\int_0^1[F'(x^\star)^{-1}(F'(x^\star+\theta(x_1-x^\star)) \\ &\quad -F'(\tfrac{x_0+y_0}{2}))](x_1-x^\star)d\theta. \end{aligned} \tag{10.53}$$

Next, using (10.43), (10.46), (10.50) and (10.53) we obtain in turn that

$$\begin{aligned} \|y_1 - x^\star\| &\leq \|F'(\tfrac{x_0+y_0}{2})^{-1}F'(x^\star)\| \int_0^1 \|F'(x^\star)^{-1}(F'(x^\star+\theta(x_1-x^\star)) \\ &\quad -F'(\tfrac{x_0+y_0}{2}))\|\|x_1-x^\star\|d\theta \\ &\leq \frac{l}{1-\frac{l_0}{2}(\|x_0-x^\star\|+\|y_0-x^\star\|)}\left(\frac{\|x^\star-x_0\|}{2}+\frac{\|x^\star-y_0\|}{2}+\frac{\|x^\star-x_1\|}{2}\right)\|x_1-x^\star\| \\ &\leq \frac{3l\rho}{2(1-l_0\rho)}\|x_1-x^\star\| \leq \|x_1-x^\star\| < \rho, \end{aligned} \tag{10.54}$$

which shows $y_1 \in U(x^\star,\rho)$. We also have that $\frac{x_1+y_1}{2} \in U(x^\star,\rho)$. If we replace the role of x_0, y_0 by x_k, y_k in the preceding estimates we obtain as in (10.52) and (10.54) that

$$\|x_{k+1} - x^\star\| < \|x_k - x^\star\| < \rho \tag{10.55}$$

and

$$\|y_{k+1}-x^\star\| < \|x_{k+1}-x^\star\| < \rho, \tag{10.56}$$

respectively. By letting $k \to \infty$ in (10.55), we obtain $\lim_{k\to\infty} x_k = x^\star$. □

Remark 10.3.6 Notice that the radius of convergence for Newton's method due to Traub [12] or Rheinboldt [11] is given by

$$\xi = \frac{2}{3l}. \tag{10.57}$$

The corresponding radius due to us in [2,5,6] is given by

$$\xi_1 = \frac{2}{2l_0+l}. \tag{10.58}$$

Comparing (10.46), (10.57) and (10.58), we see that

$$\rho < \xi < \xi_1. \tag{10.59}$$

Notice however that King-Werner-type method (10.4) is faster than Newton's method. Finally notice that

$$\frac{\rho}{\xi} \to 1, \quad \frac{\rho}{\xi_1} \to \frac{1}{3} \quad \textit{and} \quad \frac{\xi_1}{\xi} \to 3 \quad \textit{as} \, \frac{l_0}{l} \to 0. \tag{10.60}$$

10.4 Numerical examples

Example 10.4.1 *Let $X = Y = \mathbb{R}$, $D = U(0,1)$ and $x^\star = 0$. Define mapping F on D by*

$$F(x) = e^x - 1. \tag{10.61}$$

Then, the Fréchet-derivatives of F are given by

$$F'(x) = e^x, \quad F''(x) = e^x.$$

Notice that $F(x^\star) = 0$, $F'(x^\star) = F'(x^\star)^{-1} = 1$, $a = 1$, $L_{2,1} = e$, $l_0 = e-1 < l = L_1 = e$ and $\Gamma = e$. Then, using Theorem 10.3.5 and Remark 10.3.6 we obtain the following radii

$$\xi = .245252961, \quad \xi_1 = .324947231 \quad \textit{and} \quad \rho = .172541576.$$

Then, the system in (H_5), since $A = \frac{e^2}{2}$ and $B = \frac{e^2}{24}$ becomes

$$\frac{e^2}{24}v^2 + \frac{e^2}{2}w = 1$$

$$e^2v^2 + \frac{e^2}{2}vw = w.$$

Hence, we have $\rho_0 = .1350812188$, $\rho_1 = .2691499885$, $b_0 = .270670566$ and $b = .1350812188$. Notice that under our approach and under weaker hypotheses

$$b < \rho < \xi < \xi_1.$$

Next we present an example when $X = Y = C[0,1]$.

Example 10.4.2 *Let $X = Y = C[0,1]$, the space of continuous functions defined on $[0,1]$ equipped with the max norm and $D = \overline{U}(0,1)$. Define operator F on D by*

$$F(x)(t) = x(t) - \int_0^1 t\theta x^3(\theta)d\theta.$$

Then, we have

$$F'(x)(w)(t) = w(t) - 3\int_0^1 t\theta x^2(\theta)w(\theta)d\theta \quad for\ each\ w \in D.$$

Then, we have for $x^\star(t) = 0$ $(t \in [0,1])$ that $l = L_1 = 3$ and $l_0 = \frac{3}{2}$. Then, using (10.46) we obtain that

$$\rho = \frac{1}{6}.$$

Example 10.4.3 *Let also $X - Y = C[0,1]$ equipped with the max norm and $D = U(0,r)$ for some $r > 1$. Define F on D by*

$$F(x)(s) = x(s) - y(s) - \mu \int_0^1 G(s,t)x^3(t)dt, \quad x \in C[0,1],\ s \in [0,1].$$

$y \in C[0,1]$ is given, μ is a real parameter and the Kernel G is the Green's function defined by

$$G(s,t) = \begin{cases} (1-s)t & if\ t \le s \\ s(1-t) & if\ s \le t. \end{cases}$$

Then, the Fréchet derivative of F is defined by

$$(F'(x)(w))(s) = w(s) - 3\mu \int_0^1 G(s,t)x^2(t)w(t)dt, \quad w \in C[0,1],\ s \in [0,1].$$

Let us choose $x_0(s) = y(s) = 1$ and $|\mu| < \frac{8}{3}$. Then, we have that

$$\begin{array}{l} \|I - F'(x_0)\| \le \frac{3}{8}\mu, \quad F'(x_0)^{-1} \in L(Y,X), \\ \|F'(x_0)^{-1}\| \le \frac{8}{8-3|\mu|}, \quad \eta = \frac{|\mu|}{8-3|\mu|}, \quad L_0 = \frac{3(1+r)|\mu|}{8-3|\mu|}, \end{array}$$

and

$$L = \frac{6r|\mu|}{8-3|\mu|}.$$

Let us simple choose $y_0(s) = 1$, $r = 3$ and $\mu = \frac{1}{2}$. Then, we have that

$$s = 0, \quad \eta = .076923077, \quad L_0 = .923076923, \quad L = 1.384615385$$

and

$$\frac{L(s+\eta)}{2(1-\frac{L_0}{2}(2+\frac{L(s+\eta)}{2-L_0 s})\eta)} = 0.057441746, \quad \alpha = 0.711345739,$$

$$1 - L_0\eta = 0.928994083.$$

That is, condition (10.6) *is satisfied and Theorem 10.3.1 applies.*

References

[1] S. Amat, S. Busquier and M. Negra, Adaptive approximation of nonlinear operators, *Numer. Funct. Anal. Optim.*, 25, 397–405, 2004.

[2] I. K. Argyros, Computational theory of iterative methods, Series: Studies in Computational Mathematics 15, Editors, C. K. Chui and L. Wuytack, Elservier Publ. Co. New York, U.S.A., 2007.

[3] I. K. Argyros, A semilocal convergence analysis for directional Newton methods, *Math. Comput.*, 80 (273), 327–343, 2011.

[4] I. K. Argyros and Á. A. Magreñán, On the convergence of an optimal fourth-order family of methods and its dynamics, *Appl. Math. Comput.* 252, 336–346, 2015.

[5] I. K. Argyros, A. Cordero, Á. A. Magreñán and J. R. Torregrosa, On the convergence of a higher order family of methods and its dynamics. *J. Comput. Appl. Math.*, 80, 327–343, 2016.

[6] I. K. Argyros and S. Hilout, Estimating upper bounds on the limit points of majorizing sequences for Newton's method, *Numer. Algorithms*, 62, 115–132, 2013.

[7] I. K. Argyros and S. Hilout, Computational methods in nonlinear analysis. Efficient algorithms, fixed point theory and applications, World Scientific, 2013.

[8] I. K. Argyros and H. M. Ren, Ball convergence theorems for Halley's method in Banach spaces. *J. Appl. Math. Comput.*, 38, 453–465, 2012.

[9] A. Cordero, L. Feng, Á. A. Magreñán and J. R. Torregrosa, A new fourth-order family for solving nonlinear problems and its dynamics, *J. Math. Chemistry*, 53 (3), 893–910, 2015.

[10] Y. H. Geum, Y. I., Kim and Á. A., Magreñán, A biparametric extension of King's fourth-order methods and their dynamics, *Appl. Math. Comput.*, 282, 254–275, 2016.

[11] L. V. Kantorovich and G. P. Akilov, Functional Analysis, Pergamon Press, Oxford, 1982.

[12] R. F. King, Tangent methods for nonlinear equations, *Numer. Math.*, 18, 298–304, 1972.

[13] Á. A. Magreñán, Different anomalies in a Jarratt family of iterative root-finding methods, *Appl. Math. Comput.*, 233, 29–38, 2014.

[14] Á. A. Magreñán, A new tool to study real dynamics: The convergence plane, *Appl. Math. Comput.*, 248, 215–224, 2014.

[15] T. J. McDougall and S. J. Wotherspoon, A simple modification of Newton's method to achieve convergence of order $1+\sqrt{2}$, *Appl. Math. Let.*, 29, 20–25, 2014.

[16] F. A. Potra and V. Ptak, Nondiscrete induction and iterative processes [J]. *Research Notes in Mathematics*, 103, Pitman, Boston, 5, 112–119, 1984.

[17] H. M. Ren, Q. B. Wu and W. H. Bi, On convergence of a new secant like method for solving nonlinear equations, *Appl. Math. Comput.*, 217, 583–589, 2010.

[18] W. C. Rheinboldt, An adaptive continuation process for solving systems of nonlinear equations, Polish Academy of Science, *Banach Ctr. Publ.*, 3, 129–142, 1977.

[19] J. F. Traub, Iterative Methods for the Solution of Equations, Englewood Cliffs, Prentice Hull, 1984.

[20] W. Werner, Uber ein Verfahren der Ordnung $1+\sqrt{2}$ zur Nullstellenbestimmung, *Numer. Math.*, 32, 333–342, 1979.

[21] W. Werner, Some supplementary results on the $1+\sqrt{2}$ order method for the solution of nonlinear equations, *Numer. Math.*, 38, 383–392, 1982.

Chapter 11

Generalized equations and Newton's method

Ioannis K. Argyros

Department of Mathematical Sciences, Cameron University, Lawton, OK 73505, USA, Email: iargyros@cameron.edu

Á. Alberto Magreñán

Universidad Internacional de La Rioja, Escuela Superior de Ingeniería y Tecnología, 26002 Logroño, La Rioja, Spain, Email: alberto.magrenan@unir.net

CONTENTS

11.1 Introduction

In [18], G. S. Silva considered the problem of approximating the solution of the generalized equation

$$F(x)+Q(x)\ni 0, \tag{11.1}$$

where $F : D \longrightarrow H$ is a Fréchet differentiable function, H is a Hilbert space with inner product $\langle .,. \rangle$ and corresponding norm $\|.\|$, $D \subseteq H$ an open set and $T : H \rightrightarrows H$ is set-valued and maximal monotone. It is well known that the system of nonlinear equations and abstract inequality system can be modelled as equation of the form (11.1) [17]. If $\psi : H \longrightarrow (-\infty, +\infty]$ is a proper lower semicontinuous convex function and

$$Q(x) = \partial\psi(x) = \{u \in H : \psi(y) \geq \psi(x) + \langle u, y-x\rangle\}, \text{ for all } y \in H \tag{11.2}$$

then (11.1) becomes the variational inequality problem

$$F(x) + \partial\psi(x) \ni 0,$$

including linear and nonlinear complementary problems. Newton's method for solving (11.1) for an initial guess x_0 is defined by

$$F(x_k) + F'(x_k)(x_{k+1} - x_k) + Q(x_{k+1}) \ni 0, \; k = 0, 1, 2\ldots \tag{11.3}$$

has been studied by several authors [1]– [24].

In [13], Kantorovich obtained a convergence result for Newton's method for solving the equation $F(x) = 0$ under some assumptions on the derivative $F'(x_0)$ and $\|F'(x_0)^{-1}F(x_0)\|$. Kantorovich, used the majorization principle to prove his results. Later in [16], Robinson considered generalization of the Kantorovich theorem of the type $F(x) \in K$, where K is a nonempty closed and convex cone, and obtained convergence results and error bounds for this method. Josephy [12], considered a semilocal Newton's method of the kind (11.3) in order to solving (11.1) with $F = N_C$ the normal cone mapping of a convex set $C \subset \mathbb{R}^2$.

In this chapter, we expand the applicability of Newton's method for solving (11.1) using the same functions as in [18].

The rest of this chapter is organized as follows. In Section 11.2, we present some preliminaries and in Section 11.3 we present the main results are presented.

11.2 Preliminaries

Let $U(x,\rho)$ and $\bar{U}(x,\rho)$ stand respectively for open and closed balls in H with center $x \in H$ and radius $\rho > 0$.

The following items are stated briefly here in order to make the chapter as selfcontains as possible. More details can be found in [18].

Definition 11.2.1 *Let $D \subseteq H$ be an open nonempty subset of H, $h : D \longrightarrow H$ be a Fréchet differentiable function with Fréchet derivative h' and $Q : H \rightrightarrows H$ be a set mapping. The partial linearization of the mapping $h + Q$ at $x \in H$ is the set-valued mapping $L_h(x,.) : H \rightrightarrows H$ given by*

$$L_h(x,y) := h(x) + h'(x)(y-x) + Q(y). \tag{11.4}$$

For each $x \in H$, the inverse $L_h(x,.)^{-1} : H \rightrightarrows H$ of the mapping $L_h(x,.)$ at $z \in H$ is defined by

$$L_h(x,.)^{-1} := \{y \in H : z \in h(x) + h'(x)(y-x) + Q(y)\}. \tag{11.5}$$

Definition 11.2.2 *Let $Q : H \rightrightarrows H$ be a set-valued operator. Q is said to be monotone if for any $x, y \in domQ$ and $u \in Q(y), v \in Q(x)$ implies that the following inequality holds:*

$$\langle u - v, y - x \rangle \geq 0.$$

A subset of $H \times H$ is monotone if it is the graph of a monotone operator. If $\varphi : H \longrightarrow (-\infty, +\infty]$ is a proper function then the subgradient of φ is monotone.

Definition 11.2.3 *Let $Q : H \rightrightarrows H$ be monotone. Then Q is maximal monotone if the following implication holds for all $x, u \in H$:*

$$\langle u - v, y - x \rangle \geq 0 \text{ for each } y \in domQ \text{ and } v \in Q(y) \Rightarrow x \in domQ \text{ and } v \in Q(y). \tag{11.6}$$

Lemma 11.2.4 *([22]) Let G be a positive operator (i.e.,$\langle G(x), x \rangle > 0$). The following statements about G hold:*

- $\|G^2\| = \|G\|^2$;
- *If G^{-1} exists, then G^{-1} is a positive operator.*

Lemma 11.2.5 *([21, Lemma 2.2]) Let G be a positive operator. Suppose G^{-1} exists, then for each $x \in H$ we have*

$$\langle G(x), x \rangle \geq \frac{\|x\|^2}{\|G^{-1}\|}.$$

11.3 Semilocal convergence

In this section we use some more flexible scalar majorant functions than in [18] in order to obtain our main convergence result.

Theorem 11.3.1 *Let $F : D \subseteq H \longrightarrow H$ be continuous with Fréchet derivative F' continuous on D. Let also $Q : H \rightrightarrows H$ be a set-valued operator. Suppose that there exists $x_0 \in D$ such that $F'(x_0)$ is a positive operator and $\hat{F}'(x_0)^{-1}$ exists. Let $R > 0$ and $\rho := \sup\{t \in [0, R) : U(x_0, t) \subseteq D\}$. Suppose that there exists $f_0 : [0, R) \longrightarrow \mathbb{R}$ twice continuously differentiable such that for each $x \in U(x_0, \rho)$*

$$\|\hat{F}'(x_0)^{-1}\| \|F'(x) - F'(x_0)\| \leq f_0'(\|x - x_0\|) - f_0'(0). \tag{11.7}$$

Moreover, suppose that

(h_1^0) $f_0(0) > 0$ *and* $f_0'(0) = -1$.

(h_2^0) f_0' *is convex and strictly increasing.*

(h_3^0) $f_0(t) = 0$ *for some* $t \in (0, R)$.

Then, sequence $\{t_n^0\}$ *generated by* $t_0^0 = 0$,

$$t_{n+1}^0 = t_n^0 - \frac{f_0(t_n^0)}{f_0'(t_n^0)}, \quad n = 0, 1, 2, \ldots$$

is strictly increasing, remains in $(0, t_0^*)$ *and converges to* t_0^*, *where* t_0^* *is the smallest zero of function* f_0 *in* $(0, R)$. *Furthermore, suppose that for each* $x, y \in D_1 := \bar{U}(x_0, \rho) \cap U(x_0, t_0^*)$ *there exists* $f_1 : [0, \rho_1) \longrightarrow \mathbb{R}$, $\rho_1 = \min\{\rho, t_0^*\}$ *such that*

$$\|\hat{F}'(x_0)^{-1}\| \|F'(y) - F'(x)\| \leq f_1'(\|x - y\| + \|x - x_0\|) - f_1'(\|x - x_0\|) \tag{11.8}$$

$$\text{and } \|x_1 - x_0\| \leq f_1'(0)$$

Moreover, suppose that

(h_1^1) $f_1(0) > 0$ *and* $f_1'(0) = -1$.

(h_2^1) f_1' *is convex and strictly increasing.*

(h_3^1) $f_1(t) = 0$ *for some* $t \in (0, \rho_1)$.

(h_4^1) $f_0(t) \leq f_1(t)$ *and* $f_0'(t) \leq f_1'(t)$ *for each* $t \in [0, \rho_1)$.

Then, f_1 *has a smallest zero* $t_1^* \in (0, \rho_1)$, *the sequences generated by generalized Newton's method for solving the generalized equation* $F(x) + Q(x) \ni 0$ *and the scalar equation* $f_1(0) = 0$, *with initial point* x_0 *and* t_0^1 *(or* $s_0 = 0$*), respectively,*

$$0 \in F(x_n) + F'(x_n)(x_{n+1} - x_n) + Q(x_{n+1}),$$

$$t_{n+1}^1 = t_n^1 - \frac{f_1(t_n)}{f_1'(t_n)} \quad (\text{or } s_{n+1} = s_n - \frac{f_1(s_n)}{f_0'(s_n)})$$

are well defined, $\{t_n^1\}$ *(or* s_n*) is strictly increasing, remians in* $(0, t_1^*)$ *and converges to* t_1^*. *Moreover, sequence* $\{x_n\}$ *generated by generalized Newton's method (11.3) is well defined, remains in* $U(x_0, t_1^*)$ *and converges to a point* $x^* \in \bar{U}(x_0, t_1^*)$, *which is the unique solution of generalized equation* $F(x) + Q(x) \ni 0$ *in* $\bar{U}(x_0, t_1^*)$. *Furthermore, the following estimates hold:*

$$\|x_n - x^*\| \leq t_1^* - t_n^1, \quad \|x_n - x^*\| \leq t_1^* - s_n,$$

$$\|x_{n+1} - x^*\| \leq \frac{t_1^* - t_{n+1}^1}{(t_1^* - t_n^1)^2} \|x_n - x^*\|^2$$

$$\|x_{n+1} - x^*\| \leq \frac{t_1^* - s_{n+1}^1}{(t_1^* - s_n^1)^2} \|x_n - x^*\|^2$$

$$s_{n+1} - s_n \leq t_{n+1}^1 - t_n^1,$$

and sequences $\{t_n^1\}$, $\{s_n\}$ *and* $\{x_n\}$ *converge* $Q-$*linearly as follows:*

$$\|x_{n+1}-x^*\| \leq \frac{1}{2}\|x_n-x^*\|,$$

$$t_1^*-t_{n+1}^1 \leq \frac{1}{2}(t_1^*-t_n),$$

and

$$t_1^*-s_{n+1}^1 \leq \frac{1}{2}(t_1^*-s_n).$$

Finally, if
(h_5^1*)* $f_1'(t_1^*)<0,$
then the sequences $\{t_n^1\}$, $\{s_n\}$ *and* $\{x_n\}$ *converge* $Q-$*quadratically as follows:*

$$\|x_{n+1}-x^*\| \leq \frac{D^- f_1'(t_1^*)}{-2f_1'(t_1^*)}\|x_n-x^*\|^2,$$

$$\|x_{n+1}-x^*\| \leq \frac{D^- f_1'(t_1^*)}{-2f_0'(t_1^*)}\|x_n-x^*\|^2,$$

$$t_1^*-t_{n+1} \leq \frac{D^- f_1'(t_1^*)}{-2f_1'(t_1^*)}(t_1^*-t_n)^2,$$

and

$$t_1^*-s_{n+1} \leq \frac{D^- f_1'(t_1^*)}{-2f_0'(t_1^*)}(t_1^*-s_n)^2,$$

where D^- *stands for the left directional derivative of function* f_1*.*

Remark 11.3.2 *(a) Suppose that there exists* $f:[0,R) \longrightarrow \mathbb{R}$ *twice continuously differentiable such that for each* $x,y \in U(x_0,\rho)$

$$\|\hat{F}'(x_0)^{-1}\|\|F'(y)-F'(x)\| \leq f'(\|x-y\|+\|x-x_0\|)-f'(\|x-x_0\|). \tag{11.9}$$

If $f_1(t)=f_0(t)=f(t)$ *for each* $t\in[0,R)$*, then Theorem 11.3.1 specializes to Theorem 4 in [18]. On the other hand, if*

$$f_0(t) \leq f_1(t) \leq f(t) \text{ for each } t\in[0,\rho_1), \tag{11.10}$$

then, our Theorem is an improvement of Theorem 4 under the same computational cost, since in practice the computation of function f *requires the computation of functions* f_0 *or* f_1 *as special cases. Moreover, we have that for each* $t\in[0,\rho_1)$

$$f_0(t) \leq f(t) \tag{11.11}$$

and

$$f_1(t) \leq f(t) \tag{11.12}$$

leading to $t_n^1 \leq t_n$, $s_n \leq t_n$,

$$t_{n+1}^1 - t_n^1 \leq t_{n+1} - t_n \tag{11.13}$$

$$s_{n+1}^1 - s_n^1 \leq s_{n+1} - s_n \tag{11.14}$$

and

$$t_1^* \leq t^*, \tag{11.15}$$

where $\{t_n\}$ *is defined by*

$$t_0 = 0,\, t_{n+1} = t_n - \frac{f(t_n)}{f'(t_n)},$$

$t^* = \lim_{n \longrightarrow \infty} t_n$ *and* t^* *is the smallest zero of function* f *in* $(0,R)$ *(provided that the "h" conditions hold for function* f *replacing* f_1 *and* f_0*). If*

$$-\frac{f_1(t)}{f_1'(t)} \leq -\frac{f(s)}{f'(s)} \tag{11.16}$$

or

$$-\frac{f_1(t)}{f_0'(t)} \leq -\frac{f(s)}{f'(s)}, \tag{11.17}$$

respectively for each $t \leq s$. *Estimates (11.13) and (11.14) can be strict if (11.16) and (11.17) hold as strict inequalities.*

(b) *We have improved the error bounds and the location of the solution* x^* *but not necessarily the convergence domain of the generalized Newton's method (11.3). We can also show that convergence domain can be improved in some interesting special cases. Let* $F \equiv \{0\}$,

$$f(t) = \frac{L}{2}t^2 - t + \eta,$$

$$f_0(t) = \frac{L_0}{2}t^2 - t + \eta$$

and

$$f_1(t) = \frac{L_1}{2}t^2 - t + \eta,$$

where $||x_1 - x_0|| \leq \eta$ *and* L, L_0 *and* L_1 *are Lipschitz constants satisfying:*

$$||\hat{F}'(x_0)^{-1}|| ||F'(y) - F'(x)|| \leq L||y - x||$$

$$||\hat{F}'(x_0)^{-1}|| ||F'(x) - F'(x_0)|| \leq L_0||x - x_0||$$

and

$$||\hat{F}'(x_0)^{-1}|| ||F'(y) - F'(x)|| \leq L_1||y - x||,$$

on the corresponding balls. Then, we have that

$$L_0 \leq L$$

and

$$L_1 \leq L.$$

The corresponding majorizing sequences are

$$t_0 = 0,\, t_{n+1} = t_n - \frac{f(t_n)}{f'(t_n)} = t_n + \frac{L(t_n - t_{n-1})^2}{2(1 - Lt_n)}$$

$$t_0^1 = 0,\, t_{n+1}^1 = t_n^1 - \frac{f_1(t_n^1)}{f_1'(t_n^1)} = t_n^1 + \frac{L_1(t_n^1 - t_{n-1}^1)^2}{2(1 - Lt_n^1)}$$

$$s_0 = 0,\, s_{n+1} = s_n - \frac{f_1(s_n) - f_1(s_{n-1}) - f_1'(s_{n-1})(s_n - s_{n-1})}{f_0'(s_n)} = s_n + \frac{L_1(s_n - s_{n-1})^2}{2(1 - L_0 s_n)}.$$

Then, sequences converge provided, respectively that

$$h = L\eta \leq \frac{1}{2} \tag{11.18}$$

and for the last two

$$h_1 = L_1\eta \leq \frac{1}{2},$$

so

$$h \leq \frac{1}{2} \Longrightarrow h_1 \leq \frac{1}{2}.$$

It turns out from the proof of Theorem 11.3.1 that sequence $\{r_n\}$ *defined by [6]*

$$r_0 = 0,\, r_1 = \eta,\, r_2 = r_1 + \frac{L_0(r_1 - r_0)^2}{2(1 - L_0 r_1)},$$

$$r_{n+2} = r_{n+1} + \frac{L_1(r_{n+1} - r_n)^2}{2(1 - L_0 r_{n+1})}$$

is also a tighter majorizing sequence than the preceding ones for $\{x_n\}$*. The sufficient convergence condition for* $\{r_n\}$ *is given by [6]:*

$$h_2 = K\eta \leq \frac{1}{2},$$

where

$$K = \frac{1}{8}(4L_0 + \sqrt{L_1 L_0 + 8L_0^2} + \sqrt{L_0 L_1}).$$

Then, we have that

$$h_1 \leq \frac{1}{2} \Longrightarrow h_2 \leq \frac{1}{2}.$$

Therefore, the old results in [18] have been also improved. Similar improvements can follow for the Smale's alpha theory [2,6] or Wang's theory [18,22,24]. Examples where $L_0 < L$ *or* $L_1 < L$ *or* $L_0 < L_1$ *can be found in [6]. Notice that (11.18) is the celebrated Newton-Kantorovich hypothesis for solving nonlinear equations using Newton's method [13] employed as a sufficient convergence condition in all earlier studies other than ours.*

(c) *The introduction of (11.8) depends on (11.7) (i.e.,* f_1 *depends on* f_0*). Such an introduction was not possible before (i.e., when* f *was used instead of* f_1*).*

Now we can prove the Theorem 11.3.1.

Proof of Theorem 11.3.1 It is clear that the iterates $\{x_n\} \in D_1$ which is a more precise location than $\bar{U}(x_0,\rho)$ used in [18], since $D_1 \subseteq \bar{U}(x_0,\rho)$. Then, the definition of function f_1 becomes possible and replaces f in the proof [18], whereas for the computation on the upper bounds $\|\hat{F}'(x)^{-1}\|$ we use the more precise f_0 than f as it is shown in the next perturbation Banach lemma [13].

□

Lemma 11.3.3 *Let* $x_0 \in D$ *be such that* $\bar{F}'(x_0)$ *is a positive operator and* $\bar{F}'(x_0)^{-1}$ *exists. If* $\|x-x_0\| \leq t < t^*$*, then* $\bar{F}'(x)$ *is a positive operator and* $\bar{F}'(x)^{-1}$ *exists. Furthermore,*

$$\|\bar{F}'(x)^{-1}\| \leq \frac{\bar{F}'(x_0)^{-1}\|}{f_0'(t)}. \tag{11.19}$$

Proof: It is easy to see that

$$\|\bar{F}'(x)-\bar{F}'(x_0)\| \leq \frac{1}{2}\|\bar{F}'(x)-\bar{F}'(x_0)\|+\frac{1}{2}\|(\bar{F}'(x)-\bar{F}'(x_0))^*\| = \|\bar{F}'(x)-\bar{F}'(x_0)\|. \tag{11.20}$$

Let $x \in \bar{U}(x_0,t)$, $0 \leq t < t^*$. Thus $f'(t) < 0$. Using (h_1^1) and (h_2^1), we get in turn that

$$\begin{aligned} \|\bar{F}'(x_0)^{-1}\|\|\bar{F}'(x)-\bar{F}'(x_0)\| &\leq \|\bar{F}'(x_0)^{-1}\|\|\bar{F}'(x)-\bar{F}'(x_0)\| \\ &\leq f_0'(\|x-x_0\|)-f_0(0) \\ &< f_0'(t)+1 < 1. \end{aligned} \tag{11.21}$$

Now, using Banach's Lemma on invertible operators, we have $\bar{F}'(x)^{-1}$ exists. Moreover, using the above inequality,

$$\|\bar{F}'(x)^{-1}\| \leq \frac{\|\bar{F}'(x_0)^{-1}\|}{1-\|\bar{F}'(x_0)^{-1}\|\|F'(x)-F'(x_0)\|} \leq \frac{\|\bar{F}'(x_0)^{-1}\|}{1-(f_0'(t)+1)} = -\frac{\|F'(x_0)^{-1}\|}{f_0'(t)}.$$

Now, using (11.21) we obtain in turn that

$$\|\bar{F}'(x)-\bar{F}'(x_0)\| \leq \frac{1}{\|\bar{F}'(x_0)^{-1}\|}. \tag{11.22}$$

Consequently, we get

$$\langle(\bar{F}'(x_0)-\bar{F}'(x))y,y\rangle \leq \|\bar{F}'(x_0)-\bar{F}'(x)\|\|y\|^2 \leq \frac{\|y\|^2}{\|\bar{F}'(x_0)^{-1}\|},$$

which implies,

$$\langle\bar{F}'(x_0)y,y\rangle - \frac{\|y\|^2}{\|\bar{F}'(x_0)^{-1}\|} \leq \langle\bar{F}'(x)y,y\rangle. \tag{11.23}$$

Now since $\bar{F}'(x_0)$ is a positive operator and $\bar{F}'(x_0)^{-1}$ exists by assumption, we get

$$\langle\bar{F}'(x_0)y,y\rangle \geq \frac{\|y\|^2}{\|\bar{F}'(x_0)^{-1}\|}. \tag{11.24}$$

The result now follows from (11.23) and (11.24).

□

Remark 11.3.4 *This result improves the corresponding one in [18, Lemma 8] (using function f instead of f_0 or f_1) leading to more precise estimates on the distances $\|x_{n+1}-x^*\|$ which together with idea of restricted convergence domains lead to the aforementioned advantages stated in Remark 11.3.2.*

References

[1] I. K. Argyros, Concerning the convergence of Newton's method and quadratic majorants, *J. Appl. Math. Computing*, 29, 391–400, 2009.

[2] I. K. Argyros, A Kantorovich-type convergence analysis of the Newton-Josephy method for solving variational inequalities, *Numer. Algorithms*, 55, 447–466, 2010.

[3] I. K. Argyros, Variational inequalities problems and fixed point problems, *Computers Math. Appl.*, 60, 2292–2301, 2010.

[4] I. K. Argyros, Improved local convergence of Newton's method under weak majorant condition, *J. Comput. Appl. Math.*, 236, 1892–1902, 2012.

[5] I. K. Argyros, Improved local converge analysis of inexact Gauss-Newton like methods under the majorant condition, *J. Franklin Inst.*, 350 (6), 1531–1544, (2013),

[6] I. K. Argyros and S. Hilout, Weaker conditions for the convergence of Newton's method, *J. Complexity*, 28, 364–387, 2012.

[7] A. I. Dontchev and R. T. Rockafellar, Implicit functions and solution mappings, Springer Monographs in Mathematics, Springer, Dordrecht (2009).

[8] O. Ferreira, A robust semi-local convergence analysis of Newtons method for cone inclusion problems in Banach spaces under affine invariant majorant condition, *Journal of Comput. Appl. Math.*, 279, 318–335, 2015.

[9] O. P. Ferreira, M. L. N. Goncalves and P. R. Oliveria, Convergence of the Gauss-Newton method for convex composite optimization under a majorant condition, *SIAM J. Optim.*, 23 (3), 1757–1783, 2013.

[10] O. P. Ferreira and G. N. Silva, Inexact Newton's method to nonlinear functions with values in a cone, arXiv: 1510.01947 (2015).

[11] O. P. Ferreira and B. F. Svaiter, Kantorovich's majorants principle for Newton's method, *Comput. Optim. Appl.*, 42 (2), 213–229, 2009.

[12] N. Josephy, Newton's method for generalized equations and the PIES energy model, University of Wisconsin-Madison, 1979.

[13] L. V. Kantorovič, On Newton's method for functional equations, *Doklady Akad Nauk SSSR (N.S.)*, 59, 1237–1240, 1948.

[14] A. Pietrus and C. Jean-Alexis, Newton-secant method for functions with values in a cone, *Serdica Math. J.*, 39 (3-4), 271–286, 2013.

[15] F. A. Potra, The Kantorovich theorem and interior point methods, *Mathematical Programming*, 102 (1), 47–70, 2005.

[16] S. M. Robinson, Strongly regular generalized equations, *Math. Oper. Res.*, 5 (1), 43–62, 1980.

[17] R. T. Rochafellar, Convex analysis, Princeton Mathematical Series, No. 28, Princeton University Press, Princeton, N. J., 1970.

[18] G. N. Silva, On the Kantorovich's theorem for Newton's method for solving generalized equations under the majorant condition, *Appl. Math. Comput.*, 286, 178–188, 2016.

[19] S. Smale, Newtons method estimates from data at one point. In R. Ewing, K. Gross and C. Martin, editors, The Merging of Disciplines: New Directions in pure, Applied and Computational Mathematics, pages 185–196, Springer New York, 1986.

[20] J. F. Traub and H. Woźniakowski, Convergence and complexity of Newton iteration for operator equations, *J. Assoc. Comput. Mach.*, 26 (2), 250–258, 1979.

[21] L. U. Uko and I. K. Argyros, Generalized equation, variational inequalities and a weak Kantorivich theorem, *Numer. Algorithms*, 52 (3), 321–333, 2009.

[22] J. Wang, Convergence ball of Newton's method for generalized equation and uniqueness of the solution, *J. Nonlinear Convex Anal.*, 16 (9), 1847–1859, 2015.

[23] P. P. Zabrejko and D. F. Nguen, The majorant method in the theory of Newton-Kantorivich approximations and the Pták error estimates. *Numer. Funct. Anal. Optim.*, 9 (5-6), 671–684, 1987.

[24] Y. Zhang, J. Wang and S. M. Gau, Convergence criteria of the generalized Newton method and uniqueness of solution for generalized equations, *J. Nonlinear Convex. Anal.*, 16 (7), 1485–1499, 2015.

Chapter 12

Newton's method for generalized equations using restricted domains

Ioannis K. Argyros

Department of Mathematical Sciences, Cameron University, Lawton, OK 73505, USA, Email: iargyros@cameron.edu

Á. Alberto Magreñán

Universidad Internacional de La Rioja, Escuela Superior de Ingeniería y Tecnología, 26002 Logroño, La Rioja, Spain, Email: alberto.magrenan@unir.net

CONTENTS

12.1 Introduction

In this chapter we are concerned with the study of the generalized equation

$$F(x) + Q(x) \ni 0, \tag{12.1}$$

where $F : D \longrightarrow H$ is a nonlinear Fréchet differentiable defined on the open subset D of the Hilbert space H, and $Q : H \rightrightarrows H$ is set-valued and maximal monotone. Many problems from Applied Sciences can be solved finding the solutions of equations in a form like (12.1) [17–22,27,28,30,39]. If $\psi : H \longrightarrow (-\infty, +\infty]$ is a strict lower semicontinuous convex function and

$$Q(x) = \partial\psi(x) = \{u \in H : \psi(y) \geq \psi(x) + \langle u, y - x\rangle\}, \text{ for all } y \in H,$$

then (12.1) becomes the variational inequality problem

$$F(x) + \partial\psi(x) \ni 0,$$

including linear and nonlinear complementary problems, additional comments about such problems can be found in [1–32,36–40].

In the present chapter, we consider the well-known Newton's method defined as

$$F(x_k) + F'(x_k)(x_{k+1} - x_k) + F(x_{k+1}) \ni 0 \quad \text{for each } n = 0, 1, 2, \ldots \tag{12.2}$$

for solving (12.1) in an approximate way. We will use the idea of restricted convergence domains to present a convergence analysis of (12.2). In our analysis we relax the Lipschitz type continuity of the derivative of the operator involved. Our main goal in this chapter is to find larger convergence domain for the method (12.2). By means of using the restricted convergence domains, we obtained a finer convergence analysis, with the advantages (**A**):

- tighter error estimates on the distances involved and
- the information on the location of the solution is at least as precise.

These advantages were obtained (under the same computational cost) using the same or weaker hypotheses as in [36].

The rest of the chapter is organized as follows. In Section 12.2 we present some preliminaries, then in Section 12.3 we present the local convergence study of method (12.2). Finally, in Section 12.4 we provide the special cases and numerical examples.

12.2 Preliminaries

In order to make the chapter as self contained as possible we reintroduce some standard notations and auxiliary results for the monotonicity of set valued operators presented in [18,22,27].

Throughout this chapter we will denote by $U(w, \rho), \overline{U}(w, \rho)$, the open and closed balls in $\mathbb{B}_1$, respectively, with center $w \in H$ and of radius $\rho > 0$.

Definition 12.2.1 *Let $Q : H \rightrightarrows H$ be a set-valued operator. Q is said to be monotone if for any $x, y \in domQ$ and, $u \in Q(y), v \in Q(x)$ implies that the following inequality holds:*

$$\langle u - v, y - x \rangle \geq 0.$$

A subset of $H \times H$ is monotone if it is the graph of a monotone operator. If $\varphi : H \longrightarrow (-\infty, +\infty]$ is a proper function then the subgradient of φ is monotone.

Definition 12.2.2 *Let $Q : H \rightrightarrows H$ be monotone. Then Q is maximal monotone if the following holds for all $x, u \in H$:*

$$\langle u - v, y - x \rangle \geq 0 \text{ for each } y \in domQ \text{ and } v \in Q(y) \Longrightarrow x \in domQ \text{ and } v \in Q(x).$$

We will be using the following results for proving our results.

Lemma 12.2.3 *Let G be a positive operator (i.e., $\langle G(x), x \rangle \geq 0$). The following statements about G hold:*

- $\|G^2\| = \|G\|^2$;
- *If G^{-1} exists, then G^{-1} is a positive operator.*

Lemma 12.2.4 *Let G be a positive operator. Suppose that G^{-1} exists, then for each $x \in H$ we have*

$$\langle G(x), x \rangle \geq \frac{\|x\|^2}{\|G^{-1}\|}.$$

Lemma 12.2.5 *Let $B : H \longrightarrow H$ be a bounded linear operator and $I : H \longrightarrow H$ the identity operator. If $\|B - I\| < 1$ then B is invertible and $\|B^{-1}\| \leq \frac{1}{(1 - \|B - I\|)}$.*

Let $G : H \longrightarrow H$ be a bounded linear operator. Then $\widehat{G} := \frac{1}{2}(G + G^*)$ where G^* is the adjoint of G. Hereafter, we assume that $Q : H \rightrightarrows H$ is a set valued maximal monotone operator and $F : H \longrightarrow H$ is a Fréchet differentiable operator.

12.3 Local convergence

In this section, we present our main convergence result for Newton's method (12.2),

Theorem 12.3.1 *Let $F : D \subset H \longrightarrow H$ be nonlinear operator with a continuous Fréchet derivative F', where D is an open subset of H. Let $Q : H \rightrightarrows H$ be a set-valued operator and $x^* \in D$. Suppose that $0 \in F(x^*) + Q(x^*), F'(x^*)$ is a positive operator and $\widehat{F'(x^*)}^{-1}$ exists. Let $R > 0$ and suppose that there exist $f_0, f : [0, R) \longrightarrow \mathbb{R}$ twice continuously differentiable such that*

(h₀)

$$\|\widehat{F'(x^*)}^{-1}\|\|F'(x)-F'(x^*)\| \leq f_0'(\|x-x^*\|)-f_0'(0), \tag{12.3}$$

$x \in D$ and

$$\|\widehat{F'(x^*)}^{-1}\|\|F'(x)-F'(x^*+\theta(x-x^*))\| \leq f'(\|x-x^*\|)-f'(\theta\|x-x^*\|) \tag{12.4}$$

for each $x \in D_0 = D \cap U(x^*,R)$, $\theta \in [0,1]$.

(h_1) $f(0)=f_0(0)$ *and* $f'(0)=f_0'(0)=-1, f_0(t)\leq f(t), f_0'(t)\leq f'(t)$ *for each* $t \in [0,R)$.

(h_2) f_0', f' *are convex and strictly increasing.*

Let $\nu := \sup\{t \in [0,R) : f'(t) < 0\}$ *and* $r := \sup\{t \in (0,\nu) : \frac{f(t)}{tf'(t)} - 1 < 1\}$. *Then, the sequence with starting point* $x_0 \in B(x^*,r)/\{x^*\}$ *and* $t_0 = \|x^* - x_0\|$, *respectively,*

$$0 \in F(x_k)+F'(x_k)(x_{k+1}-x_k)+Q(x_{k+1}),\ t_{k+1}=|t_k-\frac{f(t_k)}{f'(t_k)}|,\ k=0,1,\ldots, \tag{12.5}$$

are well defined, $\{t_k\}$ *is strictly decreasing, is contained in* $(0,r)$ *and converges to* $0, \{x_k\}$ *is contained in* $U(x^*,r)$ *and converges to the point* x^* *which is the unique solution of the generalized equation* $F(x)+Q(x) \ni 0$ *in* $U(x^*,\bar{\sigma})$, *where* $\bar{\sigma}=\min\{r,\sigma\}$ *and* $\sigma := \sup\{0<t<R : f(t)<0\}$. *Moreover, the sequence* $\{\frac{t_{k+1}}{t_k^2}\}$ *is strictly decreasing,*

$$\|x^*-x_{k+1}\| \leq [\frac{t_{k+1}}{t_k^2}]\|x_k-x^*\|^2,\ \frac{t_{k+1}}{t_k^2} \leq \frac{f''(t_0)}{2|f'(t_0)|},\ k=0,1,\ldots. \tag{12.6}$$

If, additionally $\frac{\rho f'(\rho)-f(\rho)}{\rho f'(\rho)} = 1$ *and* $\rho < R$ *then* $r=\rho$ *is the optimal convergence radius. Furthermore, for* $t \in (0,r)$ *and* $x \in \bar{U}(x^*,t)$

$$\begin{aligned}\|x_{k+1}-x^*\| &\leq \frac{e_f(\|x_k-x^*\|,0)}{|f_0'(\|x_k^*-x^*\|)|}\\ &:= \alpha_k \leq \frac{e_f(\|x_k-x^*\|,0)}{|f'(\|x_k^*-x^*\|)|} \leq \frac{|\eta_f(t)|}{t^2}\|x_k-x^*\|^2, \end{aligned} \tag{12.7}$$

where

$$e_f(s,t) := f(t)-(f(s)+f'(s)(t-s)) \text{ for each } s,t \in [0,R)$$

and

$$\eta_f(t) := t-\frac{f(t)}{f'(t)} \text{ for each } s,t \in [0,\nu).$$

Finally, by the second inequality in (12.7) there exists $r^* \geq r$ *such that* $\lim_{k\longrightarrow\infty} x_k = x^*$, *if* $x_0 \in U(x^*,r^*)-\{x^*\}$.

Before proving the results of previous Theorem we must consider the following results.

From now on we assume that the hypotheses of Theorem 12.3.1 hold.

Remark 12.3.2 *The introduction of the center-Lipschitz-type condition (12.3) (i.e., function f_0) lends to the introduction of restricted Lipschitz-type condition (12.4). The condition used in earlier studies such as [36] is given by*

$$\|\widehat{F'(x^*)}^{-1}\|\|F'(x) - F'(x^* + \theta(x - x^*))\| \leq f_1'(\|x - x^*\|) - f_1'(\theta\|x - x^*\|) \tag{12.8}$$

for each $x \in D, \theta \in [0,1]$, where $f_1 : [0, +\infty)\mathbb{R}$ is also twice continuously differentiable. It follows from (12.3), (12.4) and (12.8) that

$$f_0'(t) \leq f_1'(t) \tag{12.9}$$

$$f(t) \leq f'(t) \tag{12.10}$$

for each $t \in [o, \nu)$ since $D_0 \subseteq D$. If $f_0'(t) = f_1'(t) = f'(t)$ for each $t \in [0, \nu)$, then our results reduce to the corresponding ones in [36]. Otherwise, (i.e., if strict inequality holds in (12.9) or (12.10)) then the new results improve the old ones. Indeed, let

$$r_1 := \sup\{t \in (0, \bar{\nu}) : -\frac{tf_1'(t) - f_1(t)}{tf_1'(t)} < 1\},$$

where $\nu_1 := \sup\{t \in [0, +\infty) : f_1'(t) < 0\}$. Then, the error bounds are (corresponding to (12.7)):

$$\begin{aligned} \|x_{k+1} - x^*\| &\leq \frac{e_{f_1}(\|x_k - x^*\|, 0)}{|f_1'(\|x_k^* - x^*\|)|} \\ &:= \beta_k \leq \frac{|\eta_{f_1}(t)|}{t^2}\|x_k - x^*\|^2. \end{aligned} \tag{12.11}$$

In view of the definition of r, r_1 and estimates (12.7), (12.9), (12.10) and (12.11), it is clear that

$$r_1 \leq r \tag{12.12}$$

and

$$\alpha_k \leq \beta_k, \; k = 1, 2, \ldots. \tag{12.13}$$

Hence, we obtain a larger radius of convergence and tighter error estimates on the distances involved, leading to a wider choice of initial guesses x^* and fewer computations of iterates x_k in order to achieve a desired error tolerance. It is also worth noticing:

The advantages are obtained under the same computational cost due to the fact in practice the computation of function f_1 requires the computation, as special cases, of functions f_0 and f. The introduction of function f was not possible before (when only function f_1 was used). This inclusion is possible using function f_0 (i.e., f is a function of f_0).

Next, we present an auxiliary Banach Lemma relating the operator F with the majorant function f_0.

Lemma 12.3.3 *Let $x^* \in H$ be such that $\widehat{F'(x^*)}$ is a positive operator and $\widehat{F'(x^*)}^{-1}$ exists. If $\|x - x^*\| \leq \min\{R, \nu\}$, then $\widehat{F'(x^*)}$ is a positive operator and $\widehat{F'(x^*)}^{-1}$ exists. Moreover,*

$$\|\widehat{F'(x)}^{-1}\| \leq \frac{\|\widehat{F'(x^*)}^{-1}\|}{|f_0'(\|x - x^*\|)|}. \tag{12.14}$$

Proof: Firstly note that

$$\|\widehat{F'(x)} - \widehat{F'(x^*)}\| \leq \frac{1}{2}\|F'(x) - F'(x^*)\| + \frac{1}{2}\|(F'(x) - F'(x^*))^*\| = \|F'(x) - F'(x^*)\|. \tag{12.15}$$

Take $x \in U(x^*, r)$. Since $r < \nu$ we have $\|x - x^*\| < \nu$. Thus, $f'(\|x - x^*\|) < 0$ which, together with (12.15), imply that for all $x \subset U(x^*, r)$

$$\|\widehat{F'(x^*)}^{-1}\|\|\widehat{F'(x)} - \widehat{F'(x^*)}\| \leq \|\widehat{F'(x^*)}^{-1}\|\|F'(x) - F'(x^*)\| \leq f_0'(\|x - x^*\|) - f_0'(0) < 1. \tag{12.16}$$

By Banach Lemma, we conclude that $\widehat{F'(x^*)}^{-1}$ exists. Moreover taking into account the last inequality

$$\|\widehat{F'(x)}^{-1}\| \leq \frac{\|\widehat{F'(x^*)}^{-1}\|}{1 - \|\widehat{F'(x^*)}^{-1}\|\|F'(x) - F'(x^*)\|} \leq \frac{\|\widehat{F'(x^*)}^{-1}\|}{1 - (f_0'(\|x - x^*\|) - f_0(0))} = \frac{\|\widehat{F'(x^*)}^{-1}\|}{|f_0'(\|x - x^*\|)|}.$$

The last step follows from the fact that $r = \min\{R, \nu\}$. Moreover, using (12.16) we have

$$\|\widehat{F'(x)} - \widehat{F'(x^*)}\| \leq \frac{1}{\|\widehat{F'(x^*)}^{-1}\|}. \tag{12.17}$$

Choose $y \in H$. Then, from the last inequality

$$\langle(\widehat{F'(x^*)} - \widehat{F'(x)})y, y\rangle \leq \|\widehat{F'(x^*)} - \widehat{F'(x)}\|\|y\|^2 \leq \frac{\|y\|^2}{\|\widehat{F'(x^*)}^{-1}\|},$$

which implies

$$\widehat{F'(x^*)}y, y\rangle - \frac{\|y\|^2}{\|\widehat{F'(x^*)}^{-1}\|} \leq \langle\widehat{F'(x)}y, y\rangle.$$

Since $\widehat{F'(x^*)}$ is a positive operator and $\widehat{F'(x^*)}^{-1}$ exists by assumption, we obtain by Lemma 12.2.5 that

$$\langle \widehat{F'(x^*)}y, y\rangle \geq \frac{\|y\|^2}{\|\widehat{F'(x^*)}^{-1}\|}.$$

Therefore, combining the two last inequalities we conclude that $\langle \widehat{F'(x)}y, y\rangle \geq 0$, i.e., $\widehat{F'(x)}$ is a positive operator.

Lemma 12.2.4 shows that $\widehat{F'(x)}$ is a positive operator and $\widehat{F'(x)}^{-1}$ exists, thus by Lemma 12.2.3 we have that for any $y \in H$

$$\langle \widehat{F'(x)}y, y\rangle \geq \frac{\|y\|^2}{\|\widehat{F'(x)}^{-1}\|}.$$

□

It is worth noticing that $\langle \widehat{F'(x)}y, y\rangle = \langle F'(x)y, y\rangle$. Thus by the second part of Lemma 12.2.4 we conclude that the Newton iteration mapping is well-defined. Now, denoting N_{F+Q}, the Newton iteration mapping for $f+F$ in that region, namely, $N_{F+Q} : U(x^*, r) \longrightarrow H$ is defined by

$$0 \in F(x) + F'(x)(N_{F+Q}(x) - x) + Q(N_{F+Q}(x)), \text{ for all } x \in U(x^*, r). \quad (12.18)$$

Remark 12.3.4 *Under condition (12.8) it was shown in [36] that*

$$\|\widehat{F'(x)}^{-1}\| \leq \frac{\|\widehat{F'(x^*)}^{-1}\|}{|f_1'(\|x - x^*\|)|} \quad (12.19)$$

instead of (12.14). However, we have that (12.14) gives a tighter error estimate than (12.15), since $|f_1'(t) \leq |f_0'(t)|$. This is really important in the proof of Theorem 12.3.1.

Proof of Theorem 12.3.1 Simply follow the proof of Theorem 4 in [36] but realize that the iterates x_k lie in D_0 which is a more precise location than D (used in [36]) allowing the usage of tighter function f than f_1 and also the usage of tighter function f_0 that f_1 for the computation of the upper bounds of the inverses $\|\widehat{F'(x)}^{-1}\|$ (i.e., we use (12.14) instead of (12.19)).

□

12.4 Special cases

First of all, let functions f_0, f, f_1 be defined by

$$f_0'(t) = \frac{L_0}{2}t^2 - t$$

$$f'(t) = \frac{L}{2}t^2 - t$$

and

$$f_1'(t) = \frac{L_1}{2}t^2 - t$$

for some positive constants L_0, L and L_1 to be determined using a specialized operator F.

Example 12.4.1 *Let $X = Y = \mathbb{R}^3, D = \bar{U}(0,1), x^* = (0,0,0)^T$. Define function F on D for $w = (x,y,z)^T$ by*

$$F(w) = (e^x - 1, \frac{e-1}{2}y^2 + y, z)^T.$$

Then, the Fréchet-derivative is given by

$$F'(v) = \begin{bmatrix} e^x & 0 & 0 \\ 0 & (e-1)y+1 & 0 \\ 0 & 0 & 1 \end{bmatrix}.$$

Notice that $L_0 = e - 1, L = e^{\frac{1}{L_0}}, L_1 = e$. Hence $f_0(t) < f(t) < f_1(t)$. Therefore, we have that the conditions of Theorem 12.3.1 hold. Moreover, we get that

$$r_1 := \frac{2}{3L_1} < r := \frac{2}{3L} < r^* := \frac{2}{2L_0 + L}.$$

Furthermore, the corresponding error bounds are:

$$\|x_{k+1} - x^*\| \leq \frac{L\|x_k - x^*\|^2}{2(1 - L_0\|x_k - x^*\|)}$$

$$\|x_{k+1} - x^*\| \leq \frac{L\|x_k - x^*\|^2}{2(1 - L\|x_k - x^*\|)}$$

and

$$\|x_{k+1} - x^*\| \leq \frac{L_1\|x_k - x^*\|^2}{2(1 - L_1\|x_k - x^*\|)}.$$

References

[1] I. K. Argyros, A semilocal convergence analysis for directional Newton methods, *Math. Comput. AMS*, 80, 327–343, 2011.

[2] I. K. Argyros, Computational theory of iterative methods, Studies in Computational Mathematics, 15, Editors: K. Chui and L. Wuytack. Elsevier New York, U.S.A., 2007.

[3] I. K. Argyros, Concerning the convergence of Newton's method and quadratic majorants, *J. Appl. Math. Comput.*, 29, 391–400, 2009.

[4] I. K. Argyros and D. González, Local convergence analysis of inexact Gauss-Newton method for singular systems of equations under majorant and center-majorant condition, *SeMA*, 69 (1), 37–51, 2015.

[5] I. K. Argyros and S. Hilout, On the Gauss-Newton method, *J. Appl. Math.*, 1–14, 2010.

[6] I. K. Argyros and S. Hilout, Extending the applicability of the Gauss-Newton method under average Lipschitz-conditions, *Numer. Algorithms*, 58, 23–52, 2011.

[7] I. K. Argyros and S. Hilout, On the solution of systems of equations with constant rank derivatives, *Numer. Algorithms*, 57, 235–253, 2011.

[8] I. K. Argyros, Y. J. Cho and S. Hilout, Numerical methods for equations and its applications, CRC Press/Taylor and Francis Group, New York, 2012.

[9] I. K. Argyros and S. Hilout, Improved local convergence of Newton's method under weaker majorant condition, *J. Comput. Appl. Math.*, 236 (7), 1892–1902, 2012.

[10] I. K. Argyros and S. Hilout, Weaker conditions for the convergence of Newton's method, *J. Complexity*, 28, 364–387, 2012.

[11] I. K. Argyros and S. Hilout, Computational Methods in Nonlinear Analysis, World Scientific Publ. Comp., New Jersey (2013).

[12] L. Blum, F. Cucker, M. Shub and S. Smale. Complexity and real computation. Springer-Verlag, New York, 1998.

[13] L. Blum, F. Cucker, M. Shub and S. Smale, Complexity and Real Computation, Springer-Verlag, New York, 1997.

[14] D. C. Chang, J. Wang and J. C. Yao, Newtons method for variational inequality problems: Smales point estimate theory under the condition, *Applicable Anal.*, 94 (1), 44–55, 2015.

[15] J. E. Dennis Jr. and R. B. Schnabel, Numerical Methods for Unconstrained Optimization and Nonlinear Equations, in: Classics in Applied Mathematics, SIAM, Philadelphia, 1996.

[16] P. Deuflhard and G. Heindl, Affine invariant convergence for Newton's method and extensions to related methods, *SIAM J. Numer. Anal.*, 16 (1), 1–10, 1979.

[17] A. L. Dontchev and R. T. Rockafellar, Implicit Functions and Solution Mappings. A View From Variational Analysis, in: Springer Monographs in Mathematics, Springer, Dordrecht, 2009.

[18] S. P. Dokov and A. L. Dontchev, Robinson's strong regularity implies robust local convergence of Newton's method Optimal control (Gainesville, FL, 1997), Vol. 15 of Appl. Optim., pages 116–129 Kluwer Acad. Publ., Dordrecht (1998).

[19] A. L. Dontchev, Local analysis of a Newton-type method based on partial linearization. In The mathematics of numerical analysis (Park City, UT, 1995). Amer. Math. Soc., Providence, RI, 1996.

[20] A. L. Dontchev, Local convergence of the Newton method for generalized equations, *C. R. Acad. Sci. Paris Ser. I Math.*, 322 (4), 327–331, 1996.

[21] A. L. Dontchev and R. T. Rockafellar, Implicit functions and solution mappings. Springer Monographs in Mathematics. Springer, Dordrecht, 2009.

[22] A. L. Dontchev and R. T. Rockafellar, Newton's method for generalized equations: a sequential implicit function theorem, *Math. Program.*, 123 (1, Ser. B), 139–159, 2010.

[23] O. Ferreira, Local convergence of Newton's method in Banach space from the viewpoint of the majorant principle, *IMA J. Numer. Anal.*, 29 (3), 746–759, 2009.

[24] O. P. Ferreira, M. L. N. Goncalves and P. R. Oliveira, Local convergence analysis of the Gauss-Newton method under a majorant condition, *J. Complexity*, 27 (1), 111–125, 2011.

[25] O. P. Ferreira and B. F. Svaiter, Kantorovich's majorants principle for Newton's method. *Comput. Optim. Appl.*, 42 (2), 213–229, 2009.

[26] O. P. Ferreira, A robust semi-local convergence analysis of Newton's method for cone inclusion problems in Banach spaces under affine invariant majorant condition, *J. Comput. Appl. Math.*, 279, 318–335, 2015.

[27] M. Josephy, Newton's Method for Generalized Equations and the PIES Energy Model. University of Wisconsin Madison, 1979.

[28] A. Pietrus and C. Jean-Alexis, Newton-secant method for functions with values in a cone, *Serdica Math. J.*, 39 (3-4), 271–286, 2013.

[29] L. B. Rall, A Note on the Convergence of Newton's Method, *SIAM J. Numer. Anal.*, 11 (1), 34–36, 1974.

[30] S. M. Robinson, Extension of Newton's method to nonlinear functions with values in a cone, *Numer. Math.*, 19, 341–347, 1972.

[31] S. M. Robinson, Strongly regular generalized equations, *Math. Oper. Res.*, 5 (1), 43–62, 1980.

[32] S. M. Robinson, Normed convex processes, *Trans. Amer. Math. Soc.*, 174, 127–140, 1972.

[33] B. Royo, J. A. Sicilia, M. J. Oliveros and E. Larrodé, Solving a long-distance routing problem using ant colony optimization. *Appl. Math.*, 9 (2L), 415–421, 2015.

[34] J. A. Sicilia, C. Quemada, B. Royo and D. Escuín, An optimization algorithm for solving the rich vehicle routing problem based on variable neighborhood search and tabu search metaheuristics. *J. Comput. Appl. Math.*, 291, 468–477, 2016.

[35] J. A. Sicilia, D. Escuín, B. Royo, E. Larrodé and J. Medrano, A hybrid algorithm for solving the general vehicle routing problem in the case of the urban freight distribution. In Computer-based Modelling and Optimization in Transportation (pp. 463–475). Springer International Publishing, 2014.

[36] G. N. Silva, Convergence of the Newton's method for generalized equations under the majorant condition, arXiv:1603.05280v1[math.NA] 16 March 2016.

[37] S. Smale, Newton method estimates from data at one point, in: R. Ewing, K. Gross and C. Martin (Eds.), The Merging of Disciplines: New Directions in Pure, Applied and Computational Mathematics, pp. 185–196. Springer-Verlag, New York, 1986.

[38] J. F. Traub and H. Wozniakowski, Convergence and complexity of Newton iteration for operator equations, *J. Assoc. Comput. Mach.*, 26 (2), 250–258, 1979.

[39] L. U. Uko, Generalized equations and the generalized Newton method, *Math. Programming*, 73 (3, Ser. A), 251–268, 1996.

[40] X. Wang, Convergence of Newton's method and uniqueness of the solution of equation in Banach space, *IMA J. Numer. Anal.*, 20, 123–134, 2000.

Chapter 13

Secant-like methods

Ioannis K. Argyros

Department of Mathematical Sciences, Cameron University, Lawton, OK 73505, USA, Email: iargyros@cameron.edu

Á. Alberto Magreñán

Universidad Internacional de La Rioja, Escuela Superior de Ingeniería y Tecnología, 26002 Logroño, La Rioja, Spain, Email: alberto.magrenan@unir.net

CONTENTS

13.1 Introduction

In this chapter we study the problem of finding a locally unique solution $x^\star$ of equation

$$F(x) = 0, \tag{13.1}$$

where F is a Fréchet-differentiable operator defined on a convex subset $\mathcal{D}$ of a Banach space $\mathcal{X}$ with values in a Banach space $\mathcal{Y}$.

Many problems from Applied Sciences can be solved finding the solutions of equations in a form like (13.1) [6, 7, 17, 22].

In this chapter, we present a both semilocal/local convergence study for the secant-like method defined for each $n = 0, 1, 2, \ldots$ by

$$x_{n+1} = x_n - A_n^{-1} F(x_n), \quad A_n = \delta F(y_n, z_n), \tag{13.2}$$

where $x_{-1}, x_0 \in \mathcal{D}$ are initial points, $y_n = g_1(x_{n-1}, x_n)$, $z_n = g_2(x_{n-1}, x_n)$, $\delta F : \mathcal{D} \times \mathcal{D} \to \mathcal{L}(\mathcal{X}, \mathcal{Y})$ and $g_1, g_2 : \mathcal{D} \times \mathcal{D} \to \mathcal{D}$. Mapping δF may be a divided difference of order one at the points $(x, y) \in \mathcal{D} \to \mathcal{D}$ with $x \neq y$ satisfying

$$\delta F(x, y)(x - y) = F(x) - F(y) \tag{13.3}$$

and if operator F is Fréchet-differentiable, then $F'(x) = \delta F(x, x)$ for each $x \in \mathcal{D}$. Many methods involving divided differences are special cases of method (13.2):

- **Amat et al. method** ($\mathcal{X} = \mathcal{Y} = \mathbb{R}^m$)

  $$x_{n+1} = x_n - \delta F(y_n, z_n)^{-1} F(x_n) \tag{13.4}$$

  Choosing $g_1(x, y) = x - aF(x)$ and $g_2(x, y) = x + bF(x)$ for $a, b \in [0, +\infty)$.

- **Ezquerro et al. method** [12]

  $$x_{n+1} = x_n - \delta F(y_n, x_n)^{-1} F(x_n) \tag{13.5}$$

  Choosing $g_1(x, y) = \lambda y + (1 - \lambda)x$ and $g_2(x, y) = y$ for all $\lambda \in [0, 1]$.

- **Kurchatov method** [6, 19, 26]

  $$x_{n+1} = x_n - \delta F(2x_n - x_{n-1}, x_{n-1})^{-1} F(x_n) \tag{13.6}$$

  Choosing $g_1(x, y) = 2y - x$ and $g_2(x, y) = x$.

- **Secant method** [5–7, 11, 17, 20–22]

  $$x_{n+1} = x_n - \delta F(x_{n-1}, x_n)^{-1} F(x_n) \tag{13.7}$$

  Choosing $g_1(x, y) = x$ and $g_2(x, y) = y$.

If $x_{-1} = x_0$ and $z_n = y_n = x_n$, for each $n = 0, 1, 2 \ldots$, then we obtain Newton-like methods [4–9, 11, 22].

Many other choices are also possible [5–7, 11, 16, 22].

A very important problem in the study of iterative procedures is the convergence domain. In general the convergence domain is small. Therefore, it is important to enlarge the convergence domain without additional hypotheses. Another important problem is to find more precise error estimates on the distances $\|x_{n+1}-x_n\|$, $\|x_n-x^\star\|$. These are our objectives in this chapter.

The simplified secant method defined for each $n=0,1,2,\ldots$ by

$$x_{n+1}=x_n-A_0^{-1}F(x_n), \quad (x_{-1},x_0\in D)$$

was first studied by S. Ulm [30].

The first semilocal convergence analysis was given by P. Laasonen [19]. His results were improved by F. A. Potra and V. Pták [20–22]. A semilocal convergence analysis for general secant-type methods was given in general by Argyros [5–9], J. E. Dennis [11], Hernández et al. [16, 17], Potra [20–22] and many others [4, 12, 14], have provided sufficient convergence conditions for the secant method based on Lipschitz–type conditions on δF.

The use of Lipschitz and center–Lipschitz conditions is one way used to enlarge the convergence domain of different methods. This technique consists on using both conditions together instead of using only the Lipschitz one which allow us to find a finer majorizing sequence, that is, a larger convergence domain. It has been used in order to find weaker convergence criteria for Newton's method by Amat et al. in [1–4], Argyros in [5–9]. On the other hand, Gutiérrez et al. in [14, 15] give sufficient conditions for Newton's method using both Lipschitz and center-Lipschitz conditions.

Here using Lipschitz and center–Lipschitz conditions, we present semilocal convergence results for method (13.2). Notice that our error bounds and the information on the location of the solution are, under the same convergence condition, more precise than the old ones given in earlier studies.

The rest of the chapter is organized as follows: In Sections 13.2 and 13.3, we present the semilocal convergence study. Then, in Sections 13.4 and 13.5 we present the local convergence analysis of the secant method. Finally, some numerical examples are presented in the concluding Section 13.6.

13.2 Semilocal convergence analysis of the secant method I

We shall show the semilocal convergence of method under a set of conditions. The conditions (C) are:

(C_1) F is a Fréchet-differentiable operator defined on a convex subset $\mathcal{D}$ of a Banach space $\mathcal{X}$ with values in a Banach space $\mathcal{Y}$.

(C_2) x_{-1} and x_0 are two points belonging in $\mathcal{D}$ satisfying the inequality

$$\|x_{-1}-x_0\|\leq c.$$

(C_3) There exist continuous and nondecreasing functions $h_1,h_2:[0,+\infty)\times$

$[0,+\infty) \to [0,+\infty)$, functions $g_1, g_2 : \mathcal{D} \times \mathcal{D} \to \mathcal{D}$ such that for each $u, v \in \mathcal{D}$ there exist $y = g_1(u,v) \in \mathcal{D}$ and $z = g_2(u,v) \in \mathcal{D}$

$$\|y - x_0\| \le h_1(\|u - x_0\|, \|v - x_0\|)$$

and

$$\|z - x_0\| \le h_2(\|u - x_0\|, \|v - x_0\|).$$

(C_4) There exist a mapping $\delta F :\to \times\mathcal{D} \to \mathcal{L}(\mathcal{X},\mathcal{Y})$, continuous, nondecreasing functions $f_1, f_2 : [0,+\infty) \times [0,+\infty) \to [0,+\infty)$ with $f_1(0,0) = f_2(0,0) = 0$, $n \ge 0$ and $x_{-1}, x_0, y_0 = g_1(x_{-1}, x_0), z_0 = g_2(x_{-1}, x_0) \in \mathcal{D}$ such that $F'(x_0)^{-1}, A_0^{-1} \in \mathcal{L}(\mathcal{Y},\mathcal{X})$,

$$\|A_0^{-1} F(x_0)\| \le \eta$$

and for each $x, y, z \in \mathcal{D}$

$$\|F'(x_0)^{-1}(\delta F(x,y) - F'(x_0))\| \le f_0(\|x - x_0\|, \|y - x_0\|)$$

and

$$\|F'(x_0)^{-1}(\delta F(x,y) - F'(z))\| \le f_1(\|x - z\|, \|y - z\|).$$

(C_5) Define functions $\varphi, \psi : [0,+\infty) \to [0,+\infty)$ by

$$\varphi(t) = f_0(h_1(t,t), h_2(t,t)),$$

and

$$\psi(t) = \int_0^1 f_1(t + \theta\eta + h_1(t+c,t), t + \theta\eta + h_2(t+c,t))d\theta.$$

Moreover, define function q by

$$q(t) = \frac{\psi(t)}{1 - \varphi(t)}.$$

Suppose that equation

$$t(1 - q(t)) - \eta = 0 \tag{13.8}$$

has at least one positive zero and the smallest positive zero, denoted by r, satisfies

$$\varphi(r) < 1,$$

$$h_1(c,0) \le \bar{r}, \quad h_1(r,r) \le \bar{r}$$

and

$$h_2(c,0) \le \bar{r}, \quad h_2(r,r) \le \bar{r}, \text{ for some } \bar{r} \ge r.$$

Set $q = q(r)$.

(C_6) $\bar{U}(x_0,\bar{r}) \subseteq \mathcal{D}$.

(C_7) There exists $r_1 \geq r$ such that

$$\int_0^1 f_0(\theta r+(1-\theta)r_1, \theta r+(1-\theta)r_1)d\theta < 1.$$

Now using the conditions (C) we present the main semilocal convergence result.

Theorem 13.2.1 *Suppose that conditions (C) hold. Then, the sequence $\{x_n\}$ generated by method (13.2) starting at x_{-1} and x_0 is well defined, remains in $U(x_0,r)$ for all $n \geq 0$ and converges to a solution $x^* \in \bar{U}(x_0,r)$ of equation $F(x)=0$, which is unique in $\bar{U}(x_0,r_1)\cap\mathcal{D}$.*

Proof: We shall first show using mathematical induction that sequence $\{x_n\}$ is well defined and belongs in $U(x_0,r)$. By (C_2)-(C_4), $y_0 = g_1(x_{-1},x_0) \in \mathcal{D}$, $z_0 = g_2(x_{-1},x_0) \in \mathcal{D}$ and $A_0^{-1} \in \mathcal{L}(\mathcal{Y},\mathcal{X})$. It follows that x_1 is well defined and $\|x_1 - x_0\| = \|A_0^{-1}F(x_0)\| < \eta < r$, since r is a solution of equation (13.8). Then, we have that $x_1 \in U(x_0,r)$. Moreover, $y_1 = g_1(x_0,x_1)$ and $z_1 = g_2(x_0,x_1)$ are well defined by the last two hypotheses in (C_5). By the second condition in (C_4), we get that

$$\begin{aligned}
\|F'(x_0)^{-1}(A_1-F'(x_0))\| &\leq f_0(\|y_1-x_0\|,\|z_1-x_0\|) \\
&\leq f_0(h_1(\|x_0-x_0\|,\|x_1-x_0\|), \\
&\qquad h_2(\|x_0-x_0\|,\|x_1-x_0\|)) \\
&\leq f_0(h_1(0,\eta),h_2(0,\eta)) \leq \varphi(r) < 1.
\end{aligned} \tag{13.9}$$

It follows from (13.9) and the Banach lemma on invertible operators [7, 18, 22] that $\mathcal{A}_1^{-1} \in \mathcal{L}(\mathcal{Y},\mathcal{X})$ and

$$\|A_1^{-1}F'(x_0)\| \leq \frac{1}{1-f_0(h_1(0,\eta),h_2(0,\eta))} \leq \frac{1}{1-\varphi(r)}. \tag{13.10}$$

Therefore, x_2 is well defined.

We can write using method (13.2) for $n=0$ and (C_1) that

$$\begin{aligned}
F(x_1) &= F(x_1)-F(x_0)-A_0(x_1-x_0) \\
&= \textstyle\int_0^1 [F'(x_0+\theta(x_1-x_0))-A_0]\,d\theta(x_1-x_0).
\end{aligned} \tag{13.11}$$

Then, using the first and second condition in (C_4), (C_3) and (13.11) we get in turn that

$$\begin{aligned}\|F'(x_0)^{-1}F(x_1)\| &= \|\int_0^1 F'(x_0)^{-1}[F'(x_0+\theta(x_1-x_0))-A_0]\,d\theta(x_1-x_0)\| \\ &\leq \int_0^1 f_1(\|x_0-x_0\|+\theta\|x_1-x_0\|+\|y_0-x_0\|, \\ &\quad \|x_0-x_0\|+\theta\|x_1-x_0\|+\|z_0-x_0\|)d\theta\|x_1-x_0\| \\ &\leq \int_0^1 f_1(\theta\eta+h_1(\|x_{-1}-x_0\|,\|x_0-x_0\|)\|x_1-x_0\| \\ &\quad +\theta\eta+h_2(\|x_{-1}-x_0\|,\|x_0-x_0\|))d\theta\|x_1-x_0\| \\ &\leq \int_0^1 f_1(\theta\eta+h_1(c,0),\theta\eta+h_2(c,0))d\theta\|x_1-x_0\| \\ &\leq \psi(r)\|x_1-x_0\|. \end{aligned} \tag{13.12}$$

By (13.2), (13.10), (C_5) and (13.12), we obtain

$$\begin{aligned}\|x_2-x_1\| &= \|A_1^{-1}F(x_1)\| \leq \|A_1^{-1}F'(x_0)\|\|F'(x_0)^{-1}F(x_1)\| \\ &\leq \frac{\psi(r)\|x_1-x_0\|}{1-\varphi(r)} \leq q\|x_1-x_0\| < \eta \end{aligned} \tag{13.13}$$

and

$$\|x_2-x_0\| = \|x_2-x_1\|+\|x_1-x_0\| \leq \frac{1-q^2}{1-q}\|x_1-x_0\| < \frac{\eta}{1-q} = r. \tag{13.14}$$

That is $x_2 \in U(x_0,r)$. On the other hand, $y_2 = g_1(x_1,x_2)$ and $z_2 = g_2(x_1,x_2)$ are well defined by the last two hypotheses in (C_5). Then, as in (13.9), we have that

$$\|F'(x_0)^{-1}(A_2-F'(x_0))\| \leq f_0(h_1(\eta,r),h_2(\eta,r)) \leq \varphi(r) < 1.$$

Hence, we have that

$$\|A_2^{-1}F'(x_0)\| \leq \frac{1}{1-f_0(h_1(\eta,r),h_2(\eta,r))} \leq \frac{1}{1-\varphi(r)}. \tag{13.15}$$

That is x_3 is well defined. Furthermore, $y_2 = g_1(x_1,x_2)$ and $z_2 = g_2(x_1,x_2)$ are well defined. Then, we have by method (13.2) for $n=0$ and (C_1) that

$$F(x_2) = F(x_2)-F(x_1)-A_1(x_2-x_1) = \int_0^1 (F'(x_1+\theta(x_2-x_1))-A_1)d\theta(x_2-x_1). \tag{13.16}$$

Then, from the first, second hypotheses in (C_4), (C_3) and (13.11) we have as in (13.12) that

$$\begin{aligned}
\|F'(x_0)^{-1}F(x_2)\| &\leq \int_0^1 f_1(\|x_1-x_0\|+\theta\|x_2-x_1\|+\|y_1-x_0\|,\\
&\quad \|x_1-x_0\|+\theta\|x_2-x_1\|+\|z_1-x_0\|)d\theta\|x_2-x_1\|\\
&\leq \int_0^1 f_1(\eta+\theta\eta+h_1(\|x_0-x_0\|,\|x_1-x_0\|),\\
&\quad \eta+\theta\eta+h_2(\|x_0-x_0\|,\|x_1-x_0\|))d\theta\|x_2-x_1\|\\
&\leq \int_0^1 f_1((1+\theta)\eta+h_1(c,\eta),(1+\theta)\eta+\\
&\quad h_2(c,\eta))d\theta\|x_2-x_1\|\\
&\leq \psi(r)\|x_2-x_1\|.
\end{aligned} \tag{13.17}$$

We get

$$\begin{aligned}
\|x_3-x_2\| &= \|A_2^{-1}F(x_0)\|\|F'(x_0)^{-1}F(x_2)\|\\
&\leq \frac{\psi(r)}{1-\varphi(r)}\|x_2-x_1\| \leq q\|x_2-x_1\| < q^2\|x_1-x_0\| < \eta
\end{aligned}$$

and

$$\|x_3-x_0\| = \|x_3-x_2\|+\|x_2-x_1\|+\|x_1-x_0\| \leq \frac{1-q^3}{1-q}\eta < \frac{\eta}{1-q} = r.$$

Then, similarly, we have that y_3, z_3 are well defined and

$$\begin{aligned}
\|A_3^{-1}F(x_0)\| &\leq \frac{1}{1-(f_0(h_1(\|x_2-x_0\|,\|x_3-x_0\|),h_2(\|x_2-x_0\|,\|x_3-x_0\|)))}\\
&\leq \frac{1}{1-f_0(h_1(r,r),h_2(r,r))} \leq \frac{1}{1-\varphi(r)}.
\end{aligned} \tag{13.18}$$

We also have that

$$\begin{aligned}\|F'(x_0)^{-1}F(x_3)\| &\le \int_0^1 f_1(\|x_2-x_0\|+\theta\|x_3-x_2\|+\|y_2-x_0\|,\\ &\quad \|x_2-x_0\|+\theta\|x_3-x_2\|+\|z_2-x_0\|)d\theta\|x_3-x_2\|\\ &\le \int_0^1 f_1(r+\theta\eta+h_1(\eta,r),\\ &\quad r+\theta\eta+h_2(\eta,r))d\theta\|x_3-x_2\|\\ &\le \psi(r)\|x_3-x_2\|.\end{aligned} \tag{13.19}$$

In view of (13.18) and (13.19) we get that

$$\begin{aligned}\|x_4-x_3\| &= \|A_3^{-1}F'(x_0)\|\|F'(x_0)^{-1}F(x_3)\|\\ &\le \frac{\psi(r)}{1-\varphi(r)}\|x_3-x_2\| \le q\|x_3-x_2\| < q^3\|x_1-x_0\| < \eta\end{aligned}$$

and

$$\begin{aligned}\|x_4-x_0\| &= \|x_4-x_3\|+\|x_3-x_2\|+\|x_2-x_1\|+\|x_1-x_0\|\\ &\le \frac{1-q^4}{1-q}\eta < \frac{\eta}{1-q} = r.\end{aligned}$$

Then, we get, as in (13.18) that y_4, z_4 are well defined and

$$\|A_4^{-1}F'(x_0)\| \le \frac{1}{1-\varphi(r)} \tag{13.20}$$

and

$$\begin{aligned}\|F'(x_0)^{-1}F(x_4)\| &\le \int_0^1 f_1(r+\theta\eta+h_1(r,r),r+\theta\eta+h_2(r,r))d\theta\|x_4-x_3\|\\ &\le \psi(r)\|x_3-x_2\|.\end{aligned} \tag{13.21}$$

Then, we have from (13.21) that

$$\begin{aligned}\|x_5-x_4\| &= \|A_4^{-1}F'(x_0)\|\|F'(x_0)^{-1}F(x_4)\|\\ &\le \frac{\psi(r)}{1-\varphi(r)}\|x_4-x_3\| \le q\|x_4-x_3\| < q^4\|x_1-x_0\| < \eta\end{aligned}$$

and

$$\begin{aligned}\|x_5-x_0\| &= \|x_5-x_4\|+\|x_4-x_3\|+\|x_3-x_2\|+\|x_2-x_1\|+\|x_1-x_0\|\\ &\le \frac{1-q^5}{1-q}\eta < \frac{\eta}{1-q} = r.\end{aligned}$$

After that, supposing that for $i = 1, 2, \ldots, k-1$

- The operator $A_k^{-1} \in \mathcal{L}(\mathcal{Y}, \mathcal{X})$ and

$$\|A_k^{-1} F'(x_0)\| \leq \frac{1}{1-\varphi(r)},$$

- $\|x_{k+1} - x_k\| \leq q \|x_k - x_{k-1}\|$
- $\|x_{k+1} - x_0\| \leq \dfrac{1-q^{k+1}}{1-q} \eta < r.$

Analogously, we get

$$\|A_{k+1}^{-1} F'(x_0)\| \leq \frac{1}{1-\varphi(r)},$$

$$\|x_{k+2} - x_{k+1}\| \leq q^{k+1} \eta \tag{13.22}$$

and

$$\|x_{k+2} - x_0\| \leq \frac{1-q^{k+2}}{1-q} \eta < r.$$

Thus, we get that $x_{k+2} \in U(x_0, r)$. That is sequence $\{x_n\}$ is well defined. Using (13.22) we have

$$\begin{aligned} \|x_{k+j} - x_k\| &= \|x_{k+j} - x_{k+j-1}\| + \|x_{k+j-1} - x_{k+j-2}\| + \cdots + \|x_{k+1} - x_k\| \\ &\leq (q^{j-1} + q^{j-2} + \ldots + q + 1) \|x_{k+1} - x_k\| \\ &= \frac{1-q^j}{1-q} \|x_{k+1} - x_k\| < \frac{q^k}{1-q} \|x_1 - x_0\|. \end{aligned} \tag{13.23}$$

for $j = 1, 2, \ldots$ and $q < 1$. It follows from (13.23) that sequence $\{x_k\}$ is complete in a Banach space $\mathcal{X}$ and as such it converges to some $x^* \in \bar{U}(x_0, r)$ (since $\bar{U}(x_0, r)$ is a closed set). From the estimate

$$\|F'(x_0)^{-1} F(x_k)\| \leq \psi(r) \|x_k - x_{k-1}\|$$

and letting $k \to \infty$ we deduce that $F(x^*) = 0$. Next, we will study the uniqueness part. Let $y^* \in \bar{U}(x_0, r)$ be such that $F(y^*) = 0$. Then, using the last condition in (C_4) and (C_7), we get in turn that for $Q = \int_0^1 F'(y^* + \theta(x^* - y^*)) d\theta$

$$\begin{aligned}\|F'(x_0)^{-1}(F'(x_0)-Q)\| &\leq \int_0^1 f_0(\|y^*+\theta(x^*-y^*)-x_0\|,\\ &\quad \|y^*+\theta(x^*-y^*)-x_0\|)d\theta\\ &\leq \int_0^1 f_0(\theta\|x^*-x_0\|+(1-\theta)\|y^*-x_0\|,\\ &\quad \theta\|x^*-x_0\|+(1-\theta)\|y^*-x_0\|)d\theta\\ &\leq \int_0^1 f_0(\theta r+(1-\theta)r_1,\theta r+(1-\theta)r_1)d\theta<1.\end{aligned} \tag{13.24}$$

If follows from (13.24) and the Banach's Lemma on invertible operators that $Q^{-1}\in\mathcal{L}(\mathcal{Y},\mathcal{X})$. Then, using the identity $0=F(x^*)-F(y^*)=Q(y^*-x^*)$, we conclude that $y^*=x^*$. ⊠.

Remark 13.2.2 **(a)** *The condition $A_0^{-1}\in\mathcal{L}(\mathcal{Y},\mathcal{X})$ can be dropped as follows. Suppose that*

$$f_0(h_1(c,0),h_2(c,0))<1. \tag{13.25}$$

Then, we have by the third condition in (C_4) that

$$\begin{aligned}\|F'(x_0)^{-1}(A_0-F'(x_0))\| &\leq f_0(\|y_0-x_0\|,\|z_0-x_0\|)\\ &\leq f_0(h_1(\|x_{-1}-x_0\|,\|x_0-x_0\|),\\ &\quad h_2(\|x_{-1}-x_0\|,\|x_0-x_0\|))\\ &\leq f_0(h_1(c,0),h_2(c,0))<1.\end{aligned}$$

It follows from (13.25) that $A_0^{-1}\in\mathcal{L}(\mathcal{Y},\mathcal{X})$ and

$$\|A_0F'(x_0)^{-1}\|\leq\frac{1}{1-f_0(h_1(c,0),h_2(c,0))}. \tag{13.26}$$

Then, due to the estimate

$$\|x_1-x_0\|\leq\|A_0^{-1}F'(x_0)\|\|F'(x_0)^{-1}F(x_0)\|\leq\frac{\|F'(x_0)^{-1}F(x_0)\|}{1-f_0(h_1(c,0),h_2(c,0))},$$

we can define

$$\eta=\frac{\|F'(x_0)^{-1}F(x_0)\|}{1-f_0(h_1(c,0),h_2(c,0))}. \tag{13.27}$$

Then, the conclusions of Theorem 2.1 hold with (13.25) replacing condition $A_0^{-1}\in\mathcal{L}(\mathcal{Y},\mathcal{X})$ and with η given by (13.27).

(b) *In view of the second and third condition in (C_4) we have that*

$$f_0(s_1,s_2)\leq f_1(s_3,s_4)\quad \textit{for each } s_i\geq 0, i=1,2,3,4 \textit{ with } s_1\leq s_3 \textit{ and } s_2\leq s_4.$$

13.3 Semilocal convergence analysis of the secant method II

Next, we show the semilocal convergence of method (13.2) under different conditions. The conditions (H) are:

(H_1) F is a continuous operator defined on a convex subset $\mathcal{D}$ of a Banach space $\mathcal{X}$ with values in a Banach space $\mathcal{Y}$.

(H_2) x_{-1} and x_0 are two points belonging in $\mathcal{D}$ satisfying the inequality

$$\|x_{-1} - x_0\| \leq c.$$

(H_3) There exist continuous and nondecreasing functions $h_1, h_2 : [0,+\infty) \times [0,+\infty) \to [0,+\infty)$, functions $g_1, g_2 : \mathcal{D} \times \mathcal{D} \to \mathcal{D}$ such that for each $u, v \in \mathcal{D}$ there exist $y = g_1(u,v) \in \mathcal{D}$ and $z = g_2(u,v) \in \mathcal{D}$

$$\|y - x_0\| \leq h_1(\|u - x_0\|, \|v - x_0\|)$$

and

$$\|z - x_0\| \leq h_2(\|u - x_0\|, \|v - x_0\|)$$

(H_4) There exist a divided difference of order one δF, continuous, nondecreasing functions $f_1, f_2 : [0,+\infty) \times [0,+\infty) \to [0,+\infty)$, $n \geq 0$ and $x_{-1}, x_0, y_0 = g_1(x_{-1}, x_0), z_0 = g_2(x_{-1}, x_0) \in \mathcal{D}$ such that $A_0^{-1} \in \mathcal{L}(\mathcal{Y}, \mathcal{X})$,

$$\|A_0^{-1} F(x_0)\| \leq \eta$$

and for each $x, y, u, v \in \mathcal{D}$

$$\|A_0^{-1}(\delta F(x,y) - A_0)\| \leq f_0(\|x - y_0\|, \|y - z_0\|)$$

and

$$\|A_0^{-1}(\delta F(x,y) - \delta F(u,v))\| \leq f_1(\|x - u\|, \|y - v\|)$$

(H_5) Define functions $\varphi, \psi : [0,+\infty) \to [0,+\infty)$ by

$$\varphi(t) = f_0(h_1(t,t) + h_1(c,0), h_2(t,t) + h_2(c,0)),$$

$$\psi(t) = f_1(t + h_1(t+c,t), t + h_2(t+c,t))$$

and

$$q(t) = \frac{\psi(t)}{1 - \varphi(t)}.$$

Suppose that equation

$$t(1 - q(t)) - \eta = 0$$

has at least one positive zero and the smallest positive zero, denoted by R, satisfies

$$\varphi(R) < 1,$$

$$h_1(c,0) \leq \bar{R}, \quad h_1(R,R) \leq \bar{R}$$

and

$$h_2(c,0) \leq \bar{R}, \quad h_2(R,R) \leq \bar{R}, \text{ for some } \bar{R} \geq R.$$

Set $q = q(R)$.

(H_6) $\bar{U}(x_0,\bar{R}) \subseteq \mathcal{D}$.

(H_7) There exists $R_1 \geq R$ such that

$$f_0(R_1 + h_1(c,0), R_1 + h_2(c,0)) < 1.$$

Using the (H) conditions we present the following semilocal convergence result.

Theorem 13.3.1 *Suppose that the conditions (H) hold. Then, the sequence $\{x_n\}$ generated by method (13.2) starting at x_{-1} and x_0 is well defined, remains in $U(x_0,R)$ for all $n \geq 0$ and converges to a solution $x^* \in \bar{U}(x_0,R)$ of equation $F(x) = 0$, which is unique in $\bar{U}(x_0,R_1) \cap \mathcal{D}$.*

It is worth noticing that the second and third conditions in (H_4) do not necessarily imply the Fréchet-differentiability of F.

13.4 Local convergence analysis of the secant method I

Next, we shall show the local convergence of method under a set of conditions (C^*). The conditions (C^*) are:

(C_1^*) F is a Fréchet-differentiable operator defined on a convex subset $\mathcal{D}$ of a Banach space $\mathcal{X}$ with values in a Banach space $\mathcal{Y}$.

(C_2^*) There exist $x^* \in \mathcal{D}$, such that $F(x^*) = 0$ and $F'(x^*)^{-1} \in \mathcal{L}(\mathcal{Y},\mathcal{X})$.

(C_3^*) There exist continuous and nondecreasing functions $h_1, h_2 : [0,+\infty) \times [0,+\infty) \to [0,+\infty)$, functions $g_1, g_2 : \mathcal{D} \times \mathcal{D} \to \mathcal{D}$ such that for each $u, v \in \mathcal{D}$ there exist $y = g_1(u,v) \in \mathcal{D}$ and $z = g_2(u,v) \in \mathcal{D}$

$$\|y - x^*\| \leq h_1(\|u - x^*\|, \|v - x^*\|)$$

and

$$\|z - x^*\| \leq h_2(\|u - x^*\|, \|v - x^*\|)$$

(C_4^*) There exist a mapping $\delta F : \mathcal{D} \times \mathcal{D} \to \mathcal{L}(\mathcal{X}, \mathcal{Y})$, continuous, nondecreasing functions $f_1, f_2 : [0, +\infty) \times [0, +\infty) \to [0, +\infty)$ with $f_1(0,0) = f_2(0,0) = 0$, $n \geq 0$ and $x_{-1}, x_0, y_0 = g_1(x_{-1}, x_0), z_0 = g_2(x_{-1}, x_0) \in \mathcal{D}$ such that for each $x, y, z \in \mathcal{D}$

$$\|F'(x^*)^{-1}(\delta F(x,y) - F'(x^*))\| \leq f_0(\|x - x^*\|, \|y - x^*\|)$$

and

$$\|F'(x^*)^{-1}(\delta F(x,y) - F'(z))\| \leq f_1(\|x - z\|, \|y - z\|)$$

(C_5^*) Define functions $\varphi, \psi : [0, +\infty) \to [0, +\infty)$ by

$$\varphi(t) = f_0(h_1(t,t), h_2(t,t)),$$

and

$$\psi(t) = \int_0^1 f_1(\theta t + h_1(t,t), \theta t + h_2(t,t)) d\theta.$$

Moreover, define functions q and p by

$$q(t) = \frac{\psi(t)}{1 - \varphi(t)}.$$

and

$$p(t) = \int_0^1 f_1(\theta t + h_1(t,t), \theta t + h_2(t,t)) d\theta + f_0(h_1(t,t), h_2(t,t)) - 1.$$

Suppose that equation

$$p(t) = 0 \tag{13.28}$$

has at least one positive zero and the smallest positive zero, denoted by r, satisfies

$$\varphi(r) < 1,$$

$$h_1(r,r) \leq \bar{r}$$

and

$$h_2(r,r) \leq \bar{r}, \text{ for some } \bar{r} \geq r.$$

Set $q = q(r)$.

(C_6^*) $\bar{U}(x_0, \bar{r}) \subseteq \mathcal{D}$.

(C_7^*) There exists $r_1 \geq r$ such that

$$\int_0^1 f_0((1-\theta)r_1, (1-\theta)r_1) d\theta < 1.$$

Theorem 13.4.1 *Suppose that the conditions (C^*) hold. Then, the sequence $\{x_n\}$ generated for $x_{-1}, x_0 \in U(x^*, r) \setminus \{x^*\}$ by method (13.2) is well defined, remains in $U(x^*, r)$ and converges to the solution x^* of equation $F(x) = 0$, which is unique in $\bar{U}(x_0, r_1) \cap \mathcal{D}$.*

Proof: The proof is similar to the proof of Theorem 13.2.1. We shall first show using mathematical induction that sequence $\{x_n\}$ is well defined and belongs in $U(x^*, r)$. We have by the first condition in (C_4^*) that

$$\begin{aligned} \|F'(x^*)^{-1}(A_0 - F'(x_1))\| &\le f_0(\|y_0 - x^*\|, \|z_0 - x^*\|) \\ &\le f_0(h_1(\|x_{-1} - x^*\|, \|x_0 - x^*\|), \\ &\quad h_2(\|x_{-1} - x^*\|, \|x_0 - x^*\|)) \\ &\le f_0(h_1(r, r), h_2(r, r)) \le \varphi(r) < 1. \end{aligned}$$

Hence, $A_0^{-1} \in \mathcal{L}(\mathcal{Y}, \mathcal{X})$ and

$$\|A_0^{-1}F'(x_0)\| \le \frac{1}{1 - f_0(h_1(\|x_{-1} - x^*\|, \|x_0 - x^*\|), h_2(\|x_{-1} - x^*\|, \|x_0 - x^*\|))} \le \frac{1}{1 - \varphi(r)}.$$

Thus, x_1 is well defined. Then, we have by the second substep of condition (C_4^*) that

$$\|F'(x^*)^{-1}(\int_0^1 F'(x^*) + \theta(x_0 - x^*)) - A_0)d\theta(x_0 - x^*)\|$$

$$\le \int_0^1 f_1(\|x^* + \theta(x_0 - x^*) - y_0\|, \|x^* + \theta(x_0 - x^*) - z_0\|)d\theta\|x_0 - x^*\|$$

$$\le \int_0^1 f_1(\theta\|x_0 - x^*\| + \|y_0 - x^*\|, \theta\|x_0 - x^*\| + \|z_0 - x^*\|)d\theta\|x_0 - x^*\|$$

$$\le \int_0^1 f_1(\theta\|x_0 - x^*\| + h_1(\|x_{-1} - x^*\|, \|x_0 - x^*\|), \theta\|x_0 - x^*\|$$

$$+ h_2(\|x_{-1} - x^*\|, \|x_0 - x^*\|)d\theta\|x_0 - x^*\|$$

$$\le f_1(\theta r + h_1(r, r), \theta r + h_2(r, r))d\theta\|x_0 - x^*\|.$$

In view of the approximation

$$\begin{aligned} x_1 - x^* &= x_0 - x^* - A_0^{-1}F(x_0) \\ &= -A_0^{-1}(F(x_0) - F(x^*) - A_0(x_0 - x^*)) \\ &= -A_0^{-1}\left[\int_0^1 F'(x^* + \theta(x_0 - x^*) - A_0)\right] d\theta(x_0 - x^*) \end{aligned}$$

we obtain

$$\begin{aligned}\|x_1 - x^*\| &\leq \|A_0^{-1}F(x^*)\|\|\int_0^1 F'(x^*)^{-1}(F'(x^* + \theta(x_0 - x^*)) - A_0)\\&\quad d\theta\|x_0 - x^*\|\\&\leq q\|x_0 - x^*\| < \|x_0 - x^*\| < r,\end{aligned}$$

which shows $x_1 \in U(x^*.r)$.

Then, we must show that $A_1^{-1} \in \mathcal{L}(\mathcal{Y}, \mathcal{X})$. We get

$$\begin{aligned}\|F'(x^*)^{-1}(A_1 - F'(x^*))\| &\leq f_0(h_1(\|x_0 - x^*\|, \|x_1 - x^*\|),\\&\quad h_2(\|x_0 - x^*\|, \|x_1 - x^*\|))\\&\leq f_0(h_1(r,r), h_2(r,r))\|x_1 - x^*\|\\&= \varphi(r) < 1,\end{aligned}$$

so

$$\|A_1^{-1}F'(x^*) \leq \frac{1}{1 - \varphi(r)}.$$

That is x_2 is well defined. Then, we obtain

$$\begin{aligned}&\|F'(x^*)^{-1}\left[\int_0^1 F'(x^* + \theta(x_1 - x^*)) - A_1\right] d\theta\|x_1 - x^*\|\\&\leq f_1(\theta\|x_1 - x^*\| + \|y_1 - x^*\|, \theta\|x_1 - x^*\| + \|z_1 - x^*\|)d\theta\|x_1 - x^*\|\\&\leq f_1(\theta r + h_1(r,r), \theta r + h_2(r,r),)d\theta\|x_1 - x^*\|.\end{aligned}$$

Using the identity

$$x_2 - x^* = -A_1^{-1}\left[\int_0^1 F'(x^* + \theta(x_1 - x^*)) - A_1\right] d\theta\|x_1 - x^*\|,$$

we get again that

$$\|x_2 - x^*\| \leq q\|x_1 - x^*\| < \|x_1 - x^*\| < r,$$

which shows that $x_2 \in U(x^*, r)$.

In a similar way, we obtain

$$\|x_{k+1} - x_k\| \leq q\|x_k - x^*\| < \|x_k - x^*\| < r,$$

which shows that $\lim_{k\to\infty} x_k = x^*$ and $x_{k+1} \in U(x^*, r)$. The part related to the uniqueness has been shown in Theorem 13.2.1. ⊠.

13.5 Local convergence analysis of the secant method II

Now we consider, as in the semilocal case, another set of conditions. The conditions (H^*) are:

(H_1^*) F is a continuous operator defined on a convex subset $\mathcal{D}$ of a Banach space $\mathcal{X}$ with values in a Banach space $\mathcal{Y}$.

(H_2^*) There exists $x^* \in \mathcal{D}$ such that $F(x^*) = 0$.

(H_3^*) There exist continuous and nondecreasing functions $h_1, h_2 : [0,+\infty) \times [0,+\infty) \to [0,+\infty)$, functions $g_1, g_2 : \mathcal{D} \times \mathcal{D} \to \mathcal{D}$ such that for each $u, v \in \mathcal{D}$ there exist $y = g_1(u,v) \in \mathcal{D}$ and $z = g_2(u,v) \in \mathcal{D}$

$$\|y - x^*\| \leq h_1(\|u - x^*\|, \|v - x^*\|)$$

and

$$\|z - x^*\| \leq h_2(\|u - x^*\|, \|v - x^*\|).$$

(H_4^*) There exist a divided difference of order one δF, a linear operator $M \in \mathcal{L}(\mathcal{X},\mathcal{Y})$, continuous, nondecreasing functions $f_1, f_2 : [0,+\infty) \times [0,+\infty) \to [0,+\infty)$, $n \geq 0$ and $x_{-1}, x_0, y_0 = g_1(x_{-1}, x_0), z_0 = g_2(x_{-1}, x_0) \in \mathcal{D}$ such that $M^{-1} \in \mathcal{L}(\mathcal{Y},\mathcal{X})$ and for each $x, y, z \in \mathcal{D}$

$$\|M^{-1}(\delta F(x,y) - M)\| \leq f_0(\|x - x^*\|, \|y - x^*\|)$$

and

$$\|M^{-1}(\delta F(x,x^*) - \delta F(u,v))\| \leq f_1(\|x - u\|, \|x^* - v\|).$$

(H_5^*) Define functions $\varphi, \psi : [0,+\infty) \to [0,+\infty)$ by

$$\varphi(t) = f_0(h_1(t,t), h_2(t,t))$$

and

$$\psi(t) = f_1(t + h_1(t,t), t + h_2(t,t)).$$

Moreover, define functions q and p by

$$q(t) = \frac{\psi(t)}{1 - \varphi(t)}$$

and

$$p(t) = f_1(t + h_1(t,t), h_2(t,t)) + f_0(h_1(t,t), h_2(t,t)) - 1.$$

Suppose that equation

$$p(t) = 0$$

has at least one positive zero and we denote R the smallest positive zero that satisfies

$$\varphi(R) < 1,$$

$$h_1(R,R) \le \bar{R}$$

and

$$h_2(R,R) \le \bar{R}, \text{ for some } \bar{R} \ge R.$$

Set $q = q(R)$.

(H_6^*) $\bar{U}(x_0,\bar{R}) \subseteq \mathcal{D}$.

(H_7^*) There exists $R_1 \ge R$ such that

$$f_0(0,R_1) < 1.$$

Theorem 13.5.1 *Suppose that the conditions* (H^*) *hold. Then, the sequence* $\{x_n\}$ *generated for* $x_{-1}, x_0 \in \bar{U}(x^*,R) \setminus x^*$ *by method* (13.2) *is well defined, remains in* $U(x^*,R)$ *and converges to the solution* x^* *of equation* $F(x) = 0$, *which is unique in* $\bar{U}(x_0,R_1) \cap \mathcal{D}$.

The results in Sections 13.4 and 13.5 hold for Newton-like methods, if we set $x_{-1} = x_0$ and $z_n = y_n = x_n$, for each $n = 0,1,2\ldots$.

13.6 Numerical examples

Example 13.6.1 *We consider*

$$x(s) = 1 + \int_0^1 G(s,t)x(t)^2 dt, \quad s \in [0,1] \tag{13.29}$$

where $x \in C[0,1]$ *and the kernel* G *is the Green's function in* $[0,1] \times [0,1]$.

We use a discretization process and transform equation (13.29) *into a finite dimensional problem. For this, we follow the process described in Section 13.2 in [12] with* $m = 8$ *and we obtain the following system of nonlinear equations:*

$$F(x) \equiv x - 1 - Av_x = 0, \quad F: \mathbb{R}^8 \to \mathbb{R}^8, \tag{13.30}$$

where

$$a = (x_1, x_2, \ldots, x_8)^T, \quad 1 = (1,1,\ldots,1)^T, \quad A = (a_{ij})_{i,j=1}^8, \quad v_x = (x_1^2, x_2^2, \ldots, x_8^2)^T.$$

We use the divided difference of first order of F *as* $[u,v;F] = I - B$, *where* $B = (b_{ij})_{i,j=1}^8$ *with* $b_{ij} = a_{ij}(u_j + v_j)$.

If we choose the starting points $x_{-1} = (\frac{7}{10}, \frac{7}{10}, \ldots, \frac{7}{10})^T$ *and* $x_0 = (\frac{18}{10}, \frac{18}{10}, \ldots, \frac{18}{10})^T$, *method* (13.5) *with* $\lambda = \frac{1}{2}$ *and the max-norm, we obtain* $c = \frac{11}{10}$, $y_0 = (\frac{5}{4}, \frac{5}{4}, \ldots, \frac{5}{4})^T$, $\beta = 1.555774\ldots$, $\eta = 0.7839875\ldots$, $m = 0.257405\ldots$ *and the polynomial*

$$p(t) = -0.789051 + t\left(1 - \frac{0.257406}{1 - 0.19223(0.55 + 2t)}\right)$$

has no real roots so the conditions in [12] are not satisfied so we cannot ensure the convergence of method (13.5).

Next, we are going to use our (C) *conditions in order to ensure the convergence of the methods. We consider* $g_1(x,y) = \frac{y+x}{2}$ *and* $g_2(x,y) = y$*, and we define the functions*

$$h_1(s,t) = \frac{s}{2} + \frac{t}{2},$$

$$h_2(s,t) = t,$$

$$f_0(s,t) = 0.0103507\ldots(s+t)$$

and

$$f_1(s,t) = 0.047827\ldots(s+t),$$

which satisfy the (C_3) *and* (C_4) *conditions. Moreover, we define*

$$\varphi(t) = 0.0207015t,$$

$$\psi(t) = 0.0640438 + 0.191311t$$

and

$$q(t) = \frac{\phi(t)}{1-\varphi(t)} = \frac{0.0640438 + 0.191311t}{1 - 0.0207015t}.$$

The solutions of equation (13.8) *are*

$$r_1 = 1.09603\ldots \quad \text{and} \quad r_2 = 3.39564\ldots.$$

Then, by denoting $r = 1.09603\ldots$ *and* $\bar{r} = r$ *it is easy to see that the following conditions are verified*

$$\varphi(r) = 0.0226895\ldots < 1,$$

$$h_1(c,0) = 0.55 \leq r, \quad h_1(r,r) = r \leq \bar{r},$$

and

$$h_2(c,0) = 0 \leq \bar{r}, \quad h_2(r,r) = r \leq \bar{r},$$

Finally to show that condition (C_7) *is verified we choose as* $r_1 = 1.25$ *and we obtain that*

$$\int_0^1 f_0(\theta r + (1-\theta)r_1, \theta r + (1-\theta)r_1)d\theta = 0.0258769\ldots < 1.$$

So, all the conditions of Theorem 13.2.1 are satisfied and a consequence we can ensure the convergence of method (13.5).

Example 13.6.2 *We consider the same problem as in Example 1 but with method* (13.4)*. If we choose the starting points* $x_{-1} = (\frac{12}{10}, \frac{12}{10}, \ldots, \frac{12}{10})^T$ *and* $x_0 = (\frac{14}{10}, \frac{14}{10}, \ldots, \frac{14}{10})^T$*, method* (13.4) *with* $a = 0.8$*,* $b = 0.75$ *and the max-norm, we obtain* $c = 0.2$*,*

$$y_0 = (1.11459, 1.16864, 1.23603, 1.28163, 1.28163, 1.23603, 1.16864, 1.11459)^T,$$

$$z_0 = (1.70443, 1.64679, 1.5749, 1.52626, 1.52626, 1.5749, 1.64679, 1.70443)^T,$$

$\beta = 1.49123\ldots$*,* $\delta = 0.380542\ldots$*,* $\eta = 0.394737\ldots$*,* $\gamma = 1.67305\ldots$ *and* $M = 0.47606\ldots$*. One of the necessary conditions given in [4] is not satisfied since*

$$M\delta\gamma^2 = 0.507089\ldots \leq 0.5.$$

so we cannot ensure the convergence of method (13.4)*.*

Next, we are going to use our (H) *conditions in order to ensure the convergence of the methods. We consider* $g_1(x,y) = x - aF(x)$ *and* $g_2(x,y) = x + bF(x)$*, and we define the functions*

$$h_1(s,t) = p_0 t + 0.5p,$$

$$h_2(s,t) = p_1 t + 0.5p,$$

$$f_0(s,t) = 0.0579754\ldots(s+t)$$

and

$$f_1(s,t) = 0.0579754\ldots(s+t),$$

where $p = \|A_0\|$*,* $p_0 = \|I - 0.75\delta F[x,x_0]\| = 1.16988\ldots$ *and* $p_1 = \|I + 0.8\delta F[x,x_0]\| = 1.03601$ *which satisfy the* (C_3) *and* (C_4) *conditions. Moreover, we define*

$$\varphi(t) = 0.0579754(2.98245 + 2.20589t),$$

$$\psi(t) = 0.10934 + 0.243838t$$

and

$$q(t) = \frac{\phi(t)}{1-\varphi(t)} = \frac{0.10934 + 0.243838t}{1 - 0.0579754(2.98245 + 2.20589t)}.$$

The solutions of equation (13.8) *are*

$$r_1 = 0.598033\ldots \quad \text{and} \quad r_2 = 1.46863\ldots.$$

Then, by denoting $r = 0.598033\ldots$ *and* $\bar{r} = 1.44524\ldots$ *it is easy to see that the following conditions are verified*

$$\varphi(r) = 0.24939\ldots < 1,$$

$$h_1(c,0) = 0.745613\ldots \leq \bar{r}, \quad h_1(r,r) = \bar{r}\ldots \leq \bar{r},$$

and

$$h_2(c,0) = 0.745613\ldots \leq \bar{r}, \quad h_2(r,r) = \bar{r}\ldots \leq \bar{r},$$

Finally to show that condition (C_7) *is verified we choose as* $r_1 = 0.6$ *and we obtain that*

$$\int_0^1 f_0(\theta r + (1-\theta)r_1, \theta r + (1-\theta)r_1)d\theta = 0.0695705\ldots < 1.$$

So, all the conditions of Theorem 13.2.2 are satisfied and a consequence we can ensure the convergence of method (13.4).

Example 13.6.3 *Let* $X = Y = \mathbb{R}^3$, $D = U(0,1)$, $x^* = (0,0,0)^T$ *and define function* F *on* D *by*

$$F(x,y,z) = (e^x - 1, y^2 + y, z)^T. \tag{13.31}$$

We have that for $u = (x,y,z)^T$

$$F'(u) = \begin{pmatrix} e^x & 0 & 0 \\ 0 & 2y+1 & 0 \\ 0 & 0 & 1 \end{pmatrix}, \tag{13.32}$$

Using the norm of the maximum of the rows and (13.31)–(13.32) we see that since $F'(x^*) = diag\{1,1,1\}$, *In this case, we take the domain* $\Omega = B(x_0, R)$, *where* $x_0 = 0.5$ *and* $R = 0.3$. *Choosing* $x_{-1} = 0.4$, *it is easy to see that now, we are going to use our* (C^*) *conditions in order to ensure the convergence of the method* (13.7). *We define the functions*

$$h_1(s,t) = t$$

$$h_2(s,t) = s,$$

$$f_0(s,t) = \frac{e}{2}(s+t)$$

and

$$f_1(s,t) = e(s+t)$$

which satisfy the (H_3^*) *and* (H_4^*) *conditions. Moreover, we define*

$$\varphi(t) = et,$$

$$\psi(t) = 3et,$$

$$q(t) = \frac{\phi(t)}{1-\varphi(t)} = \frac{3et}{1-et}$$

and

$$p(t) = \int_0^1 f_1(\theta t + h_1(t,t), \theta t + h_2(t,t))d\theta + f_0(h_1(t,t), h_2(t,t)) - 1 = 10.8731t - 1.$$

The root of $p(t)$ *is*

$$r_1 = 0.0919699\ldots$$

so denoting $r = 0.0919699\ldots$ *and* $\bar{r} = r$ *it is easy to see that the following conditions are verified*

$$\varphi(r) = 0.25 < 1,$$

$$h_1(r,r) = \bar{r} \leq \bar{r},$$

and

$$h_2(r,r) = r \leq \bar{r},$$

Finally, it is clear that $\bar{U}(x_0,\bar{r}) \subseteq U(x_0,R)$ *and to show that condition* (H_7^*) *is verified we choose as* $r_1 = 0.1$ *and we obtain that*

$$\int_0^1 f_0(\theta r + (1-\theta)r_1, \theta r + (1-\theta)r_1)d\theta = 0.135914\ldots < 1.$$

So, all the conditions of Theorem 13.2.4 are satisfied and as a consequence we can ensure the convergence of method (13.7).

Example 13.6.4 *In order to show the applicability of our theory in a real problem we are going to consider the following quartic equation that describes the fraction of the nitrogen-hydrogen feed that gets converted to ammonia, called the fractional conversion showed in [13,25]. In Figure 13.1 they process of ammonia is shown.*

For 250 *atm and* 500*C, this equation takes the form:*

$$f(x) = x^4 - 7.79075x^3 + 14.7445x^2 + 2.511x - 1.674$$

In this case, we take the domain $\Omega = B(x_0,R)$*, where* $x_0 = 0.5$ *and* $R = 0.3$*. Choose* $x_{-1} = 0.4$*. We are going to use our* (C^*) *conditions in order to ensure the convergence of the method* (13.6)*. Define the functions*

$$h_1(s,t) = s + 2t$$

$$h_2(s,t) = s,$$

$$f_0(s,t) = 0.111304(32.2642s + 20.5458t)$$

and

$$f_1(s,t) = 0.111304(40.9686s + 30.7779t),$$

which satisfy the (C_3^*) *and* (C_4^*) *conditions. Moreover, we define*

$$\varphi(t) = 13.0603t,$$

$$\psi(t) = 21.0985t,$$

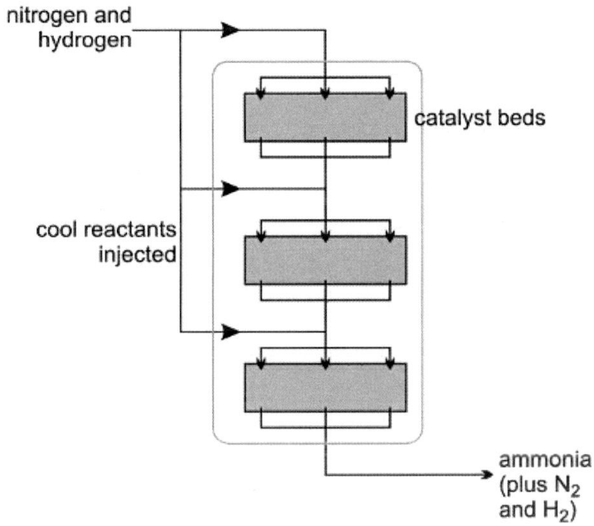

Figure 13.1: Ammonia process.Taken from www.essentialchemicalindustry.org/chemicals/ammonia.

$$q(t) = \frac{\phi(t)}{1-\varphi(t)} = \frac{-1+34.1587t}{1-13.0603t}$$

and

$$p(t) = \int_0^1 f_1(\theta t+h_1(t,t),\theta t+h_2(t,t))d\theta + f_0(h_1(t,t),h_2(t,t))-1 = 34.1587t-1.$$

The root of $p(t)$ *is*

$$r_1 = 0.0292751\ldots.$$

Then, by denoting $r = 0.0292751\ldots$ *and* $\bar{r} = 3r = 0.0878253\ldots$ *it is easy to see that the following conditions are verified*

$$\varphi(r) = 0.38234\ldots < 1,$$

$$h_1(r,r) = \bar{r} \leq \bar{r},$$

and

$$h_2(r,r) = r \leq \bar{r},$$

Finally, it is clear that $\bar{U}(x_0,\bar{r}) \subseteq U(x_0,R)$. *Next, to show that condition* (C_7^*) *is verified we choose as* $r_1 = 0.1$ *and we obtain that*

$$\int_0^1 f_0(\theta r+(1-\theta)r_1,\theta r+(1-\theta)r_1)d\theta = 0.293899\ldots < 1.$$

So, all the conditions of Theorem 13.2.3 are satisfied and as a consequence we can ensure the convergence of method (13.6).

References

[1] S. Amat, S. Busquier and J. M. Gutiérrez, On the local convergence of secant-type methods, Int. J. Comput. Math., 81(8), 1153–1161 (2004).

[2] S. Amat, S. Busquier and Á. A. Magreñán, Reducing chaos and bifurcations in Newton-type methods, *Abst. Appl. Anal.*, 2013 (2013), Article ID 726701, 10 pages, `http://dx.doi.org/10.1155/2013/726701`.

[3] S. Amat, Á. A. Magreñán and N. Romero, On a family of two step Newton-type methods, *Appl. Math. Comput.*, 219 (4), 11341–11347, 2013.

[4] S. Amat, J. A. Ezquerro and M. A. Hernández, On a Steffensen-like method for solving equations, *Calcolo*, 53 (2), 171–188, 2016.

[5] I. K. Argyros, A unifying local–semilocal convergence analysis and applications for two–point Newton–like methods in Banach space, *J. Math. Anal. Appl.*, 298, 374–397, 2004.

[6] I. K. Argyros, Computational theory of iterative methods, Elsevier Science, Editors C. K. Chui and L. Wuytack, San Diego (U.S.A.), 2007.

[7] I. K. Argyros, A semilocal convergence analysis for directional Newton methods, *Math. Comput. AMS*, 80, 327–343, 2011.

[8] I. K. Argyros, Weaker conditions for the convergence of Newton's method, *J. Comp.*, 28, 364–387, 2012.

[9] I. K. Argyros, D. González and Á. A. Magreñán, A semilocal convergence for a uniparametric family of efficient secant-Like methods, *J. Funct. Spaces*, Volume 2014 (2014), Article ID 467980, 10 pages. `http://dx.doi.org/10.1155/2014/467980`.

[10] I. K. Argyros and Á. A. Magreñán, A unified convergence analysis for Secant-type methods, *J. Korean Math. Soc.*, 51 (6), 1155-1175, 2014.

[11] J. E. Dennis, Toward a unified convergence theory for Newton-like methods, *Nonlinear Funct. Anal. App. (L. B. Rall, ed.), Academic Press*, New York, 425–472,1971.

[12] J. A. Ezquerro, M. A. Hernández and A. I. Velasco, An analysis of the semilocal convergence for secant-like methods, *Appl. Math. Comput.*, 266, 883–892, 2015.

[13] V. B. Gopalan and J. D. Seader, Application of interval Newton's method to chemical engineering problems, *Reliable computing*, 1 (3), 215–223, 1995.

[14] J. M. Gutiérrez, Á. A. Magreñán and N. Romero, On the semilocal convergence of Newton-Kantorovich method under center-Lipschitz conditions, *Appl. Math. Comput.*, 221, 79–88, 2013.

[15] J. M. Gutiérrez, Á. A. Magreñán and J. L. Varona, The "Gauss-Seidelization" of iterative methods for solving nonlinear equations in the complex plane. *Appl. Math. Comput.*, 218 (6), 2467–2479, 2011.

[16] M. A. Hernández, M. J. Rubio and J. A. Ezquerro, Solving a special case of conservative problems by secant–like method, *Appl. Math. Comput.*, 169, 926–942, 2005.

[17] M. A. Hernández, M. J. Rubio and J. A. Ezquerro, Secant-like methods for solving nonlinear integral equations of the Hammerstein type, *J. Comput. Appl. Math.*, 115, 245–254, 2000.

[18] V. A. Kurchatov, On a method of linear interpolation for the solution of functional equations, *Dokl. Akad. Nauk SSSR (Russian)*, 198 (3), 524–526, 1971. Translation in *Soviet Math. Dokl.*, 12, 835–838, 1974.

[19] P. Laarsonen, Ein überquadratisch konvergenter iterativer algorithmus, *Ann. Acad. Sci. Fenn. Ser I*, 450, 1–10, 1969.

[20] F. A. Potra, An error analysis for the secant method, *Numer. Math.*, 38, 427–445, 1982.

[21] F. A. Potra, Sharp error bounds for a class of Newton-like methods, *Lib. Math.*, 5, 71–84, 1985.

[22] F. A. Potra and V. Pták, Nondiscrete induction and iterative processes, Pitman, New York, 1984.

[23] J. W. Schmidt, Untere Fehlerschranken fur Regula–Falsi Verhafren, *Period. Hungar.*, 9, 241–247, 1978.

[24] A. S. Sergeev, On the method of Chords (in Russian), Sibirsk, *Math. J.*, 11, 282–289, 1961.

[25] M. Shacham, An improved memory method for the solution of a nonlinear equation, *Chem. Engin. Sci.*, 44, 1495–150, 1989.

[26] S. M. Shakhno, On a Kurchatov's method of linear interpolation for solving nonlinear equations, *PAMM-Proc. Appl. Math. Mech.*, 4, 650–651, 2004.

[27] B. Royo, J. A. Sicilia, M. J. Oliveros and E. Larrodé, Solving a long-distance routing problem using ant colony optimization. *Appl. Math.*, 9 (2L), 415–421, 2015.

[28] J. A. Sicilia, C. Quemada, B. Royo and D. Escuín, An optimization algorithm for solving the rich vehicle routing problem based on variable neighborhood search and tabu search metaheuristics. *J. Comput. Appl. Math.*, 291, 468–477, 2016.

[29] J. A. Sicilia, D. Escuín, B. Royo, E. Larrodé and J. Medrano, A hybrid algorithm for solving the general vehicle routing problem in the case of the urban freight distribution. In Computer-based Modelling and Optimization in Transportation (pp. 463–475). Springer International Publishing, 2014.

[30] S. Ulm, Majorant principle and the method of Chords (in Russian), *Izv. Akad. Nauk Eston. SSR, Ser. Fiz.-Mat.*, 13, 217–227, 1964.

[31] T. Yamamoto, A convergence Theorem for Newton-like methods in Banach spaces, *Numer. Math.*, 51, 545–557, 1987.

Chapter 14

King-Werner-like methods free of derivatives

Ioannis K. Argyros

Department of Mathematical Sciences, Cameron University, Lawton, OK 73505, USA, Email: iargyros@cameron.edu

Á. Alberto Magreñán

Universidad Internacional de La Rioja, Escuela Superior de Ingeniería y Tecnología, 26002 Logroño, La Rioja, Spain, Email: alberto.magrenan@unir.net

CONTENTS

14.1 Introduction

Recently, Argyros and Ren in [6] studied King-Werner-like methods for approximating a locally unique solution $x^\star$ of equation

$$F(x) = 0, \tag{14.1}$$

where F is Fréchet-differentiable operator defined on a convex subset of a Banach space $\mathbb{B}_1$ with values in a Banach space $\mathbb{B}_2$.

In particular, they studied the semilocal convergence analysis of method defined for $n = 0, 1, 2, \ldots$ by

$$\begin{aligned} x_{n+1} &= x_n - A_n^{-1}F(x_n) \\ y_{n+1} &= x_{n+1} - A_n^{-1}F(x_{n+1}), \end{aligned} \tag{14.2}$$

where x_0, y_0 are initial points, $A_n = [x_n, y_n; F]$ and $[x, y; F]$ denotes a divided difference of order one for operator F at points $x, y \in \Omega$ [2,4,7] satisfying

$$[x,y;F](x-y) = F(x) - F(y) \quad for\ each\ x, y \in \Omega\ with\ x \neq y. \tag{14.3}$$

If F is Fréchet-differentiable on Ω, then $F'(x) = [x, x; F]$ for each $x \in \Omega$.

On the other hand, the local convergence analysis of method (14.2) was given in [9] in the special case when $\mathbb{B}_1 = \mathbb{B}_2 = \mathbb{R}$. The convergence order of method (14.2) was shown to be $1 + \sqrt{2}$.

The chapter is organized as follows: In Section 14.2 the semilocal convergence study of method (14.2) is presented. Then, in Section 14.3 the local convergence study is provided. Finally, the numerical examples are presented in Section 14.4.

14.2 Semilocal convergence

First of all, we need some auxiliary results related to majorizing sequence. All the proofs of these results can be found in [6].

Lemma 14.2.1 *Let $L_0 > 0$, $L > 0$, $s_0 \geq 0$, $t_1 \geq 0$ be given parameters. Denote by α the only root in the interval $(0,1)$ of polynomial p defined by*

$$p(t) = L_0 t^3 + L_0 t^2 + 2Lt - 2L. \tag{14.4}$$

Suppose that

$$0 < \frac{L(t_1 + s_0)}{1 - L_0(t_1 + s_1 + s_0)} \leq \alpha \leq 1 - \frac{2L_0 t_1}{1 - L_0 s_0}, \tag{14.5}$$

where

$$s_1 = t_1 + L(t_1 + s_0)t_1. \tag{14.6}$$

Then, scalar sequence $\{t_n\}$ defined for each $n = 1, 2, \ldots$ by

$$\begin{aligned} &t_0 = 0, s_{n+1} = t_{n+1} + \frac{L(t_{n+1} - t_n + s_n - t_n)(t_{n+1} - t_n)}{1 - L_0(t_n - t_0 + s_n + s_0)}, \quad for\ each\ n = 1, 2, \ldots, \\ &t_{n+2} = t_{n+1} + \frac{L(t_{n+1} - t_n + s_n - t_n)(t_{n+1} - t_n)}{1 - L_0(t_{n+1} - t_0 + s_{n+1} + s_0)}, \quad for\ each\ n = 0, 1, 2, \ldots \end{aligned} \tag{14.7}$$

is well defined, increasing, bounded above by

$$t^{\star\star} = \frac{t_1}{1-\alpha} \tag{14.8}$$

and converges to its unique least upper bound $t^\star$ *which satisfies*

$$t_1 \le t^\star \le t^{\star\star}. \tag{14.9}$$

Moreover, the following estimates hold

$$s_n - t_n \le \alpha(t_n - t_{n-1}) \le \alpha^n (t_1 - t_0), \tag{14.10}$$

$$t_{n+1} - t_n \le \alpha(t_n - t_{n-1}) \le \alpha^n (t_1 - t_0) \tag{14.11}$$

and

$$t_n \le s_n \tag{14.12}$$

for each $n = 1, 2, \ldots$.

Throughout this chapter, we denote by $U(w,\rho)$, $\overline{U}(w,\rho)$, the open and closed balls in $\mathbb{B}_1$, respectively, with center $w \in \mathbb{B}_1$ and of radius $\rho > 0$.

Now, using $\{t_n\}$ as a majorizing sequence we present our main semilocal result.

Theorem 14.2.2 *Let* $F : \Omega \subset \mathbb{B}_1 \to \mathbb{B}_2$ *be a Fréchet-differentiable operator. Suppose that there exists a divided differentiable* $[.,.;F]$ *of order one for operator* F *on* $\Omega \times \Omega$. *Moreover, suppose that there exist* $x_0, y_0 \in \Omega$, $L_0 > 0$, $L > 0$, $s_0 \ge 0$, $t_1 \ge 0$ *such that*

$$A_0^{-1} \in L(\mathbb{B}_2, \mathbb{B}_1) \tag{14.13}$$

$$\|A_0^{-1} F(x_0)\| \le t_1, \tag{14.14}$$

$$\|x_0 - y_0\| \le s_0, \tag{14.15}$$

$$\|A_0^{-1}([x,y;F] - A_0)\| \le L_0(\|x - x_0\| + \|y - y_0\|), \text{ for each } x, y \in \Omega \tag{14.16}$$

$$\begin{aligned} \|A_0^{-1}([x,y;F] - [z,v;F])\| &\le L(\|x - z\| + \|y - v\|), \\ &\qquad \text{for each } x, y, z \in \Omega_1 = \Omega \cap U(x_0, \frac{1}{2L_0}) \end{aligned} \tag{14.17}$$

$$\overline{U}(x_0, t^\star) \subseteq \Omega \tag{14.18}$$

and hypotheses of Lemma 14.2.1 hold, where $A_0 = [x_0; y_0; F]$ *and* $t^\star$ *is given in Lemma 14.2.1. Then, sequence* $\{x_n\}$ *generated by method (14.2) is well defined, remains in* $\overline{U}(x_0, t^\star)$ *and converges to a unique solution* $x^\star \in \overline{U}(x_0, t^\star)$ *of equation* $F(x) = 0$. *Moreover, the following estimates hold for each* $n = 0, 1, 2, \ldots$

$$\|x_n - x^\star\| \le t^\star - t_n. \tag{14.19}$$

Furthermore, if there exists $R > t^\star$ such that

$$U(x_0,R) \subseteq \Omega \tag{14.20}$$

and

$$L_0(t^\star + R + s_0) < 1, \tag{14.21}$$

then, the point $x^\star$ is the only solution of equation $F(x) = 0$ in $U(x_0,R)$.

Proof: Simply notice that the iterates remain in Ω_1 which is a more precise location than Ω used in [6], since $\Omega_1 \subseteq \Omega$. Then, in view of this the proof follows from the corresponding one in [6].

□

Remark 14.2.3 *(a) The limit point $t^\star$ can be replaced by $t^{\star\star}$ given in closed form by (14.8) in Theorem 14.2.1.*

(b) In [6], Argyros and Ren used the stronger condition

$$\|A_0^{-1}([x,y;F] - [z,v;F])\| \leq L_1(\|x-z\| + \|y-v\|) \text{ for each } x,y,z,v \in \Omega.$$

Notice that from we have

$$L_0 \leq L_1 \text{ and } L \leq L_1$$

holds in general and $\frac{L_1}{L_0}$ can be arbitrarily large [2–6]. Moreover, it follows from the proof of Theorem 14.2.2 that hypothesis (14.17) is not needed to compute an upper bound for $\|A_0^{-1}F(x_1)\|$. Hence, we can define the more precise (than $\{t_n\}$) majorizing sequence $\{\bar{t}_n\}$ (for $\{x_n\}$) by

$$\begin{aligned} &\bar{t}_0 = 0, \bar{t}_1 = t_1, \bar{s}_0 = s_0, \bar{s}_1 = \bar{t}_1 + L_0(\bar{t}_1 + \bar{s}_0)\bar{t}_1, \\ &\bar{s}_{n+1} = \bar{t}_{n+1} + \frac{L(\bar{t}_{n+1} - \bar{t}_n + \bar{s}_n - \bar{t}_n)(\bar{t}_{n+1} - \bar{t}_n)}{1 - L_0(\bar{t}_n - \bar{t}_0 + \bar{s}_n + \bar{s}_0)} \quad for\ each\ n = 1,2,... \end{aligned} \tag{14.22}$$

and

$$\bar{t}_{n+2} = \bar{t}_{n+1} + \frac{L(\bar{t}_{n+1} - \bar{t}_n + \bar{s}_n - \bar{t}_n)(\bar{t}_{n+1} - \bar{t}_n)}{1 - L_0(\bar{t}_{n+1} - \bar{t}_0 + \bar{s}_{n+1} + \bar{s}_0)} \quad for\ each\ n = 0,1,... \tag{14.23}$$

Then, from a simple induction argument we have that

$$\bar{t}_n \leq t_n, \tag{14.24}$$

$$\bar{s}_n \leq s_n, \tag{14.25}$$

$$\bar{t}_{n+1} - \bar{t}_n \leq t_{n+1} - t_n, \tag{14.26}$$

$$\bar{s}_n - \bar{t}_n \leq s_n - t_n \tag{14.27}$$

and

$$\bar{t}^\star = \lim_{n\to\infty} \bar{t}_n \leq t^\star. \tag{14.28}$$

Furthermore, if $L_0 < L$, then (14.24)-(14.27) are strict for $n \geq 2$, $n \geq 1$, $n \geq 1$, $n \geq 1$, respectively. Clearly, sequence $\{\bar{t}_n\}$ increasing converges to $\bar{t}^\star$ under the hypotheses of Lemma 14.2.1 and can replace $\{t_n\}$ as a majorizing sequence for $\{x_n\}$ in Theorem 14.2.2. Finally, the old sequences using L_1 instead of L in [6] are less precise than the new ones.

14.3 Local convergence

Now, we present the main local convergence result.

Theorem 14.3.1 *Let $F : \Omega \subseteq \mathbb{B}_1 \to \mathbb{B}_2$ be a Fréchet-differentiable operator. Suppose that there exist $x^\star \in \Omega$, $l_0 > 0$ and $l > 0$ such that for each $x, y, z, u \in \Omega$*

$$F(x^\star) = 0,\ F'(x^\star)^{-1} \in L(\mathbb{B}_2, \mathbb{B}_1), \tag{14.29}$$

$$\|F'(x^\star)^{-1}([x,y;F] - F'(x^\star))\| \leq l_0(\|x - x^\star\| + \|y - x^\star\|) \text{ for each } x, y \in \Omega \tag{14.30}$$

$$\|F'(x^\star)^{-1}([x,y;F] - [z,u;F])\| \leq l(\|x - z\| + \|y - u\|), \tag{14.31}$$

$$\text{for each } x, y, z, u \in \Omega_2 := \Omega \cap U(x^*, \frac{1}{2l_0})$$

and

$$\overline{U}(x^\star, \rho) \subseteq \Omega, \tag{14.32}$$

where

$$\rho = \frac{1}{(1+\sqrt{2})l + 2l_0}. \tag{14.33}$$

Then, sequence $\{x_n\}$ generated by method (14.2) is well defined, remains in $\overline{U}(x^\star, \rho)$ and converges to $x^\star$ with order of $1 + \sqrt{2}$ at least, provided that $x_0, y_0 \in U(x^\star, \rho)$. Furthermore, the following estimates

$$\|x_{n+2} - x^\star\| \leq \frac{\sqrt{2}-1}{\rho^2}\|x_{n+1} - x^\star\|^2\|x_n - x^\star\| \tag{14.34}$$

and

$$\|x_n - x^\star\| \leq (\frac{\sqrt{\sqrt{2}-1}}{\rho})^{F_n - 1}\|x_1 - x^\star\|^{F_n} \tag{14.35}$$

hold for each $n = 1, 2, \ldots$, where F_n is a generalized Fibonacci sequence defined by $F_1 = F_2 = 1$ and $F_{n+2} = 2F_{n+1} + F_n$.

Proof: Similar to the Proof of Theorem 14.2.2.

□

Remark 14.3.2 *(a) For the special case $\mathbb{B}_1 = \mathbb{B}_2 = \mathbb{R}$, the radius of convergence ball for method (14.2) is given in [10] by*

$$\rho_\star = \frac{s^\star}{M}, \tag{14.36}$$

where $s^\star \approx 0.55279$ is a constant and $M > 0$ is the upper bound for $|F(x^\star)^{-1}F''(x)|$ in the given domain Ω. Using (14.31) we have

$$\|F'(x^\star)^{-1}(F'(x) - F'(y))\| \leq 2l\|x - y\| \quad \mathit{for\ any}\ x, y \in \Omega. \tag{14.37}$$

That is, we can choose $l = \frac{M}{2}$. *Simply set* $l_0 = l$, *we have from (14.33) that*

$$\rho = \frac{2}{(3+\sqrt{2})M} = \frac{2(3-\sqrt{2})}{5M} \approx \frac{0.63432}{M} > \frac{s^\star}{M} = \rho_\star. \tag{14.38}$$

Therefore, even in this special case, a bigger radius of convergence ball for method (14.2) has been given in Theorem 14.3.1.

(b) Notice that we have

$$l_0 \leq l_1 \text{ and } l \leq l_1 \tag{14.39}$$

$$\|A_0^{-1}([x,y;F]-[z,u;F])\| \leq l_1(\|x-z\|+\|y-u\|) \text{ for each } x,y,z,u \in \Omega.$$

The radius given in [6]:

$$\rho_0 = \frac{1}{(1+\sqrt{2})l_1 + l_0} \leq \rho. \tag{14.40}$$

Moreover, if $l < l_1$, *then* $\rho_0 < \rho$ *and the new error bounds (14.34) and (14.35) are tighter than the old ones in [6] using* ρ_0 *instead of* ρ.

14.4 Numerical examples

Example 14.4.1 *Let* $\mathbb{B}_1 = \mathbb{B}_2 = \mathbb{R}$, $\Omega = (-1,1)$ *and define* F *on* Ω *by*

$$F(x) = e^x - 1. \tag{14.41}$$

Then, $x^\star = 0$ *is a solution of* (14.41), *and* $F'(x^\star) = 1$. *Note that for any* $x,y,z,u \in \Omega$, *we have*

$$\begin{aligned}
&|F'(x^\star)^{-1}([x,y;F]-[z,u;F])| \\
&= |\textstyle\int_0^1 (F'(tx+(1-t)y) - F'(tz+(1-t)u))dt| \\
&= |\textstyle\int_0^1\int_0^1 (F''(\theta(tx+(1-t)y)+(1-\theta)(tz+(1-t)u))) \\
&\times (tx+(1-t)y-(tz+(1-t)u))d\theta dt| \\
&= |\textstyle\int_0^1\int_0^1 (e^{\theta(tx+(1-t)y)+(1-\theta)(tz+(1-t)u)})(tx+(1-t)y-(tz+(1-t)u))d\theta dt| \\
&\leq \textstyle\int_0^1 e|t(x-z)+(1-t)(y-u)|dt \\
&\leq \tfrac{e}{2}(|x-z|+|y-u|)
\end{aligned} \tag{14.42}$$

and

$$\begin{aligned}
&|F'(x^\star)^{-1}([x,y;F]-[x^\star,x^\star;F])| = |\textstyle\int_0^1 F'(tx+(1-t)y)dt - F'(x^\star)| \\
&= |\textstyle\int_0^1 (e^{tx+(1-t)y}-1)dt| \\
&= |\textstyle\int_0^1 (tx+(1-t)y)(1+\frac{tx+(1-t)y}{2!}+\frac{(tx+(1-t)y)^2}{3!}+\cdots)dt| \\
&\leq |\textstyle\int_0^1 (tx+(1-t)y)(1+\frac{1}{2!}+\frac{1}{3!}+\cdots)dt| \\
&\leq \tfrac{e-1}{2}(|x-x^\star|+|y-x^\star|).
\end{aligned} \tag{14.43}$$

That is to say, the Lipschitz condition (14.31) and the center-Lipschitz condition (14.30) are true for $l_1 = \frac{e}{2}, l = \frac{e^{\frac{1}{e-1}}}{2}$ *and* $l_0 = \frac{e-1}{2}$*, respectively. Using (14.33) in Theorem 14.3.1, we can deduce that the radius of convergence ball for method (14.2) is given by*

$$\rho_0 = \frac{1}{(1+\sqrt{2})l_1 + 2l_0} = \frac{2}{(3+\sqrt{2})e - 2} \approx 0.200018471, \tag{14.44}$$

which is smaller than the corresponding radius

$$\rho = \frac{1}{(1+\sqrt{2})l + 2l_0} \approx 0.2578325131698342986 \tag{14.45}$$

Let us choose $x_0 = 0.2, y_0 = 0.199$*. Suppose sequences* $\{x_n\}$ *and* $\{y_n\}$ *are generated by method (14.2). Table 14.1 gives a comparison results of error estimates, which shows that tighter error estimates can be obtained from the new (14.34) or (14.35) using* ρ *instead of* ρ_0 *used in [6].*

Table 14.1: The comparison results of error estimates.

n	using ρ new (14.35)	using ρ_0 old (14.35)
3	0.0498	0.0828
4	0.0031	0.0142
5	2.9778e-06	1.7305e-04
6	1.7108e-13	4.4047e-09
6	5.4309e-31	3.4761e-20

Hence the new results are more precise than the old ones in [6].

Example 14.4.2 *Let* $\mathbb{B}_1 = \mathbb{B}_2 = C[0,1]$*, the space of continuous functions defined on* $[0,1]$*, equipped with the max norm and* $\Omega = \overline{U}(0,1)$*. Define function F on* Ω*, given by*

$$F(x)(s) = x(s) - 5\int_0^1 stx^3(t)dt, \tag{14.46}$$

and the divided difference of F is defined by

$$[x,y;F] = \int_0^1 F'(tx + (1-t)y)dt. \tag{14.47}$$

Then, we have

$$[F'(x)y](s) = y(s) - 15\int_0^1 stx^2(t)y(t)dt, \ \ for \ all \ \ y \in \Omega. \tag{14.48}$$

We have $x^\star(s) = 0$ *for all* $s \in [0,1]$*,* $l_0 = 3.75$ *and* $l = l_1 = 7.5$*. Using Theorem*

14.3.1, we can deduce that the radius of convergence ball for method (14.2) is given by

$$\rho_0 = \rho = \frac{1}{(1+\sqrt{2})l + 2l_0} \approx 0.039052429. \tag{14.49}$$

Example 14.4.3 *Let $\mathbb{B}_1 = \mathbb{B}_2 = C[0,1]$ be equipped with the max norm and $\Omega = U(0,r)$ for some $r > 1$. Define F on Ω by*

$$F(x)(s) = x(s) - y(s) - \mu \int_0^1 G(s,t)x^3(t)dt, \quad x \in C[0,1],\ s \in [0,1].$$

$y \in C[0,1]$ is given, μ is a real parameter and the Kernel G is the Green's function defined by

$$G(s,t) = \begin{cases} (1-s)t & \text{if } t \le s \\ s(1-t) & \text{if } s \le t. \end{cases}$$

Then, the Fréchet derivative of F is defined by

$$(F'(x)(w))(s) = w(s) - 3\mu \int_0^1 G(s,t)x^2(t)w(t)dt, \quad w \in C[0,1],\ s \in [0,1].$$

Let us choose $x_0(s) = y_0(s) = y(s) = 1$ and $|\mu| < \frac{8}{3}$. Then, we have that

$$\begin{array}{l} \|I - A_0\| \le \frac{3}{8}\mu, \quad A_0^{-1} \in L(\mathbb{B}_2, \mathbb{B}_1), \\ \|A_0^{-1}\| \le \frac{8}{8-3|\mu|}, \quad s_0 = 0, \quad t_1 = \frac{|\mu|}{8-3|\mu|}, \quad L_0 = \frac{3(1+r)|\mu|}{2(8-3|\mu|)}, \end{array}$$

and

$$L = \frac{3r|\mu|}{8-3|\mu|}.$$

Let us choose $r = 3$ and $\mu = \frac{1}{2}$. Then, we have that

$$t_1 = 0.076923077, \quad L_0 \approx 0.461538462, \quad L = L_1 \approx 0.692307692$$

and

$$\frac{L(t_1 + s_0)}{1 - L_0(t_1 + s_1 + s_0)} \approx 0.057441746, \quad \alpha \approx 0.711345739,$$

$$1 - \frac{2L_0 t_1}{1 - L_0 s_0} \approx 0.928994083.$$

That is, condition (14.5) is satisfied and Theorem 14.2.2 applies.

References

[1] S. Amat, S. Busquier and M. Negra, Adaptive approximation of nonlinear operators, *Numer. Funct. Anal. Optim.*, 25, 397–405, 2004.

[2] I. K. Argyros, Computational theory of iterative methods, Series: Studies in Computational Mathematics 15, Editors, C. K. Chui and L. Wuytack, Elservier Publ. Co. New York, USA, 2007.

[3] I. K. Argyros, A semilocal convergence analysis for directional Newton methods, *Math. Comput.*, 80, 327–343, 2011.

[4] I. K. Argyros and S. Hilout, Computational methods in nonlinear analysis. Efficient algorithms, fixed point theory and applications, World Scientific, 2013.

[5] I. K. Argyros and S. Hilout, Weaker conditions for the convergence of Newton's method, *J. Complexity*, 28, 364–387, 2012.

[6] I. K. Argyros and H. Ren, On the convergence of efficient King-Werner-type methods of order $1+\sqrt{2}$ free of derivatives, *App. Math. Comput.*, 217 (2), 612–621, 2010.

[7] L. V. Kantorovich and G. P. Akilov, Functional Analysis, Pergamon Press, Oxford, 1982.

[8] R. F. King, Tangent methods for nonlinear equations, *Numer. Math.*, 18, 298–304, 1972.

[9] T. J. McDougall and S. J. Wotherspoon, A simple modification of Newton's method to achieve convergence of order $1+\sqrt{2}$, *Appl. Math. Let.*, 29, 20–25, 2014.

[10] H. Ren, Q. Wu and W. Bi, On convergence of a new secant like method for solving nonlinear equations, *Appl. Math. Comput.*, 217, 583–589, 2010.

[11] W. C. Rheinboldt, An adaptive continuation process for solving systems of nonlinear equations, *Polish Academy of Science, Banach Ctr. Publ.*, 3, 129–142, 1977.

[12] J. F. Traub, Iterative Methods for the Solution of Equations, Englewood Cliffs, Prentice Hull, 1984.

[13] W. Werner, Uber ein Verfahren der Ordnung $1+\sqrt{2}$ zur Nullstellenbestimmung, *Numer. Math.*, 32, 333–342, 1979.

[14] W. Werner, Some supplementary results on the $1+\sqrt{2}$ order method for the solution of nonlinear equations, *Numer. Math.*, 38, 383–392, 1982.

Chapter 15

Müller's method

Ioannis K. Argyros

Department of Mathematical Sciences, Cameron University, Lawton, OK 73505, USA, Email: iargyros@cameron.edu

Á. Alberto Magreñán

Universidad Internacional de La Rioja, Escuela Superior de Ingeniería y Tecnología, 26002 Logroño, La Rioja, Spain, Email: alberto.magrenan@unir.net

CONTENTS

15.1 Introduction

In this chapter we are concerned with approximating a solution of the equation

$$f(x) = 0, \tag{15.1}$$

where f is defined on an open domain or closed domain D on a real space $\mathbb{R}$.

The widely used [4] Müller's method is defined in [2] by

$$x_{n+1} = x_n - \frac{2C_n}{B_n \pm \sqrt{B_n^2 - 4A_nC_n}}, n = 0,1,2,..., \quad x_{-2}, x_{-1}, x_0 \in D, \tag{15.2}$$

where,

$$A_n = f[x_n, x_{n-1}, x_{n-2}], \quad B_n = f[x_n, x_{n-1}] + A_n(x_n - x_{n-1}), \quad C_n = f(x_n), \tag{15.3}$$

and $f[.,.], f[.,.,.]$ are divided differences of order one and two, respectively (see [3]). The sign in the denominator of (15.2) is chosen so as to give the larger value.

It is a free-derivative method and has a convergence order 1.839... under reasonable conditions [2]. Xie in [5] established a semilocal convergence Theorem of the method by means of bounding third and fourth derivatives. Bi et al. [6] presented a new semilocal convergence Theorem of the method using $\gamma-$condition. Wu et al. [7] gave the convergence ball and error analysis of the method under the hypotheses that the second-order and third-order derivative of function f are bounded.

In this chapter, we will use a weaker condition to provide a new estimate on the radius of convergence ball of Müller's method. We will assume that f is differentiable in D, $f'(x_\star) \neq 0$, and the following Lipschitz conditions are true:

$$|f'(x_\star)^{-1}(f[x,y] - f[u,v])| \leq K(|x-u| + |y-v|)\ , for\ any\ x,y,u,v \in D \quad (15.4)$$

and

$$|f'(x_\star)^{-1}(f[x,x_\star] - f[x_\star,x_\star])| \leq K_\star|x-x_\star|\ , for\ any\ x \in D, \quad (15.5)$$

where, $K > 0$ and $K_\star > 0$ are constants, and $x_\star \in D$ is a solution of (15.1).

The chapter is organized as follows: In Section 15.2 we present the convergence ball analysis of method (15.2). Finally, some numerical examples are provided in Section 15.3.

15.2 Convergence ball for method (15.1.2)

Throughout this chapter we will denote $U(x,r)$ as an open ball around x with radius r.

Next, we have the following result.

Theorem 15.2.1 *Suppose $x_\star \in D$ is a solution of equation (15.1), $f'(x_\star) \neq 0$, conditions (15.4) and (15.5) are satisfied. Denote*

$$R' = \frac{1}{5K + K_\star + 2\sqrt{4K^2 + 2KK_\star}}. \quad (15.6)$$

Assume

$$U(x_\star, R') \subseteq D. \quad (15.7)$$

Then, the sequence $\{x_n\}$ generated by Müller's method (15.2) starting from any three distinct points $x_{-2}, x_{-1}, x_0 \in U(x_\star, R')$ is well defined, and converges to $x_\star$. Moreover, the following estimates hold:

$$|x_\star - x_{n+1}| \leq (\frac{C}{R'})^{F_{n+1}-1}|x_\star - x_0|^{F_n}|x_\star - x_{-1}|^{F_{n-1}},\ for\ any\ n = 0,1,2,..., \quad (15.8)$$

where,

$$C = \frac{2K}{\sqrt{4K^2 + 2KK_\star}},$$

and $\{F_n\}$ *is Fibonacci sequence, and is defined by* $F_{-1} = 1, F_0 = 1$ *and* $F_{n+1} = F_n + F_{n-1}$ *for any* $n = 0, 1, ...$

Proof: We will use induction. First of all, denote $e_n = x_\star - x_n (n = -2, -1, ...)$. Let $x_{-2}, x_{-1}, x_0 \in U(x_\star, R')$ be distinct points. By (15.3) and (15.4), we have

$$|f'(x_\star)^{-1}A_0| = |f'(x_\star)^{-1}f[x_0, x_{-1}, x_{-2}]| = |f'(x_\star)^{-1}\tfrac{f[x_0,x_{-2}]-f[x_{-1},x_{-2}]}{x_0-x_{-1}}| \le K. \tag{15.9}$$

Now, using $x_{-1}, x_0 \in U(x_\star, R')$, (15.3), (15.5) and (15.9), we get

$$\begin{aligned}
&|1 - f'(x_\star)^{-1}B_0| = |1 - f'(x_\star)^{-1}(f[x_0, x_{-1}] + A_0(e_{-1} - e_0))| \\
&= |f'(x_\star)^{-1}(f[x_\star, x_\star] - f[x_\star, x_0] + f[x_0, x_\star] - f[x_0, x_{-1}] + A_0(e_0 - e_{-1}))| \\
&\le K_\star|e_0| + K|e_{-1}| + K(|e_0| + |e_{-1}|) = (K + K_\star)|e_0| + 2K|e_{-1}| \\
&< (3K + K_\star)R' = \tfrac{3K+K_\star}{5K+K_\star+2\sqrt{4K^2+2KK_\star}} < 1,
\end{aligned} \tag{15.10}$$

which means

$$|f'(x_\star)^{-1}B_0| > 1 - (3K + K_\star)R' = \tfrac{2K+2\sqrt{4K^2+2KK_\star}}{5K+K_\star+2\sqrt{4K^2+2KK_\star}} > 0 \tag{15.11}$$

and

$$|B_0^{-1}f'(x_\star)| < \tfrac{1}{1-(3K+K_\star)R'} = \tfrac{5K+K_\star+2\sqrt{4K^2+2KK_\star}}{2K+2\sqrt{4K^2+2KK_\star}}. \tag{15.12}$$

Using $x_0 \in U(x_\star, R')$, $f(x_\star) = 0$, (15.3) and (15.5), we have

$$\begin{aligned}
|f'(x_\star)^{-1}C_0| &= |f'(x_\star)^{-1}(f(x_0) - f(x_\star))| = |f'(x_\star)^{-1}(f[x_\star, x_0] \\
&\quad -f[x_\star, x_\star] + f[x_\star, x_\star])e_0| \\
&\le K_\star|e_0|^2 + |e_0| < K_\star R'^2 + R'.
\end{aligned} \tag{15.13}$$

In view of (15.9), (15.11), (15.13) and (15.6), we have

$$\begin{aligned}
&(f'(x_\star)^{-1})^2B_0^2 - 4(f'(x_\star)^{-1})^2A_0C_0 \ge |f'(x_\star)^{-1}B_0|^2 - 4|f'(x_\star)^{-1}A_0| \\
&|f'(x_\star)^{-1}C_0| \\
&> (1 - (3K + K_\star)R')^2 - 4K(K_\star R'^2 + R') \\
&= [(3K + K_\star)^2 - 4KK_\star]R'^2 - [2(3K + K_\star) + 4K]R' + 1 \\
&= (9K^2 + 2KK_\star + K_\star^2)R'^2 - 2(5K + K_\star)R' + 1 = 0.
\end{aligned} \tag{15.14}$$

Therefore, x_1 can be defined. Denoting $sign(s)$ as $sign(s) = 1$ if $s \ge 0$ and $sign(s) = -1$ if $s < 0$. Now, considering (15.2), we obtain

$$\begin{aligned}
e_1 &= e_0 + \tfrac{2C_0}{B_0+sign(B_0)\sqrt{B_0^2-4A_0C_0}} = e_0 + \tfrac{B_0-sign(B_0)\sqrt{B_0^2-4A_0C_0}}{2A_0} \\
&= \tfrac{2A_0e_0+B_0-sign(B_0)\sqrt{B_0^2-4A_0C_0}}{2A_0} = \tfrac{(2A_0e_0+B_0)^2-B_0^2+4A_0C_0}{2A_0(2A_0e_0+B_0+sign(B_0)\sqrt{B_0^2-4A_0C_0})} \\
&= \tfrac{2(A_0e_0^2+B_0e_0+C_0)}{2A_0e_0+B_0+sign(B_0)\sqrt{B_0^2-4A_0C_0}},
\end{aligned} \tag{15.15}$$

so we have

$$|e_1| = \frac{2|A_0e_0^2+B_0e_0+C_0|}{|2A_0e_0+B_0+sign(B_0)\sqrt{B_0^2-4A_0C_0}|}. \tag{15.16}$$

Next we shall show $2A_0e_0+B_0$ has the same sign as B_0. In fact, using (15.9), (15.6) and (15.12), we get

$$\begin{aligned}\frac{2A_0e_0+B_0}{B_0} &= 1+2f'(x_\star)^{-1}A_0f'(x_\star)B_0^{-1}e_0 \geq 1-2|f'(x_\star)^{-1}A_0||f'(x_\star)B_0^{-1}||e_0| \\ &> 1-\frac{2K(5K+K_\star+2\sqrt{4K^2+2KK_\star})}{2K+2\sqrt{4K^2+2KK_\star}}R' \\ &= \frac{\sqrt{4K^2+2KK_\star}}{K+\sqrt{4K^2+2KK_\star}} > 0.\end{aligned} \tag{15.17}$$

Hence, by (15.16), (15.3), (15.9), (15.4), (15.14), (15.11) and (15.17), we obtain

$$\begin{aligned}|e_1| &= \frac{2|A_0e_0^2+B_0e_0+C_0|}{|2A_0e_0+B_0|+\sqrt{B_0^2-4A_0C_0}} = \frac{2|(A_0e_0+B_0)e_0-f[x_\star,x_0]e_0|}{|2A_0e_0+B_0|+\sqrt{B_0^2-4A_0C_0}} \\ &= \frac{2|A_0e_{-1}+A_0(e_0-e_{-1})+f[x_0,x_{-1}]-A_0(e_0-e_{-1})-f[x_0,x_\star]|}{|2A_0e_0+B_0|+\sqrt{B_0^2-4A_0C_0}}|e_0| \\ &= \frac{2|f'(x_\star)^{-1}(A_0e_{-1}+f[x_0,x_{-1}]-f[x_0,x_\star])|}{|f'(x_\star)^{-1}B_0|\frac{2A_0e_0+B_0}{B_0}+\sqrt{(f'(x_\star)^{-1}B_0)^2-4f'(x_\star)^{-1}A_0f'(x_\star)^{-1}C_0}}|e_0| \\ &\leq \frac{4K|e_{-1}|}{\frac{2K+2\sqrt{4K^2+2KK_\star}}{5K+K_\star+2\sqrt{4K^2+2KK_\star}}\frac{\sqrt{4K^2+2KK_\star}}{K+\sqrt{4K^2+2KK_\star}}}|e_0| \\ &= \frac{2K(5K+K_\star+2\sqrt{4K^2+2KK_\star})}{\sqrt{4K^2+2KK_\star}}|e_0||e_{-1}|.\end{aligned} \tag{15.18}$$

as a consequence, from the definition of C in this Theorem, we obtain

$$\frac{|e_1|}{R'} \leq \frac{2K}{\sqrt{4K^2+2KK_\star}}\frac{|e_0|}{R'}\frac{|e_{-1}|}{R'} = C\frac{|e_0|}{R'}\frac{|e_{-1}|}{R'} < 1, \tag{15.19}$$

that is, $x_1 \in U(x_\star, R')$.

Using an inductive procedure on $n = 0,1,2,...$, we get $x_n \in U(x_\star, R')$, and

$$\frac{|e_{n+1}|}{R'} \leq C\frac{|e_n|}{R'}\frac{|e_{n-1}|}{R'},\ for\ any\ n = 0,1,2,.... \tag{15.20}$$

Denote

$$\rho_n = C\frac{|e_n|}{R'}, \quad n = 0,1,2,...., \tag{15.21}$$

we obtain

$$\rho_{n+1} \leq \rho_n\rho_{n-1}, \quad n = 0,1,2,.... \tag{15.22}$$

Clearly the following relation holds:

$$\rho_{n+1} \leq \rho_0^{F_n}\rho_{-1}^{F_{n-1}}, \quad n = 0,1,2,.... \tag{15.23}$$

Consequently, estimates (15.8) are true. □

Remark 15.2.2

(a) Condition (15.4) used in Theorem 15.2.1 is weaker than bounded conditions of the second-order and third-order derivative of function f used in [7]. In fact, suppose that f is twice differentiable on D, and

$$|f'(x_\star)^{-1}f''(x)| \leq N\,, for\,any\,x \in D, \tag{15.24}$$

then, for any $x, y, u, v \in D$, we get

$$\begin{aligned}
&|f'(x_\star)^{-1}(f[x,y]-f[u,v])| = |f'(x_\star)^{-1}\int_0^1(f'(tx+(1-t)y) \\
&-f'(tu+(1-t)v))dt| \\
&= |f'(x_\star)^{-1}\int_0^1\int_0^1 f''\big(s(tx+(1-t)y)+(1-s)(tu+(1-t)v)\big) \\
&\big(t(x-u)+(1-t)(y-v)\big)dsdt| \\
&\leq \tfrac{N}{2}(|x-u|+|y-v|),
\end{aligned} \tag{15.25}$$

which shows condition (15.4) holds for $K = \frac{N}{2}$.

(b) Notice that [6] a semilocal convergence result is given under the γ-condition of order two which is a condition of $third-order$ derivative of function f, however, we don't use any information of f''' in our Theorem.

Remark 15.2.3 It is worth noticing that from (15.4) and (15.5)

$$K_\star \leq K \tag{15.26}$$

holds in general and $\frac{K}{K_\star}$ can be arbitrarily large [1]. Therefore, the condition $K_\star \leq K$ in Theorem 15.2.1 involves no loss of generality. If we use only condition (15.4) in Theorem 15.2.1, we can establish a similar Theorem by replacing (15.6) and (15.8) by

$$\overline{R}' = \frac{1}{2(3+\sqrt{6})K} \tag{15.27}$$

and

$$|x_\star - x_{n+1}| \leq ((4+2\sqrt{6})K)^{F_{n+1}-1}|x_\star - x_0|^{F_n}|x_\star - x_{-1}|^{F_{n-1}},\ for\,any\,n = 0,1,2,..., \tag{15.28}$$

respectively. On the other hand, if (15.26) holds strictly, it is clear that the following inequalities are true:

$$\overline{R}' < R' \tag{15.29}$$

and

$$(4+2\sqrt{6})K < \frac{C}{R'} = \frac{2K(5K+K_\star)}{\sqrt{4K^2+2KK_\star}} + 4K, \tag{15.30}$$

which means that we have bigger radius of convergence ball of Müller's method and tighter errors by using conditions (15.4) and (15.5) simultaneously instead of using only condition (15.4).

15.3 Numerical examples

Example 15.3.1 *Let f be defined on $D = [-\frac{5}{2}, \frac{1}{2}]$ by*

$$f(x) = \begin{cases} x^3 \ln x^2 + x^5 - x^4 + 4x, & x \neq 0; \\ 0, & x = 0. \end{cases} \tag{15.31}$$

Then, we have

$$f'(x) = \begin{cases} 3x^2 \ln x^2 + 5x^4 - 4x^3 + 2x^2 + 4, & x \neq 0, \\ 4, & x = 0, \end{cases} \tag{15.32}$$

$$f''(x) = \begin{cases} 6x \ln x^2 + 20x^3 - 12x^2 + 10x, & x \neq 0, \\ 0, & x = 0 \end{cases} \tag{15.33}$$

and

$$f'''(x) = \begin{cases} 6 \ln x^2 + 60x^2 - 24x + 22, & x \neq 0, \\ -\infty, & x = 0, \end{cases} \tag{15.34}$$

which means f'' is continuous on D and f''' is unbounded on D. Then, Theorem 15.2.1 applies. However, the Theorem in [7] cannot apply.

Example 15.3.2 *Let f be defined on $D = [-1, 1]$ by*

$$f(x) = e^x - 1. \tag{15.35}$$

Then, we have

$$f'(x) = e^x, \quad f''(x) = e^x, \quad f'''(x) = e^x. \tag{15.36}$$

It is obvious that $x_\star = 0$, $f'(x_\star) = 1$ and for any $x, y, u, v \in D$, we have

$$\begin{aligned} &|f'(x_\star)^{-1}(f[x,y] - f[u,v])| = |\int_0^1 (f'(tx + (1-t)y) - f'(tu + (1-t)v))dt| \\ &= |\int_0^1 (e^{tx+(1-t)y} - e^{tu+(1-t)v})dt| \\ &= |\int_0^1 \int_0^1 e^{s(tx+(1-t)y)+(1-s)(tu+(1-t)v)} ds(t(x-u) + (1-t)(y-v))dt| \\ &\leq \frac{e}{2}(|x-u| + |y-v|). \end{aligned} \tag{15.37}$$

Moreover, for any $x \in D$, we have

$$\begin{aligned} &|f'(x_\star)^{-1}(f[x,x_\star] - f[x_\star,x_\star])| = |\int_0^1 (f'(tx) - f'(0))dt| \\ &= |\int_0^1 (e^{tx} - 1)dt| = |\int_0^1 (tx + \frac{(tx)^2}{2!} + \ldots)dt| \\ &\leq \int_0^1 t|x|(1 + \frac{1}{2!} + \ldots)dt| = \frac{e-1}{2}|x - x_\star|. \end{aligned} \tag{15.38}$$

Therefore, we can choose $K = \frac{e}{2}$ and $K_\star = \frac{e-1}{2}$ in (15.4) *and* (15.5)*, respectively. By* (15.6)*, we have $R' \approx 0.0720$ and* (15.7) *is true. Therefore, all conditions of Theorem 15.2.1 hold, so it applies. Moreover, if we use only condition* (15.4)*, we have only $R' \approx 0.0675$. Note also that we can choose constants N (the upper bound for $|f'(x_\star)^{-1} f''(x)|$ on D) and M (the upper bound for $|f'(x_\star)^{-1} f'''(x)|$ on D) of Theorem 1 in [7] as $N = M = e$. However, the other condition $1215N^2 \leq 32M$ of Theorem 1 in [7] is not satisfied, and so Theorem 1 in [7] cannot apply.*

Example 15.3.3 *Let f be defined on $D = [-1,1]$ by*

$$f(x) = \begin{cases} \frac{1}{2}x^2 + 3x, & 0 \le x \le 1, \\ -\frac{1}{2}x^2 + 3x, & -1 \le x < 0. \end{cases} \tag{15.39}$$

Then, we have $x_\star = 0$,

$$f'(x) = |x| + 3 = \begin{cases} x+3, & 0 \le x \le 1, \\ -x+3, & -1 \le x < 0 \end{cases} \tag{15.40}$$

and

$$f''(x) = \begin{cases} 1, & 0 < x \le 1, \\ \textit{does not exist}, & x = 0, \\ -1, & -1 \le x < 0. \end{cases} \tag{15.41}$$

Notice that, for any $x, y, u, v \in D$, we have

$$\begin{aligned} |f'(x_\star)^{-1}(f[x,y] - f[u,v])| &= \tfrac{1}{3}|\int_0^1 (f'(tx+(1-t)y) - f'(tu+(1-t)v))dt| \\ &= \tfrac{1}{3}|\int_0^1 (|tx+(1-t)y| - |tu+(1-t)v|)dt| \\ &\le \tfrac{1}{3}\int_0^1 ||tx+(1-t)y| - |tu+(1-t)v||dt \\ &\le \tfrac{1}{3}\int_0^1 |tx+(1-t)y - (tu+(1-t)v)|dt \\ &\le \tfrac{1}{3}\int_0^1 (t|x-u| + (1-t)|y-v|)dt \\ &= \tfrac{1}{6}(|x-u| + |y-v|). \end{aligned} \tag{15.42}$$

That is, condition (15.4) *holds for $K = \frac{1}{6}$ but $f''(x)$ is not differentiable at $0 \in D$.*

References

[1] I. K. Argyros, Computational theory of iterative methods, Series: Studies in Computational Mathematics 15, Editors, C. K. Chui and L. Wuytack, Elservier Publ. Co. New York, USA, 2007.

[2] D. E. Müller, A method for solving algebraic equations using an automatic computer, *Math. Tables Other Aids Comput.*, 10, 208–215, 1956.

[3] J. Stoer and R. Bulirsch, Introduction to Numerical Analysis, Springer-Verlag, New York, 1980.

[4] C. F. Gerald and P. O. Wheatley, Applied Numerical Analysis, Addison-Wesley, Reading, MA, 1994.

[5] S. Q. Xie, Convergence of Müller's method for finding roots, *Mathematics in Practice and Theory*, 2, 18–25, 1980 (in Chinese).

[6] W. H. Bi, H. M. Ren and Q. B. Wu, A new semilocal convergence Theorem of Müller's method, *Appl. Math. Comput.*, 199, 375–384, 2008.

[7] Q. B. Wu, H. M. Ren and W. H. Bi, Convergence ball and error analysis of Müller's method, *Appl. Math. Comput.*, 184, 464–470, 2007.

Chapter 16

Generalized Newton method with applications

Ioannis K. Argyros

Department of Mathematical Sciences, Cameron University, Lawton, OK 73505, USA, Email: iargyros@cameron.edu

Á. Alberto Magreñán

Universidad Internacional de La Rioja, Escuela Superior de Ingeniería y Tecnología, 26002 Logroño, La Rioja, Spain, Email: alberto.magrenan@unir.net

CONTENTS

16.1 Introduction

In this chapter we are interested in the approximately solving the generalized equation: Find $x^* \in H$ such that

$$0 \in F(x^*) + T(x^*). \tag{16.1}$$

where $F : H \longrightarrow H$ is a Fréchet differentiable function, H is a Hilbert space and $T : H \rightrightarrows H$ is a set valued maximal monotone operator.

Importance of studying problem (16.1) and its applications in the physical and engineering sciences and many other areas can be found in [1]– [33].

The Generalized Newton method for equation (16.1) is defined as

$$0 \in F(x_n) + F'(x_n)(x_{n+1} - x_n) + T(x_{n+1}), \tag{16.2}$$

for $n = 1, 2, 3, \ldots$.

Using classical Lipschitz condition Uko [26, 27] established the convergence of (16.2). The convergence (16.2) has been studied by many other authors under various conditions (see Robinson [17], Josephy [12] (also see [18, 28])).

The convergence of generalized Newton method (16.2) was shown in [29, 33] using Lipschitz continuity conditions on F'. However, there are problems where the Lipschitz continuity of F' does not hold. Motivated by this constrains, we present a convergence analysis of the generalized Newton method (16.2) using generalized continuity on F'. Our results are weaker even, if we specialize the conditions on F' to the condition given in [23].

The rest of the chapter is organized as follows: In Section 16.2 we present some needed preliminaries. Then, in Section 16.3 the convergence of generalized Newton method (16.2) is presented.

16.2 Preliminaries

Let $x \in H$ and $r > 0$. Throughout this chapter we will use $U(x, r)$ and $\overline{U(x, r)}$ to denote the open and closed metric ball, respectively, at x with radius r

$$U(x, r) := \{y \in H : \|x - y\| < r\} \text{ and } \overline{U(x, r)} := \{y \in H : \|x - y\| \leq r\}.$$

Recall that a bounded linear operator $G : H \to H$ is called a positive operator if G is self conjugate and $\langle Gx, x \rangle \geq 0$ for each $x \in H$(cf. [22], p.313).

We present the following lemma taken from [29].

Lemma 16.2.1 *Let G be a positive operator. Then the following conditions hold:*

- $\|G^2\| = \|G\|^2$
- *If G^{-1} exists, then G^{-1} is also a positive operator and*

$$\langle Gx, x \rangle \geq \frac{\|x\|^2}{\|G^{-1}\|}, \quad \text{for each } x \in H. \tag{16.3}$$

Now, let $T : H \rightrightarrows H$ be a set valued operator. The domain $domT$ of T is defined as $domT := \{x \in H : Tx \neq \emptyset\}$. Next, let us recall notions of montonicity for set-values operators taken from [32].

Definition 16.2.2 *Let $T : H \rightrightarrows H$ be a set valued operator. T is said to be*

(a) monotone if

$$\langle u - v, y - x \rangle \geq 0 \text{ for each } u \in T(y) \text{ and } v \in T(x) \tag{16.4}$$

- *maximal monotone if it is monotone and the following implications hold:*

$$\langle u-v, x-y\rangle \geq 0 \text{ for each } y \in domT \text{ and } v \in T(y) \Rightarrow x \in domT \text{ and } u \in T(x). \tag{16.5}$$

Let $G : H \longrightarrow H$ be a bounded linear operator. Then $\widehat{G} := \frac{1}{2}(G+G^*)$ where G^* is the adjoint of G. Hereafter, we assume that $T : H \rightrightarrows H$ is a set valued maximal monotone operator and $F : H \longrightarrow H$ is a Fréchet differentiable operator.

16.3 Semilocal convergence

First, we begin with some needed definitions.

Definition 16.3.1 *Let $R > 0$ and $\bar{x} \in H$ be such that $\widehat{F'(\bar{x})}^{-1}$ exists. Then, opera tor $\|\widehat{F'(\bar{x})}^{-1}\|F$ is said to satisfy the center-Lipschitz condition with $L-$ average at $\bar{x} \in U(\bar{x}, R)$, if*

$$\|\widehat{F'(\bar{x})}^{-1}\|\|F'(x) - F'(\bar{x})\| \leq \int_0^{\|x-\bar{x}\|} L_0(u)du,$$

for each $\bar{x} \in U(\bar{x}, R)$, where L_0 is a non-negative non-decreasing integrable function on the interval $[0, R]$ satisfying $\int_0^{R_0} L_0(u)du = 1$, for some $R_0 \geq 0$.

Definition 16.3.2 *Let $R > 0$ and $\bar{x} \in H$ be such that $\widehat{F'(\bar{x})}^{-1}$ exists. Then, operator $\|\widehat{F'(\bar{x})}^{-1}\|F'$ is said to satisfy the center-Lipschitz condition in the inscribed sphere with $(L_0, L)-$average at $\bar{x}$ on $U(\bar{x}, \overline{R})$, $\overline{R} = \min\{R_0, R\}$, if*

$$\|\widehat{F'(\bar{x})}^{-1}\|\|F'(y) - F'(x)\| \leq \int_{\|x-\bar{x}\|}^{\|x-\bar{x}\|+\|y-x\|} L(u)du,$$

for each $x, y \in U(\bar{x}, R_0)$, with $\|x-\bar{x}\| + \|y-x\| < \overline{R}$ and L is a non-negative non-decreasing integrable function on the interval $[0, \overline{R}]$.

Definition 16.3.3 *Let $R > 0$ and $\bar{x} \in H$ be such that $\widehat{F'(\bar{x})}^{-1}$ exists. Then, operator $\|\widehat{F'(\bar{x})}^{-1}\|F'$ is said to satisfy the center-Lipschitz condition in the inscribed sphere with L_1-average at $\bar{x}$ on $U(\bar{x}, \overline{R})$, if*

$$\|\widehat{F'(\bar{x})}^{-1}\|\|F'(y) - F'(x)\| \leq \int_{\|x-\bar{x}\|}^{\|x-\bar{x}\|+\|y-x\|} L_1(u)du$$

for each $x, y \in U(\bar{x}, R)$, with $\|x-\bar{x}\| + \|y-x\| < \overline{R}$ and L_1 is a non-negative non-decreasing integrable function on the interval $[0, R]$.

The convergence analysis of generalized Newton's method (16.2) was based on Definition 16.3.3 [33]. However, we have that

$$L_0(u) \leq L_1(u) \tag{16.6}$$

and

$$L(u) \leq L_1(u) \tag{16.7}$$

for each $u \in [0, \overline{R}]$, since $\overline{R} \leq R$.

Hence, if functions L_0 and L are used instead of L_1 we can obtain a more precise convergence analysis, which is the main objective of the present chapter.

It is worth noticing that in practice the computation of function L_1 requires, as special cases, the computation of functions L_0 and L.

Now, let $\beta > 0$. Define scalar majorizing function h by

$$h_1(t) = \beta - t + \int_0^t L_1(u)(t-u)du, \text{ for each } t \in [0, R] \tag{16.8}$$

and majorizing sequences $\{t_n\}$ by $t_0 = 0$, $t_{n+1} = t_n - h_1'(t_n)^{-1}h_1(t_n)$. Sequence $\{t_n\}$ can also be written as $t_0 = 0$, $t_1 = \beta$,

$$t_{n+1} = t_n + \frac{\int_0^1 \int_{t_{n-1}}^{t_{n-1}+\theta(t_n - t_{n-1})} L_1(u)(t_n - t_{n-1})d\theta}{1 - \int_0^{t_n} L_1(u)du}. \tag{16.9}$$

Sequence $\{t_n\}$ was used in [33]. However, we will use in this chapter the sequence $\{s_n\}$ defined by $s_0 = 0$, $s_1 = \beta$,

$$s_{n+1} = s_n + \frac{\int_0^1 \int_{s_{n-1}}^{s_{n-1}+\theta(s_n - s_{n-1})} L(u)du(s_n - s_{n-1})d\theta}{1 - \int_0^{s_n} L_0(u)du}. \tag{16.10}$$

We need an auxiliary result where we compare sequence $\{t_n\}$ to $\{s_n\}$.

Let $R_1 > 0$ and $b > 0$ be such that

$$\int_0^{R_1} L_1(u)du = 1 \text{ and } b = \int_0^{R_1} L_1(u)udu. \tag{16.11}$$

Lemma 16.3.4 *The following items hold.*

(a) The function h_1 is monotonically decreasing on $[0, R_1]$ and monotonically increasing on $[R_1, R]$. Suppose that

$$\beta \leq b. \tag{16.12}$$

Then, h has a unique zero, respectively in $[0, R_1]$ and $[R_1, R]$, which are denoted by r_1 and r_2 which satisfy

$$\beta < r_1 < \frac{R_1}{b}\beta < R_1 < r_2 < R. \tag{16.13}$$

Moreover, if $\beta \leq b$ and $r_1 = r_2$ if $\beta = b$.

(b) *Under hypothesis (16.12) sequences* $\{t_n\}$ *to* $\{s_n\}$ *are non-decreasing, converge to* r_1 *and* $s^* = \lim_{n\to\infty} s_n$, *respectively so that*

$$s_n \le t_n \le r_1 \tag{16.14}$$

$$0 \le s_{n+1} - s_n \le t_{n+1} - t_n \tag{16.15}$$

and

$$s_n^* \le r_1. \tag{16.16}$$

Proof: The proof of part (a) and that sequence $\{t_n\}$ non-decreasingly converges to r_1 can be found in [29, 33]. Using simple induction (16.6), (16.7), (16.9), (16.10) estimates (16.14) and (16.15) are obtained. Therefore, sequence $\{s_n\}$ is non-decreasing and bounded above by r_1 and as such it converges to s^* which satisfies (16.16) □

We also need the following lemma.

Lemma 16.3.5 *Let* $r < R_0$. *Let also* $\bar{x} \in H$ *be such that* $\widehat{F'(\bar{x})}$ *is a positive operator and* $\widehat{F'(\bar{x})}^{-1}$ *exists. Suppose that* $\|\widehat{F'(\bar{x})}\|F'$ *satisfies the center-Lipschitz condition with* L_0-*average at* $\bar{x} \in U(\bar{x}, r)$. *Then, for each* $x \in U(\bar{x}, r)$, $\widehat{F'(x)}$ *is a positive operator,* $\widehat{F'(\bar{x})}^{-1}$ *exists and*

$$\|\widehat{F'(x)}^{-1}\| \le \frac{\|\widehat{F'(\bar{x})}^{-1}\|}{1 - \int_0^{\|x-\bar{x}\|} L_0(u)du}. \tag{16.17}$$

Proof: Simply use L_0 instead of L_1 in the proof of corresponding result in [29, 33].

Remark 16.3.6 *The estimate corresponding to (16.17) is*

$$\|\widehat{F'(x)}^{-1}\| \le \frac{\|\widehat{F'(\bar{x})}^{-1}\|}{1 - \int_0^{\|x-\bar{x}\|} L_1(u)du}. \tag{16.18}$$

In view of (16.6), (16.17) and (16.18), estimate (16.17) is more precise than (16.18), if $L_0(u) < L_1(u)$.

Next, we present the main semilocal result of the chapter.

Theorem 16.3.7 *Suppose that there exists* $x_0 \in D$ *be such that* $\widehat{F'(x_0)}^{-1}$ *exists, (16.12) holds,* $\|\widehat{F'(x_0)}^{-1}\|F'$ *satisfies the center-Lipschitz condition in the inscribed sphere with* L_0-*average at* x_0 *on* $U(x_0, s^*)$, *the center-Lipschitz condition in the inscribed sphere with* $(L_0, L)-$*average at* $\bar{x}$ *on* $U(x_0, s^*)$, *and* $F'(x_0)$

is a positive operator. Then, sequence $\{x_n\}$ generated with initial point x_0 by Newton's method (16.1) provided that

$$\|x_1 - x_0\| \leq \beta \tag{16.19}$$

is well defined in $\overline{U}(x_0, s^*)$, remains in $\overline{U}(x_0, s^*)$ for each $n = 0, 1, 2, \cdots$ and converges to a solution x^* of (16.1) in $\overline{U}(x_0, s^*)$. Moreover, the following estimates hold

$$\|x_{n+1} - x_n\| \leq s^* - s_n \text{ for each } n = 0, 1, 2, \cdots \tag{16.20}$$

Proof: We shall show using mathematical induction that sequence $\{x_n\}$ is well defined and satisfies

$$\|x_{k+1} - x_k\| \leq s_{k+1} - s_k \text{ for each } k = 0, 1, 2, \cdots \tag{16.21}$$

Estimate (16.21) holds for $k = 0$ by (16.19). Let us suppose that (16.21) holds for $n = 0, 1, 2, \cdots, k-1$. We shall show that (16.21) holds for $n = k$. Notice that

$$\|x_k - x_0\| \leq \|x_k - x_{k+1}\| + \cdots + \|x_1 - x_0\| \leq s_k - s_{k-1} + \cdots + s_1 - s_0 = s_k - s_0 < s^*. \tag{16.22}$$

By (16.22) and Lemma 16.3.5, $F'(x_k)$ is a positive operator, $\widehat{F'(x_k)}^{-1}$ exists and

$$\|\widehat{F'(x_k)}^{-1}\| \leq \frac{\|\widehat{F'(x_0)}^{-1}\|}{1 - \int_0^{\|x_k - x_0\|} L_0(u)du}. \tag{16.23}$$

By Lemma 16.2.1, we have that

$$\frac{\|x\|^2}{\|\widehat{F'(x_k)}^{-1}\|} \leq \langle F'(x_k)x, x\rangle = \langle F'(x_k)x, x\rangle \text{ for each } x \in H, \tag{16.24}$$

so by Remark 2.5 [29, 33]:

$$0 \in F(x_k) + F'(x_k)(x_{k+1} - x_k) + T(x_{k+1}). \tag{16.25}$$

From hypotheses we obtain

$$0 \in F(x_{k-1}) + F'(x_{k-1})(x_k - x_{k-1}) + T(x_k). \tag{16.26}$$

Using (16.25), (16.26) and T being a maximal monotone operator we get that $\langle -F(x_{k-1}) - F'(x_{k-1})(x_k - x_{k-1}) + F(x_k) + F'(x_k)(x_{k+1} - x_k), x_k - x_{k+1}\rangle \geq 0$, so

$$\langle F(x_k) - F(x_{k-1}) - F'(x_{k-1})(x_k - x_{k-1}), x_k - x_{k+1}\rangle \geq \langle F'(x_k)(x_k - x_{k+1}), x_k - x_{k+1}\rangle. \tag{16.27}$$

By (16.24), we have that

$$\frac{\|x_k - x_{k+1}\|^2}{\|\widehat{F'(x_k)}^{-1}\|} \leq \langle \widehat{F'(x_k)}(x_k - x_{k+1}), x_k - x_{k+1}\rangle = \langle F'(x_k)(x_k - x_{k+1}), x_k - x_{k+1}\rangle$$

leading together with (16.27) to

$$\begin{aligned}
\|x_k - x_{k+1}\| &\leq \|\widehat{F'(x_k)}^{-1}\|\|F(x_k) - F(x_{k-1}) - F'(x_{k-1})(x_k - x_{k-1})\| \\
&\leq \|\widehat{F'(x_k)}^{-1}\|\|\int_0^1 (F'(x_{k-1} + \theta(x_k - x_{k-1}))) \\
&\quad -F'(x_{k-1})(x_k - x_{k-1})d\theta\| \\
&\leq \frac{\|\widehat{F'(x_0)}^{-1}\|}{1 - \int_0^{\|x_k - x_0\|} L_0(u)du}\|\int_0^1 (F'(x_{k-1} + \theta(x_k - x_{k-1}))) \\
&\quad -F'(x_{k-1})\| \\
&\quad \times\|(x_k - x_{k-1})d\theta\| \\
&\leq \frac{1}{1 - \int_0^{s_k - s_0} L_0(u)du}\int_0^1 \int_{\|x_{k-1} - x_0\|}^{\|x_{k-1} - x_0\| + \theta\|x_k - x_{k-1}\|} L(u)du \\
&\quad \|x_k - x_{k-1}\|d\theta \\
&\leq \frac{1}{1 - \int_0^{s_k} L_0(u)du}\int_0^1 \int_{s_{k-1}}^{s_{k-1} + \theta(s_k - s_{k-1})} L(u)du(s_k - s_{k-1})d\theta \\
&= s_{k+1} - s_k. \qquad (16.28)
\end{aligned}$$

It follows from (16.28) that sequence $\{x_k\}$ is complete in a Hilbert space H and as such it converges to some $x^* \in \overline{U}(x_0, s^*)$ (since $\overline{U}(x^*, s^*)$ is a closed set). Moreover, since T is a maximal monotone and $F \in C^1$, we deduce that x^* solves (16.1). Furthermore, (16.20) follows from (16.21) by using standard majorization techniques [1,4,5]. □

Next, we present the following result related to the uniqueness.

Proposition 16.3.8 *Suppose that the hypotheses of Theorem 16.3.7 hold except the center-Lipschitz (L_0, L) condition for $x_1 \in H$ such that*

$$0 \in F(x_0) + F'(x_0)(x_1 - x_0) + T(x_1).$$

Then, there exists a unique solution x^* of (16.1) in $\overline{U}(x_0, s^*)$.

Proof: Simply replace function L_1 by L_0 in the proof of Theorem 4.1 in [33].

Remark 16.3.9 *(a) Notice that function L_1 is not needed in the proof of Theorem 4.1 in [33], since it can be replaced by L_0 which is more precise than L_1(see (16.6)) in all stages of the proof. On the other hand, in view of (16.16), we obtain a better information on the uniqueness of x^*.*

(b) The results obtained in this chapter can be improved even further, if in Definition 16.3.2 we consider instead the center Lipschitz condition in

the inscribed sphere with (L_0,K)–average at $\bar{x}$ on $U(\bar{x},\overline{R}-\|x_1-x_0\|)$ (or $\overline{U}(\bar{x},\overline{R}-\beta)$) given by

$$\|\widehat{F'(\bar{x})}^{-1}\|\|F'(y)-F'(x)\|\leq\int_{\|x-\bar{x}\|}^{\|x-\bar{x}\|+\|y-x\|}K(u)du,$$

for each $x,y\in U(\bar{x},\overline{R}-\|x_1-x_0\|)$ (or $\overline{U}(\bar{x},\overline{R}-\beta)$). It is worth noticing that

$$K(u)\leq L(u),\quad \text{for each}\ \ u\in U(0,\overline{R}). \tag{16.29}$$

Then, by simply noticing that the iterates $\{x_n\}$ remain in $U(x_0,\overline{R}-\|x_1-x_0\|)$ which is more precise location than $U(x_0,\overline{R})$, function K can replace L in all the preceding results.

Furthermore, define sequence $\{\alpha_n\}$ by $\alpha_0=0$, $\alpha_1=\beta$,

$$\alpha_{n+1}=\alpha_n+\frac{\int_0^1\int_{\alpha_{n-1}}^{\alpha_{n-1}+\theta(\alpha_n-\alpha_{n-1})}K(u)du(\alpha_n-\alpha_{n-1})d\theta}{1-\int_0^{\alpha_n}L_0(u)du}. \tag{16.30}$$

Then, in view of (16.10), (16.29) and (16.30), we have that

$$\alpha_n\leq s_n, \tag{16.31}$$

$$0\leq\alpha_{n+1}-\alpha_n\leq s_{n+1}-s_n \tag{16.32}$$

and

$$\alpha^*=\lim_{n\to\infty}\alpha_n\leq s^* \tag{16.33}$$

hold under the hypotheses (16.12).

(c) *We have extended the applicability of Newton's method under hypotheses (16.12). At this point we are wondering, if sequences $\{\alpha_n\}$ and $\{s_n\}$ converge under a hypotheses weaker than (16.12). It turns out that this is indeed the case, when L_0,L,L_1,K are constant functions.*

It follows from (16.9), (16.10) and (16.30) that sequences $\{t_n\}$, $\{s_n\}$ and $\{\alpha_n\}$ reduce respectively to

$$t_0=0,\ \ t_1=\beta,\ \ t_{n+1}=t_n+\frac{L(t_n-t_{n-1})^2}{2(1-Lt_n)}, \tag{16.34}$$

$$s_0=0,\ \ s_1=\beta,\ \ s_{n+1}=s_n+\frac{L(s_n-s_{n-1})^2}{2(1-L_0s_n)}, \tag{16.35}$$

$$\alpha_0=0,\ \ \alpha_1=\beta,\ \ \alpha_{n+1}=\alpha_n+\frac{K(\alpha_n-\alpha_{n-1})^2}{2(1-L_0\alpha_n)}. \tag{16.36}$$

Sequence $\{t_n\}$, *converges provided that the Kantorovich condition [13, 14, 25]*

$$h_k = L\beta \leq \frac{1}{2} \tag{16.37}$$

is satisfied, whereas $\{s_n\}$ *and* $\{\alpha_n\}$ *converge [6], if*

$$h_1 = L_1\beta \leq \frac{1}{2} \tag{16.38}$$

and

$$h_2 = L_2\beta \leq \frac{1}{2} \tag{16.39}$$

are satisfied, respectively, where $L_1 = \frac{1}{8}(L+4L_0+\sqrt{L^2+8L_0L})$ *and* $L_2 = \frac{1}{8}(L+4L_0+\sqrt{L^2+8L_0K})$. *Notice that*

$$h_k \leq \frac{1}{2} \Rightarrow h_1 \leq \frac{1}{2} \Rightarrow h_2 \leq \frac{1}{2} \tag{16.40}$$

but not necessarily vice versa, unless, if $L_0 = L = K$.

Examples, where $L_0 < K < L < L_1$ *can be found in [6].*

Hence, we have also extended the convergence domain in this case. Similar advantages are obtained in the case of the Smale's $\alpha-$*theory [23, 24] of Wang's* $\gamma-$*condition [30], [33].*

References

[1] I. K. Argyros, Concerning the convergence of Newton's method and quadratic majorants, *J. Appl. Math. Computing*, 29, 391–400, 2009.

[2] I. K. Argyros, A Kantorovich-type convergence analysis of the Newton-Josephy method for solving variational inequalities, *Numer. Algorithms*, 55, 447–466, 2010.

[3] I. K. Argyros, Variational inequalities problems and fixed point problems, *Computers Math. Appl.*, 60, 2292–2301, 2010.

[4] I. K. Argyros, Improved local convergence of Newton's method under weak majorant condition, J. Comput. Appl. Math., 236, 1892–1902 (2012).

[5] I. K. Argyros and Á. A. Magreñán, Local convergence analysis of proximal Gauss-Newton method for penalized nonlinear least squares problems, *Appl. Math. Comput.*, 241, 401–408 , 2014.

[6] I. K. Argyros and S. Hilout, Weaker conditions for the convergence of Newton's method, *J. Complexity*, 28, 364–387, 2012.

[7] A. I. Dontchev and R. T. Rockafellar, Implicit functions and solution mappings, Springer Monographs in Mathematics, Springer, Dordrecht, 2009.

[8] O. Ferreira, A robust semi-local convergence analysis of Newtons method for cone inclusion problems in Banach spaces under affine invariant majorant condition, *J. Comput. Appl. Math.*, 279, 318–335, 2015.

[9] O. P. Ferreira, M. L. N. Goncalves and P. R. Oliveria, Convergence of the Gauss-Newton method for convex composite optimization under a majorant condition, *SIAM J. Optim.*, 23 (3), 1757–1783, 2013.

[10] O. P. Ferreira and G. N. Silva, Inexact Newton's method to nonlinear functions with values in a cone, arXiv: 1510.01947, 2015.

[11] O. P. Ferreira and B. F. Svaiter, Kantorovich's majorants principle for Newton's method, *Comput. Optim. Appl.*, 42 (2), 213–229, 2009.

[12] N. Josephy, Newton's method for generalized equations and the PIES energy model, University of Wisconsin-Madison, 1979.

[13] L. V. Kantorovič, On Newton's method for functional equations, *Doklady Akad Nauk SSSR (N.S.)*, 59, 1237–1240, 1948.

[14] L. V. Kantorovič and G. P. Akilov, Functional analysis, Oxford, Pergamon, 1982.

[15] A. Pietrus and C. Jean-Alexis, Newton-secant method for functions with values in a cone, *Serdica Math. J.*, 39 (3-4), 271–286, 2013.

[16] F. A. Potra, The Kantorovich theorem and interior point methods, *Mathem. Programming*, 102 (1), 47–70, 2005.

[17] S. M. Robinson, Strongly regular generalized equations, *Math. Oper. Res.*, 5 (1), 43–62, 1980.

[18] R. T. Rochafellar, Convex analysis, Princeton Mathematical Series, No. 28, Princeton University Press, Princeton, N. J., 1970.

[19] B. Royo, J. A. Sicilia, M. J. Oliveros and E. Larrodé, Solving a long-distance routing problem using ant colony optimization. *Appl. Math.*, 9 (2L), 415–421, 2015.

[20] J. A. Sicilia, C. Quemada, B. Royo and D. Escuín, An optimization algorithm for solving the rich vehicle routing problem based on variable neighborhood search and tabu search metaheuristics. *J. Comput. Appl. Math.*, 291, 468–477, 2016.

[21] J. A. Sicilia, D. Escuín, B. Royo, E. Larrodé and J. Medrano, A hybrid algorithm for solving the general vehicle routing problem in the case of the urban freight distribution. In Computer-based Modelling and Optimization in Transportation (pp. 463–475). Springer International Publishing, 2014.

[22] W. Rudin, Functional Analysis, McGraw-Hill, Inc., 1973.

[23] S. Smale, Newtons method estimates from data at one point. In R. Ewing, K. Gross and C. Martin, editors, *The Merging of Disciplines: New Directions in pure, Applied and Computational Mathematics*, 185–196, Springer New York, 1986.

[24] S. Smale, Complexity theory and numerical analysis, *Acta. Numer.*, 6, 523–551, 1997.

[25] J. F. Traub and H. Woźniakowski, Convergence and complexity of Newton iteration for operator equations, *J. Assoc. Comput. Mach.*, 26 (2), 250–258, 1979.

[26] L. U. Uko and I. K. Argyros, Generalized equation, variational inequalities and a weak Kantorivich theorem, *Numer. Algorithms*, 52 (3), 321–333, 2009.

[27] L. U. Uko, Generalized equations and the generalized Newton method, *Math. Program.*, 73, 251–268, 1996.

[28] J. Wang, Convergence ball of Newton's method for generalized equation and uniqueness of the solution, *J. Nonlinear Convex Anal.*, 16 (9), 1847–1859, 2015.

[29] J. H. Wang, Convergence ball of Newton's method for inclusion problems and uniqueness of the solution, *J. Nonlinear Convex Anal. (To Appear).*

[30] X. H. Wang, Convergence of Newton's method and uniqueness of the solution of equations in Banach space, *IMA J. Numer. Anal.*, 20, 123–134, 2000.

[31] P. P. Zabrejko and D. F. Nguen, The majorant method in the theory of Newton-Kantorivich approximations and the Pták error estimates, *Numer. Funct. Anal. Optim.*, 9 (5-6), 671–684, 1987.

[32] E. Zeidler, Non-linear Functional analysis and Applications IIB, Non-linear Monotone operators, Springer, Berlin, 1990.

[33] Y. Zhang, J. Wang and S. M. Gau, Convergence criteria of the generalized Newton method and uniqueness of solution for generalized equations, *J. Nonlinear Convex. Anal.*, 16 (7), 1485–1499, 2015.

Chapter 17

Newton-secant methods with values in a cone

Ioannis K. Argyros

Department of Mathematical Sciences, Cameron University, Lawton, OK 73505, USA, Email: iargyros@cameron.edu

Á. Alberto Magreñán

Universidad Internacional de La Rioja, Escuela Superior de Ingeniería y Tecnología, 26002 Logroño, La Rioja, Spain, Email: alberto.magrenan@unir.net

CONTENTS

17.1 Introduction

We study the variational inclusion

$$0 \in F(x) + G(x) + E(x), \tag{17.1}$$

where X, Y are Banach space $D \subset X$ is an open set $F : D \longrightarrow Y$ is a smooth operator, $G : D \longrightarrow Y$ is continuous operator, $[.,.;G]$ is a divided difference of order one for G [13] and $E : X \rightrightarrows Y$ is a set-valued operator.

Many problems from Applied Sciences can be solved finding the solutions of equations in a form like (17.1) [1]– [28].

In particular, the inclusion

$$0 \in F(x_m) + G(x_m) + (F'(x_m) + [x_{m-1}, x_m; G])(x_{m+1} - x_m) + E(x_{m+1}), \quad (17.2)$$

where $x_0 \in D$ is an initial point was studied in [19]. If $E = \{0\}$, the method was studied by Cătinas [5]. Then, assumimg $E(x) = -C$ for each $x \in X$ and $C \subseteq Y$ being a closed convex cone, Pietrus and Alexis [19] studied the algorithm

$$\text{mimimize}\{\|x - x_m\| / F(x_m) + G(x_m) + (F'(x_m) + [x_{m-1}, x_m; G])(x - x_m) \in C\} \quad (17.3)$$

Next, we shall define the Algorithm for solving (17.1). Let $u_1, u_2 \in D$. Define a set-valued operator $Q(u_1, u_2) : X \longrightarrow Y$ by

$$Q(u_1, u_2)x := (F'(u_2) + [u_1, u_2; G])x - C. \quad (17.4)$$

Then, $T(u_1, u_2)$ is a normed convex process. The inverse, defined for each $y \in Y$ by

$$Q^{-1}(u_1, u_2)y := \{z \in X : (F'(u_2) + [u_1, u_2; G])z \in y + C\}$$

is also a normed convex process. Then, we have the following [19].

Algorithm. Newton-secant-cone $(F, G, C, x_0, x_1, \varepsilon)$

1. If $Q^{-1}(x_0, x_1)[-F(x_1) - G(x_1)] = \emptyset$, stop.
2. Do while $e > \varepsilon$.
   (a) Choose x as a solution of the problem

   $$\text{mimimize}\{\|x - x_1\| / F(x_1) + G(x_1) + (F'(x_1) + [x_0, x_1; G])(x - x_1) \in C\}.$$

   (b) Compute $e : \|x - x_1\|; x_0 := x_1; x_1 := x$.
3. Return x.

Remark 17.1.1 *Notice that the continuity of the linear operator $F'(x_m)$ and G being closed and convex, imply that the feasible set of (17.3) is a closed convex set for all m. Hence, the existence of a feasible point $\bar{x}$ implies that each solution of (17.3) lies in the intersection of the feasible set of (17.3) with the closed ball of center x_m and radius $\|\bar{x} - x_m\|$. Then, by [6–10] a solution of (17.3) exists, since X is reflexive and the function $\|x - x_m\|$ is weakly lower semicontinuous.*

The convergence of Newton-secant method (17.3) was shown in [19] using Lipschitz continuity conditions on F' and divided differences of order one and two for G. However, there are problems where the Lipschitz continuity of F' does not hold or the divided differences of order two of G do not exist. Motivated by these constrains, we present a convergence analysis of the Newton-secant method (17.3) using generalized continuity on F' and hypotheses only on the divided difference of order one for G. Our results are weaker even, if we specialize the conditions on F' to the condition given in [19]. This way we expand the applicability of Newton-secant method (17.3).

The rest of the chapter is organized as follows: In Section 17.2 we present the convergence of Newton-secant method (17.3).

17.2 Convergence of the Newton-secant method

First of all, we present some needed auxiliary result on majorizing sequences.

Lemma 17.2.1 *Let $a \geq 0, b \geq 0$ and $c > 0$ be given parameters. Let also $w : \mathbb{R}_+ \longrightarrow \mathbb{R}_+$, $w_1 : \mathbb{R}_+^3 \longrightarrow \mathbb{R}_+$, $w_2 : \mathbb{R}_+^4 \longrightarrow \mathbb{R}_+$ be continuous, nondecreasing functions. Suppose that*

$$a \leq b \leq 2a \tag{17.5}$$

and equation

$$(t-a)(1-q(t))+a-b=0 \tag{17.6}$$

has zeros greater than a, where

$$q(t) = \frac{c}{1-cp(t)}\left(\int_0^1 w(a\theta)d\theta + w_1(a,a,a)\right) \tag{17.7}$$

and

$$p(t) = w(t-a) + w_2(t,t-a,t-a). \tag{17.8}$$

Denote by t^* the smallest such zero. Then, scalar sequence $\{t_n\}$ defined by

$$\begin{aligned} t_0 &= 0, t_1 = a, t_2 \geq b, \\ t_{n+1} &= t_n + \frac{c}{1-cp_n}\left[\int_0^1 w(\theta(t_n - t_{n-1}))d\theta \right. \\ &\quad \left. + w_1(t_n - t_{n-1}, t_n - t_{n-2}, t_{n-1} - t_{n-2})\right](t_n - t_{n-1}) \end{aligned} \tag{17.9}$$

is well defined, nondecreasing, bounded from above by t^{**} given by

$$t^{**} = \frac{t_2 - t_1}{1-q} + a \tag{17.10}$$

and converges to its unique least upper bound t^* which satisfies

$$b \leq t^* \leq t^{**}, \tag{17.11}$$

where

$$p_n = w(t_n - a) + w_2(t_{n-1}, t_{n-1} - a, t_n, t_{n-1} - t_1),$$

$$q = \frac{c}{1-cp}$$

and

$$p = p(t^*).$$

Even more, the following estimates hold

$$0 \leq t_2 - t_1 \leq t_1 - t_0 \tag{17.12}$$

$$0 \leq t_{n+1} - t_n \leq q(t_n - t_{n-1}) \tag{17.13}$$

and

$$0 \leq t^* - t_n \leq \frac{q^{n-1}}{1-q}(t_2 - t_1) \; n = 2,3,\ldots. \tag{17.14}$$

Proof: By (17.5), we have that $t_0 \leq t_1 \leq t_2$ and $t_2 - t_1 \leq t_1 - t_0$. It then follows by (17.9) and a simple inductive argument that

$$t_0 \leq t_1 \leq t_2 \leq \ldots \leq t_k \leq t_{k+1}$$

and

$$\begin{aligned} t_{k+1} - t_0 &= (t_{k+1} - t_k) + (t_k - t_{k-1}) + \ldots + (t_2 - t_1) + (t_1 - t_0) \\ &\leq (q^{k-1} + \ldots + 1)(t_2 - t_1) + t_1 - t_0 \\ &\leq \frac{t_2 - t_1}{1-q} + t_1 - t_0 = t^{**}. \end{aligned}$$

Therefore, sequence $\{t_k\}$ is nondecreasing, bounded from above by t^{**} and as such it converges to t^* which satisfies (17.11). Let $m \geq 1$. We can write

$$\begin{aligned} t_{n+m} - t_n &= (t_{n+m} - t_{n+m-1}) + \ldots + (t_{n+1} - t_n) \\ &\leq (q^{n+m-2} + \ldots q^{n-1})(t_2 - t_1) \\ &= \frac{1-q^m}{1-q} q^{n-1}(t_2 - t_1). \end{aligned}$$

By letting $m \longrightarrow \infty$ in the preceding estimate we show (17.14).

□

Now, we present the main convergence result of the chapter.

Theorem 17.2.2 *Let D, X, Y, F, G, Q be as defined above. Suppose:*

(i) There exist points $x_0, x_1 \in D$ such that $Q(x_0, x_1)$ carries X onto Y.

(ii) There exist $c > 0$ and $a \geq 0$ such that

$$\|Q^{-1}(x_0, x_1)\| \leq c$$

and

$$\|x_1 - x_0\| \leq a.$$

(iii) *There exist* $w:\mathbb{R}_+\longrightarrow\mathbb{R}_+, w_1:\mathbb{R}_+^3\longrightarrow\mathbb{R}_+, w_2:\mathbb{R}_+^4\longrightarrow\mathbb{R}_+$ *continuous, nondecreasing functions such that for each* $x,y,z\in D$

$$\|F'(x)-F'(y)\|\leq w(\|x-y\|),$$

$$\|[x,y;G]-[z,y;G]\|\leq w_1(\|x-z\|,\|x-y\|,\|z-y\|)$$

and

$$\|[x,y;G]-[x_0,x_1;G]\|\leq w_2(\|x-x_0\|,\|y-x_1\|,\|x-x_1\|,\|y-x_0\|).$$

(iv) $\|x_2-x_1\|\leq b-a$, *where* x_2 *defined by the algorithm and Remark 17.1.1.*

(v) *Hypotheses of Lemma 17.2.1 hold.*

(vi) $\bar{U}(x_0,t^*)\subset D$.

(vii) $w(0)=w_1(0,0,0)=w_2(0,0,0,0)=0$.

Then, there exists at least a sequence $\{x_n\}$ *generated by method (17.3) which is well defined in* $U(x_0,t^*)$, *remains in* $U(x_0,t^*)$ *for each* $n=0,1,2,\ldots$ *and converges to some* $x^*\in\bar{U}(x_0,t^*)$ *such that* $F(x^*)+G(x^*)\in C$. *Moreover, the following estimates hold*

$$\|x_n-x^*\|\leq t^*-t_n, \tag{17.15}$$

where sequence $\{t_n\}$ *is defined by (17.9) and* $t^*=\lim_{n\longrightarrow\infty}t_n$.

Proof: We begin by defining operator A_m for each $m=1,2,\ldots$ as

$$\begin{aligned} A_m \;=\;& F(x_{m+1})+G(x_{m+1})-F(x^*)-G(x^*)\\ &-[F(x_{m+1})-F(x_m)-(F'(x_m)+[x_{m-1},x_m;G])(x_{m+1}-x_m).\\ &+G(x_{m+1})-G(x_m)]. \end{aligned} \tag{17.16}$$

Then, we obtain $A_m\in C-F(x^*)-G(x^*)$ and by continuity $\lim_{n\longrightarrow\infty}A_m=0$. It follows that $F(x^*)+G(x^*)\in C$, since $C-F(x^*)-G(x^*)$ is closed. Therefore, the point x^* solves (17.1). Using induction, we shall show that sequence $\{x_m\}$ is well defined and satisfies

$$\|x_{m+1}-x_m\|\leq t_{m+1}-t_m \text{ for each } m=0,1,2,\ldots, \tag{17.17}$$

where sequence $\{t_m\}$ is defined by (17.9). By the second hypothesis in (ii), (iv) and (17.9), we obtain the existence of x_2 which solves (17.1) for $m=1$, $\|x_1-x_0\|\leq a=t_1-t_0$, $\|x_2-x_1\|\leq b-a=t_2-t_1$, which shows (17.17) for $m=0,1$. Suppose that (17.17) holds for $i=3,\ldots m$, where $x_1,x_2,\ldots x_m$ are defined by (17.3). Then, we get that $\|x_i-x_0\|\leq t_i-t_0<t^*$. That is $x_i\in U(x_0,t^*)$.

Now, it is easy that

$$B_i(x)=-(F'(x_i)-F'(x_1)+[x_{i-1},x_i;G]-[x_0,x_1;G])x \tag{17.18}$$

that

$$Q(x_{i-1},x_i)x=(Q(x_0,x_1)-B_i)x. \tag{17.19}$$

By (ii), (iii), (17.19), Lemma 17.2.1 and the induction hypotheses we get

$$\begin{aligned}\|Q^{-1}(x_0,x_1)\|\|B_i\| &\leq c[w(\|x_i-x_1\|)+w_2(\|x_{i-1}-x_0\|,\|x_{i-1}-x_1\|,\\ &\quad \|x_i-x_0\|,\|x_{i-1}-x_1\|)] \qquad (17.20)\\ &\leq c[w(t_i-t_1)+w_2(t_{i-1}-t_0,t_{i-1}-t_1,t_i-t_0,t_{i-1}-t_1)]\\ &\leq c[w(t^*-a)+w_2(t^*,t^*-a,t^*,t^*-a)]<1.\end{aligned}$$

It follows from (17.21) and the Banach perturbation Lemma [13] that

$$Q^{-1}(x_{i-1},x_i)\in L(Y,X) \tag{17.21}$$

and

$$\begin{aligned}\|Q^{-1}(x_{i-1},x_i)\| &\leq \frac{\|Q^{-1}(x_0,x_1)\|}{1-\|A^{-1}(x_0,x_1)\|\|B_i\|}\\ &\leq \frac{c}{1-cp_i}.\end{aligned} \tag{17.22}$$

The existence of x_i solving (17.3) with $i=k$ follows from the fact that $Q(x_{i-1},x_i):X\longrightarrow Y$. Next, we must find a solution of

$$\begin{aligned}F(x_m)+G(x_m)+[x_{m-1},x_m;G](x-x_m) &\in\\ F(x_{m-1})+(F'(x_{m-1})+[x_{m-2},x_{m-1};G](x_m-x_{m-1})+G(x_{m-1}))+C.\end{aligned} \tag{17.23}$$

The right hand side of (17.23) is contained in the cone C, since x_m solves (ii). That is any x satisfying (17.23) is feasible for (17.3). Using (17.23), we can get x as the solution of

$$\begin{aligned}x-x_m &\in Q^{-1}(x_{m-1},x_m)(-F(x_m)-G(x_m)+F(x_{m-1}))\\ &\quad +G(x_{m-1})+(F'(x_{m-1})+[x_{m-2},x_{m-1};G](x_m-x_{m-1})).\end{aligned} \tag{17.24}$$

The right hand side of (17.24) contains an element of least norm, so there exists $\bar{x}$ satisfying (17.22) and (17.23) such that

$$\begin{aligned}\|\bar{x}-x_m\| &\leq \|Q^{-1}(x_{m-1},x_m)\|(\|-F(x_m)+F(x_{m-1})+F'(x_{m-1})(x_m-x_{m-1})\|\\ &\quad +\|F(x_m)-G(x_{m-1}-[x_{m-2},x_{m-1};G](x_m-x_{m-1})\|).\end{aligned} \tag{17.25}$$

In view of (17.9), (iii), (17.22), (17.25) and the induction hypotheses we obtain

$$\begin{aligned}
\|\bar{x}-x_m\| &\leq \frac{c}{1-cp_m}\Big[\int_0^1 w(\theta\|x_m-x_{m-1}\|)d\theta\|x_m-x_{m-1}\| \\
&\quad +\|[x_m,x_{m-1};G]-[x_{m-2},x_{m-1};G]\|\|x_m-x_{m-1}\|\Big] \\
&\leq \frac{c}{1-cp_m}\Big[\int_0^1 w(\theta\|x_m-x_{m-1}\|)d\theta \\
&\quad +w_1(\|x_m-x_{m-1}\|,\|x_m-x_{m-2}\|,\|x_{m-1}-x_{m-2}\|)\Big]\|x_m-x_{m-1}\| \\
&\leq \frac{c}{1-cp_m}\Big[\int_0^1 w(\theta(t_m-t_{m-1}))d\theta \\
&\quad +w_1(t_m-t_{m-1},t_m-t_{m-2},t_{m-1}-t_{m-2})\Big]|t_m-t_{m-1}| \\
&= t_{m+1}-t_m.
\end{aligned} \tag{17.26}$$

By Lemma 17.2.1, sequence $\{t_m\}$ is complete. In view of (17.26), sequence $\{x_m\}$ is complete in a Banach space X and

$$\|x_{m+1}-x_m\| \leq \|\bar{x}-x_m\| \leq t_{m+1}-t_m.$$

□

Remark 17.2.3 *(a) Our results can specialize to the corresponding ones in [19]. Choose $w(t)=Lt$, $w_1(t)=L_1$ and $w_2(t,t,t,t)=L_2$. But even in this case, our results are weaker, since we do not use the hypothesis on the divided difference of order two*

$$\|[x,y,z;G]\| \leq K$$

but use instead the second hypothesis in (iii) which involves only the divided difference of order one. Moreover, the results in [19] cannot be used to solve the numerical examples at the end of the chapter, since F' is not Lipschitz continuous. However, our results can apply to solve inclusion problems. Hence, we have expanded the applicability of Newton-secant method (17.3).

(b) Our results can be improved even further as follows: Let r_0 be defined as the smallest positive zero of equation

$$w_2(t,t,t,t)=1.$$

Define $D_0=D\cap U(x_0,r_0)$. Suppose that the first and second hypotheses in (iii) are replaced by

$$\|F'(x)-F'(y)\| \leq \bar{w}(\|x-y\|)$$

and

$$\|[x,y;G]-[z,y;G]\| \leq \bar{w}_1(\|x-z\|,\|x-y\|,\|z-y\|)$$

for each $x,y,z \in D_0$. *Denote the new conditions by (iii)'. Then, clearly condition (iii)' can replace (iii) in Theorem 17.2.2, since the iterates* $\{x_m\}$ *lie in* D_0 *which is a more accurate location than* D. *Moreover, since* $D_0 \subseteq D$, *we have that*

$$\bar{w}(t) \leq w(t)$$

and

$$\bar{w}_1(t,t,t) \leq w_1(t,t,t)$$

for each $t \in [0, r_0)$ *hold. Notice that the definition of functions* $\bar{w}$ *and* $\bar{w}_1$ *depends on* w *and* r_0. *Another way of extending our results is to consider instead of* D_0 *the set* $D_1 = D \cap U(x_1, r_0 - \|x_1 - x_0\|)$. *Then, corresponding functions* $\bar{\bar{w}}$ *and* $\bar{\bar{w}}_1$ *will be at least as small as* $\bar{w}, \bar{w}_1$, *respectively, since* $D_1 \subseteq D_0$. *Notice that the construction of the* $\|w\|$ *functions is based on the initial data* F, G, C, x_0, x_1.

References

[1] I. K. Argyros, S. George and Á. A. Magreñán, Local convergence for multi-point parametric Chebyshev-Halley-type methods of high convergence order, *J. Comput. Appl. Math.*, 282, 215–224, 2015.

[2] I. K. Argyros and S. George, Ball comparison between two optimal eight-order methods under weak conditions, *SeMA*, 73 (1), 1–1, 2015.

[3] J. P. Aubin, Lipschitz behaviour of solutions to convex minimization problems, *Math. Oper. Res.*, 9, 87–111, 1984.

[4] J. P. Aubin and H. Frankowska, Set-Valued Analysis, Systems Control Found. Appl. Boston, MA, Birkhauser Boston, Inc.,1990.

[5] E. Catinas, On some iterative methods for solving nonlinear equations, *Rev. Anal. Nwner. Thdor. Approx.*, 23, 1, 47–53, 1994.

[6] A. L. Dontchev and W. W. Hager, An inverse mapping theorem for set valued maps, *Proc. Amer. Math. Soc.*, 121, 2, 481–489, 1994.

[7] A. L. Dontchev, Local convergence of the Newton method for generalized equation, *G. R. Acad. Sci., Paris, Ser. I*, 322, 4, 327–331, 1996.

[8] A. L. Dontchev, Uniform convergence of the Newton method for Aubin continuous maps, *Serdica Math. J.*, 22, 3, 385–398, 1996.

[9] A. L. Dontchev, M. Quincampoix and N. Zlateva, Aubin Criterion for Metric Regularity, *J. Convex Anal.*, 13, 2, 281–297 (2006).

[10] A. L. Dontchev and R. T. Rockafellar, Implicit functions and solution mappings. Springer Monographs in Mathematics. New York, Springer (2009).

[11] M. H. Geoffroy and A. Pietrus, Local convergence of some iterative methods for generalized equations, *J. Math. Anal Appl.*, 290, 2, 497–505, 2004.

[12] M. Grau-Sanchez, M. Noguera and S. Amat, On the approximation of derivatives using divided difference operators preserving the local convergence order of iterative methods, *J. Comput. Appl. Math.,* 237, 1, 363–372, 2013.

[13] L. V. Kantorovich, On Newton's method, *Tr. Mat. Inst Steklova*, 28, 104–144, 1949.

[14] E. S. Levitin and B. T. Polyak, Constrained minimization methods. *Zh. Vychisl. Mat. Mat. Fiz.*, 6, 5, 787–823, 1966 (in Russian).

[15] A. S. Lewis, Ill-conditioned convex processes and Conic linear systems, *Math. Oper. Res.*, 24, 4, 829–834, 1999.

[16] A. S. Lewis, Ill-Conditioned Inclusions, *Set-Valued Anal.*, 9, 4, 375–381, 2001.

[17] B. S. Mordukhovich, Complete characterization of openess metric regularity and lipschitzian properties of multifunctions,*Trans. Amer. Math. Soc.*, 340, 1, 1–35, 1993.

[18] B. S. Mordukhovich, Variational analysis and generalized differentiation I: Basic theory. Grundlehren der Mathematischen Wissenschaften [Fundamental Principles of Mathematical Sciences] Vol. 330, Berlin, Springer, 2006.

[19] A. Pietrus and C. J. Alexis, Newton-secant method for functions with values in a cone, *Serdica Math.*, 39, 271–296, 2013.

[20] W. C. Rheinboldt, A unified convergence theory for a class of iterative processes, *SIAM J. Numer. Anal.*, 5, 1, 42–63, 1968.

[21] S. M. Robinson, Normed convex processes, *Trans. Amer. Math. Soc.*, 174, 127–140, 1972.

[22] S. M. Robinson, Extension of Newton's method to non linear functions with values in a cone, *Numer. Math.*, 19, 341–347, 1972.

[23] S. M. Robinson, Generalized equations and their solutions, part I: basic theory, *Math. Program. Study*, 10, 128–141, 1979.

[24] S. M. Robinson. Generalized equations and their solutions, part II: application to nonlinear programming, *Math. Program. Study*, 19, 200–221, 1982.

[25] R. T. Rockafellar, Lipsehitzian properties of multifunctions. *Nonlinear Anal., Theory Methods Appl.*, 9, 867–885, 1985.

[26] R. T. Rockafellar and R. J. B. Wets, Variational analysis. A Series of Comprehensives Studies in Mathematics Vol. 317, Berlin, Springer, 1998.

[27] R. T. Rockafellar, Monotone processes of convex and concave type, *Mem. Am. Math. Soc.*, 77, 1967.

[28] R. T. Rockafellar, Convex Analysis. Princeton, Princeton University Press, XVIII, 1970.

Chapter 18

Gauss-Newton method with applications to convex optimization

Ioannis K. Argyros
Department of Mathematical Sciences, Cameron University, Lawton, OK 73505, USA, Email: iargyros@cameron.edu

Á. Alberto Magreñán
Universidad Internacional de La Rioja, Escuela Superior de Ingeniería y Tecnología, 26002 Logroño, La Rioja, Spain, Email: alberto.magrenan@unir.net

CONTENTS

18.1 Introduction

In this chapter we will study the convex composite optimizations problem.

We present a convergence analysis based on the majorant function presented in [17] of Gauss-Newton method (defined by Algorithm (GNA) in Sec. 18.2). We will use the same formulation using the majorant function provided in [23]. In [3,5,8], a convergence study in a Banach space setting was shown for (GNM) defined by

$$x_{k+1} = x_k - \left[F'(x_k)^+ F'(x_k)\right]^{-1} F'(x_k)^+ F(x_k) \text{ for each } k = 0,1,2,\ldots,$$

where x_0 is an initial point and $F'(x)^+$ in the Moore-Penrose inverse [11–13, 19, 26] of operator $F'(x)$ with $F : \mathbb{R}^n \to \mathbb{R}^m$ being continuously differentiable. In [23], a semilocal convergence analysis using a combination of a majorant and a center majorant function was given with the advantages $(\mathcal{A})$:

- tighter error estimates on the distances involved and
- the information on the location of the solution is at least as precise.

These advantages were obtained, under the same computational cost, using same or weaker sufficient convergence hypotheses. In this chapter, we even extend the same advantages $(\mathcal{A})$ for GNA method.

The chapter is organized as follows: In Section 18.2 the definition of GNA method is presented. In Section 18.3, we provide the semilocal convergence analysis of GNA. Finally, some Numerical examples and applications are provided in the concluding Section 18.4.

18.2 Gauss-Newton Algorithm and Quasi-Regularity condition

18.2.1 Gauss-Newton Algorithm GNA

From the idea of restricted convergence domains, we analyze the convex composite optimization problem defined as follows

$$\min_{x \in \mathbb{R}^n} p(x) := h(F(x)), \tag{18.1}$$

where $h : \mathbb{R}^m \longrightarrow \mathbb{R}$ is convex, $F : \mathbb{R}^n \longrightarrow \mathbb{R}^m$ is Fréchet-differentiable operator and $m, l \in \mathbb{N}^\star$. The importance of (18.1) has been show in [2, 10, 12, 19, 21–23, 25, 27]. We assume that the minimum h_{min} of the function h is attained. Problem (18.1) is related to

$$F(x) \in \mathcal{C}, \tag{18.2}$$

where

$$\mathcal{C} = \operatorname{argmin} h \tag{18.3}$$

is the set of all minimum points of h.

Now, let $\rho \in [1,\infty[$, $\Delta \in]0,\infty]$ and for each $x \in I\!R^n$, define $\mathcal{D}_\Delta(x)$ by

$$\begin{aligned}\mathcal{D}_\Delta(x) &= \{d \in I\!R^n : \| d \| \leq \Delta,\ h(F(x)+F'(x)d) \leq h(F(x)+F'(x)d') \\ &\quad \text{for all } d' \in I\!R^n \text{ with } \| d' \| \leq \Delta\}.\end{aligned} \tag{18.4}$$

Let $x_0 \in I\!R^n$ be an initial point. The GNA associated with (ρ,Δ,x_0) as defined in [12, 17] is :

Algorithm GNA : (ρ,Δ,x_0)

INICIALIZATION. Take $\rho \in [1,\infty)$, $\Delta \in (0,\infty]$ and $x_0 \in \mathbb{R}^n$, set $k=0$.
STOP CRITERION. Compute $D_\Delta(x_k)$. If $0 \in \mathcal{D}_\Delta(x_k)$, STOP. Otherwise.
ITERATIVE STEP. Compute d_k satisfying $d_k \in D_\Delta(x_k)$, $\|d_k\| \leq \rho d(0, D\Delta(x_k))$,

Then, set $x_{k+1} = x_k + d_k$, $k = k+1$ and GO TO STOP CRITERION.

Where $d(x,W)$ denotes the distance from x to W in the finite dimensional Banach space containing W. It is worth noticing that the set $\mathcal{D}_\Delta(x)$ ($x \in I\!R^n$) is nonempty and is the solution of the following convex optimization problem

$$\min_{d \in I\!R^n,\, \|d\| \leq \Delta} h(F(x)+F'(x)d), \tag{18.5}$$

which can be solved by well known methods such as the subgradient or cutting plane or bundle methods (see, e.g., [12, 19, 25–27]). Throughout the chapter we will denote $U(x,r)$ as the open ball in $\mathbb{R}^n$ (or $\mathbb{R}^m$) centered at x and of radius $r > 0$ and by $\overline{U}(x,r)$ its closure. Let W be a closed convex subset of $\mathbb{R}^n$ (or $\mathbb{R}^m$). The negative polar of W denoted by $W^\ominus$ is defined as

$$W^\ominus = \{z : < z, w > \leq 0 \quad \text{for each} \quad w \in W\}. \tag{18.6}$$

18.2.2 Quasi Regularity

In this section, we mention some concepts and results on regularities which can be found in [12] (see also, e.g., [17, 21–23, 25]).

For a set-valued mapping $T : \mathbb{R}^n \rightrightarrows \mathbb{R}^m$ and for a set A in $\mathbb{R}^n$ or $\mathbb{R}^m$, we denote by

$$D(T) = \{x \in \mathbb{R}^n : Tx \neq \emptyset\}, \quad R(T) = \bigcup_{x \in D(T)} Tx, \tag{18.7}$$

$$T^{-1}y = \{x \in \mathbb{R}^n : y \in Tx\} \quad \text{and} \quad \| A \| = \inf_{a \in A} \| a \| .$$

Consider the inclusion

$$F(x) \in C, \tag{18.8}$$

where C is a closed convex set in $\mathbb{R}^m$. Let $x \in \mathbb{R}^n$ and

$$\mathcal{D}(x) = \{d \in \mathbb{R}^n : F(x) + F'(x)d \in C\}. \tag{18.9}$$

Definition 18.2.1 Let $x_0 \in \mathbb{R}^n$.

(a) x_0 is quasi-regular point of (18.8) if there exist $R_0 \in]0, +\infty[$ and an increasing positive function β on $[0, R_0[$ such that

$$\mathcal{D}(x) \neq \emptyset \text{ and } d(0, \mathcal{D}(x)) \leq \beta(\| x - x_0 \|)\, d(F(x), C) \text{ for all } x \in U(x_0, R_0). \tag{18.10}$$

$\beta(\| x - x_0 \|)$ is an "error bound" in determining how for the origin is away from the solution set of (18.8).

(b) x_0 is a regular point of (18.8) if

$$ker(F'(x_0)^T) \cap (C - F(x_0))^{\ominus} = \{0\}. \tag{18.11}$$

In order for us to make the chapter as self contained as possible, the notion of quasi-regularity is also re-introduced (see, e.g., [12, 17, 21]).

Proposition 18.2.2 (see, e.g., [12,17,21,25]) *Let x_0 be a regular point of (18.8). Then, there are constants $R_0 > 0$ and $\beta > 0$ such that (18.10) holds for R_0 and $\beta(\cdot) = \beta$. Therefore, x_0 is a quasi-regular point with the quasi-regular radius $R_{x_0} \geq R_0$ and the quasi-regular bound function $\beta_{x_0} \leq \beta$ on $[0, R_0]$.*

Remark 18.2.3 (a) $\mathcal{D}(x)$ can be considered as the solution set of the linearized problem associated to (18.8)

$$F(x) + F'(x)d \in C. \tag{18.12}$$

(b) If C defined in (18.8) is the set of all minimum points of h and if there exists $d_0 \in \mathcal{D}(x)$ with $\| d_0 \| \leq \Delta$, then $d_0 \in \mathcal{D}_\Delta(x)$ and for each $d \in \mathbb{R}^n$, we have the following equivalence

$$d \in \mathcal{D}_\Delta(x) \Longleftrightarrow d \in \mathcal{D}(x) \Longleftrightarrow d \in \mathcal{D}_\infty(x). \tag{18.13}$$

(c) Let R_{x_0} denote the supremum of R_0 such that (18.10) holds for some function β defined in Definition 18.2.1. Let $R_0 \in [0, R_{x_0}]$ and $\mathcal{B}_R(x_0)$ denotes the set of function β defined on $[0, R_0)$ such that (18.10) holds. Define

$$\beta_{x_0}(t) = \inf\{\beta(t) : \beta \in \mathcal{B}_{R_{x_0}}(x_0)\} \quad \text{for each} \quad t \in [0, R_{x_0}). \tag{18.14}$$

All function $\beta \in \mathcal{B}_R(x_0)$ with $\lim\limits_{t \to R^-} \beta(t) < +\infty$ can be extended to an element of $\mathcal{B}_{R_{x_0}}(x_0)$ and we have that

$$\beta_{x_0}(t) = \inf\{\beta(t) : \beta \in \mathcal{B}_R(x_0)\} \quad \text{for each} \quad t \in [0, R_0). \tag{18.15}$$

R_{x_0} and β_{x_0} are called the quasi-regular radius and the quasi-regular function of the quasi-regular point x_0, respectively.

Definition 18.2.4 (a) *A set-valued mapping $T : \mathbb{R}^n \rightrightarrows \mathbb{R}^m$ is convex if the following items hold*

(i) $Tx + Ty \subseteq T(x+y)$ *for all* $x, y \in \mathbb{R}^n$.

(ii) $T\lambda x = \lambda Tx$ *for all* $\lambda > 0$ *and* $x \in \mathbb{R}^n$.

(iii) $0 \in T0$.

18.3 Semilocal convergence for GNA

In this section we provide the main semilocal convergence result for GNA. First, we need some auxiliary definitions.

Definition 18.3.1 *Let $R > 0$, $x_0 \in \mathbb{R}^n$ and $F : \mathbb{R}^n \to \mathbb{R}^m$ be continuously Fréchet-differentiable. A twice-differentiable function $f_0 : [0, R) \to \mathbb{R}$ is called a center-majorant function for F on $U(x_0, R)$, if for each $x \in U(x_0, R)$,*

(h_0^0) $\|F'(x) - F'(x_0)\| \leq f_0'(\|x - x_0\|) - f_0'(0)$;

(h_1^0) $f_0(0) = 0$, $f_0'(0) = -1$;

and

(h_2^0) f_0' *is convex and strictly increasing.*

Definition 18.3.2 *[5,8,17] Let $x_0 \in \mathbb{R}^n$ and $F : \mathbb{R}^n \to \mathbb{R}^m$ be continuously differentiable. Define $R_0 = \sup\{t \in [0, R) : f_0'(t) < 0\}$. A twice-differentiable function $f : [0, R_0) \to \mathbb{R}$ is called a majorant function for F on $U(x_0, R_0)$, if for each $x, y \in U(x_0, R_0)$, $\|x - x_0\| + \|y - x\| < R_0$,*

(h_0) $\|F'(y) - F'(x)\| \leq f'(\|y - x\| + \|x - x_0\|) - f'(\|x - x_0\|)$;

(h_1) $f(0) = 0$, $f'(0) = -1$;

and

(h_2) f' *is convex and strictly increasing.*

On the other hand, assume that

(h_3) $f_0(t) \leq f(t)$ and $f_0'(t) \leq f'(t)$ for each $t \in [0, R_0)$.

Remark 18.3.3 *Suppose that $R_0 < R$. If $R_0 \geq R$, then we do not need to introduce Definition 18.3.2.*

Let $\rho > 0$ and $\alpha > 0$ be fixed and define auxiliary function $\varphi : [0, R_0) \to \mathbb{R}$ by

$$\varphi(t) = \rho + (\alpha - 1)t + \alpha f(t). \tag{18.16}$$

We shall use the following hypotheses

(h_4) there exists $s^* \in (0,R)$ such that for each $t \in (0,s^*)$, $\varphi(t) > 0$ and $\varphi(s^*) = 0$;

(h_5) $\varphi(s^*) < 0$.

From now on we assume the hypotheses $(h_0) - (h_4)$ and $(h_0^0) - (h_2^0)$ which will be called the hypotheses (H).

Now, we provide the main convergence result of the Gauss-Newton method generated by the Algorithm GNA for solving (18.1).

Theorem 18.3.4 *Suppose that the (H) conditions are satisfied. Then,*

(i) *sequence $\{s_k\}$ generated by the Gauss-Newton method for $s_0 = 0, s_{k+1} = s_k - \frac{\varphi(s_k)}{\varphi'(s_k)}$ for solving equation $\psi(t) = 0$ is: well defined; strictly increasing; remains in $[0,s^*)$ and converges Q-linearly to s^*.*

Let $\eta \in [1,\infty]$, $\Delta \in (0,\infty]$ and $h : \mathbb{R}^m \to \mathbb{R}$ be real-valued convex with minimizer set C such that $C \neq \emptyset$.

(ii) *Suppose that $x_0 \in \mathbb{R}^n$ is a quasi-regular point of the inclusion*

$$F(x) \in C,$$

with the quasi-regular radius r_{x_0} and the quasi-regular bound function β_{x_0} defined by (18.2.14) and (18.2.15), respectively. If $d(F(x_0),C) > 0$, $s^* \leq r_{x_0}$, $\Delta \geq \rho \geq \eta\beta_{x_0}(0)d(F(x_0),C)$,

$$\alpha \geq \sup\left\{\frac{\eta\beta_{x_0}(t)}{\eta\beta_{x_0}(t)(1+f'(t))+1} : \rho \leq t < s^*\right\}$$

then, sequence $\{x_k\}$ generated by GNA is well defined, remains in $U(x_0,s^*)$ for each $k = 0,1,2,\ldots$, such that

$$F(x_k) + F'(x_k)(x_{k+1} - x_k) \in C \text{ for each } k = 0,1,2\ldots. \tag{18.17}$$

Moreover, the following estimates hold

$$\|x_{k+1} - x_k\| \leq s_{k+1} - s_k, \tag{18.18}$$

$$\|x_{k+1} - x_k\| \leq \frac{s_{k+1} - s_k}{(s_k - s_{k-1})^2}\|x_k - x_{k-1}\|^2, \tag{18.19}$$

for each $k = 0,1,2\ldots$, and $k = 1,2,\ldots$, respectively and converges to a point $x^* \in U(x_0,s^*)$ satisfying $F(x^*) \in C$ and

$$\|x^* - x_k\| \leq t^* - s_k \text{ for each } k = 0,1,2,\ldots. \tag{18.20}$$

The convergence is R-linear. If hypothesis (h_5) holds, then the sequences

$\{s_k\}$ and $\{x_k\}$ converge Q-quadratically and R-quadratically to s^* and x^*, respectively. Furthermore, if

$$\alpha > \overline{\alpha} := \sup\left\{ \frac{\eta_{\beta_{x_0}}(t)}{\eta_{\beta_{x_0}}(t)(1+f'(t))+1} : \rho \leq t < s^* \right\},$$

then, the sequence $\{x_k\}$ converges R-quadratically to x^*.

Proof: Simply replace function g in [23] (see also [17]) by function f in the proof, where g is a majorant function for F on $U(x_0,R)$. That is we have instead of (h_0):
(h_0')

$$\|F'(y)-F'(x)\| \leq g'(\|y-x\|+\|x-x_0\|)-g'(\|x-x_0\|) \qquad (18.21)$$

for each $x,y \in U(x_0,R)$ with $\|x-x_0\|+\|y-x\| < R$. The iterates x_n lie in $U(x_0,R_0)$ which is a more precise location than $U(x_0,R)$.

Remark 18.3.5 *(a) If $f(t) = g(t) = f_0(t)$ for each $t \in [0,R_0)$ and $R_0 = R$, then Theorem 18.2.1 reduces to the corresponding Theorem in [17]. Furthermore, if $f_0(t) \leq f(t) = g(t)$ we obtain the results in [23]. It is worth noticing that, we have that*

$$f_0'(t) \leq g'(t) \text{ for each } t \in [0,R) \qquad (18.22)$$

and

$$f'(t) \leq g'(t) \text{ for each } t \in [0,R_0). \qquad (18.23)$$

Hence, if

$$f_0'(t) \leq f'(t) < g'(t) \text{ for each } t \in [0,R_0), \qquad (18.24)$$

then the following advantages denoted by ($\mathcal{A}$) are obtained: weaker sufficient convergence criteria, tighter error bounds on the distances $\|x_n - x^*\|$, $\|x_{n+1}-x_n\|$ and an at least as precise information on the location of the solution x^*. These advantages are obtained using less computational cost, since in practice the computation of function g requires the computation of functions f_0 and f as special cases. Realize that under (h_0^0) function f_0' is defined and therefore R_0 which is at least as small as R. Therefore the majorant function to satisfy (i.e., f') is at least as small as the majorant function satisfying (h_0') (i.e., g') leading to the advantages of the new approach over the approach in [17] or [23]. Indeed, we have that if function ψ has a solution t^*, then, since $\varphi(t^*) \leq \psi(t^*) = 0$ and $\varphi(0) = \psi(0) = \rho > 0$, we get that function φ has a solution r^* such that

$$r^* \leq t^*. \qquad (18.25)$$

but not necessarily vice versa. If also follows from (18.25) that the new information about the location of the solution x^* is at least as precise as the one given in [18].

(b) *Let us specialize conditions (18.8)–(18.10) even further in the case when L_0, K and L are constant functions and $\alpha = 1$. Then, function corresponding to (18.16) reduce to*

$$\psi(t) = \frac{L}{2}t^2 - t + \rho \tag{18.26}$$

[17, 23] and

$$\varphi(t) = \frac{K}{2}t^2 - t + \rho, \tag{18.27}$$

respectively. In this case the convergence criteria become, respectively

$$h = L\rho \leq \frac{1}{2} \tag{18.28}$$

and

$$h_1 = K\rho \leq \frac{1}{2}. \tag{18.29}$$

Notice that

$$h \leq \frac{1}{2} \Longrightarrow h_1 \leq \frac{1}{2} \tag{18.30}$$

but not necessarily vice versa. Unless, if $K = L$. Criterion (18.28) is the celebrated Kantorovich hypotheses for the semilocal convergence of Newton's method [20]. In the case of Wang's condition [29] we have

$$\varphi(t) = \frac{\gamma t^2}{1 - \gamma t} - t + \rho,$$

$$\psi(t) = \frac{\beta t^2}{1 - \beta t} - t + \rho,$$

$$L(u) = \frac{2\gamma}{(1 - \gamma u)^3}, \ \gamma > 0, \ 0 \leq t \leq \frac{1}{\gamma}$$

and

$$K(u) = \frac{2\beta}{(1 - \beta u)^3}, \ \beta > 0, \ 0 \leq t \leq \frac{1}{\beta}$$

with convergence criteria, given respectively by

$$H = \gamma\rho \leq 3 - 2\sqrt{2} \tag{18.31}$$

$$H_1 = \beta\rho \leq 3 - 2\sqrt{2}. \tag{18.32}$$

Then, we also have

$$H \leq 3 - 2\sqrt{2} \Longrightarrow H_1 \leq 3 - 2\sqrt{2}$$

but not necessarily vice versa, unless, if $\beta = \gamma$.

(c) *Concerning the error bounds and the limit of majorizing sequence, let us define majorizing sequence* $\{r_{\alpha,k}\}$ *by*

$$r_{\alpha,0} = 0; r_{\alpha,k+1} = r_{\alpha,k} - \frac{\varphi(r_{\alpha,k})}{\varphi'_{\alpha,0}(r_{\alpha,k})}$$

for each $k = 0, 1, 2, \ldots$, *where*

$$\varphi_{\alpha,0}(t) = \rho - t + \alpha \int_0^t L_0(t)(t-u)du.$$

Suppose that

$$-\frac{\varphi(r)}{\varphi'_{\alpha,0}(r)} \leq -\frac{\varphi(s)}{\varphi'(s)}$$

for each $r, s \in [0, R_0]$ *with* $r \leq s$. *According to the proof of Theorem 18.3.1 sequence* $\{r_{\alpha,k}\}$ *is also a majorizing sequence for GNA. On the other hand, a simple inductive argument shows that*

$$r_k \leq s_k,$$

$$r_{k+1} - r_k \leq s_{k+1} - s_k$$

and

$$r^* = \lim_{k \longrightarrow \infty} r_k \leq s^*.$$

Moreover, the first two preceding inequality are strict for $n \geq 2$, *if*

$$L_0(u) < K(u) \text{ for each } u \in [0, R_0].$$

Similarly, suppose that

$$-\frac{\varphi(s)}{\varphi'(s)} \leq -\frac{\psi(t)}{\psi'(t)} \tag{18.33}$$

for each $s, t \in [0, R_0]$ *with* $s \leq t$. *Then, we have that*

$$s_{\alpha,k} \leq t_{\alpha,k}$$

$$s_{\alpha,k+1} - s_{\alpha,k} \leq t_{\alpha,k+1} - t_{\alpha,k}$$

and

$$s^* \leq t^*.$$

The first two preceding inequalities are also strict for $k \geq 2$, *if strict inequality holds in (18.33).*

18.4 Specializations and numerical examples

Specializations of Theorem 18.3.3 to some interesting cases such as Smale's $\alpha-$theory (see also Wang's $\gamma-$theory) and Kantorovich theory have been reported in [17,23,25], if $f_0'(t) = f'(t) = g'(t)$ for each $t \in [0,R)$ with $R_0 = R$ and in [23], if $f_0'(t) < f'(t) = g'(t)$ for each $t \in [0,R_0)$. Next, we present examples where $f_0'(t) < f'(t) < g'(t)$ for each $t \in [0,R_0)$ to show the advantages of the new approach over the ones in [17,24,26]. We choose for simplicity $m = n = \alpha = 1$.

Example 18.4.1 *Let $x_0 = 1, D = U(1, 1-q), q \in [0, \frac{1}{2})$ and define function F on D by*

$$F(x) = x^3 - q. \tag{18.34}$$

Then, we have that $\rho = \frac{1}{3}(1-q), L_0 = 3-q, L = 2(2-q)$ and $K = 2(1+\frac{1}{L_0})$. The Newton-Kantorovich condition (18.28) is not satisfied, since

$$h > \frac{1}{2} \text{ for each } q \in [0, \frac{1}{2}). \tag{18.35}$$

Hence, there is no guarantee by the Newton-Kantorovich Theorem [17] that Newton's method (18.1) converges to a zero of operator F. Let us see what gives: We have by (18.29) that

$$h_1 \leq \frac{1}{2}, \text{ if } 0.461983163 < q < \frac{1}{2} \tag{18.36}$$

References

[1] S. Amat, S. Busquier and J. M. Gutiérrez, Geometric constructions of iterative functions to solve nonlinear equations, *J. Comput. Appl. Math.*, 157, 197–205, 2003.

[2] I. K. Argyros, Computational theory of iterative methods, Studies in Computational Mathematics, 15, Editors: K. Chui and L. Wuytack. Elsevier, New York, U.S.A., 2007.

[3] I. K. Argyros, Concerning the convergence of Newton's method and quadratic majorants, *J. Appl. Math. Comput.*, 29, 391–400, 2009.

[4] I. K. Argyros and S. Hilout, On the Gauss-Newton method, *J. Appl. Math.*, 1–14, 2010.

[5] I. K. Argyros and S. Hilout, Extending the applicability of the Gauss-Newton method under average Lipschitz-conditions, *Numer. Algorithms*, 58, 23–52, 2011.

[6] I. K. Argyros, A semilocal convergence analysis for directional Newton methods, *Math. Comput., AMS*, 80, 327–343, 2011.

[7] I. K. Argyros, Y. J. Cho and S. Hilout, Numerical methods for equations and its applications, CRC Press/Taylor and Francis Group, New York, 2012.

[8] I. K. Argyros and S. Hilout, Improved local convergence of Newton's method under weaker majorant condition, *J. Comput. Appl. Math.*, 236 (7), 1892–1902, 2012.

[9] I. K. Argyros and S. Hilout, Weaker conditions for the convergence of Newton's method, *J. Complexity*, 28, 364–387 (2012).

[10] I. K. Argyros and S. Hilout, Computational Methods in Nonlinear Analysis, World Scientific Publ. Comp., New Jersey, 2013.

[11] A. Ben-Israel and T. N. E. Greville, Generalized inverses. CMS Books in Mathematics/Ouvrages de Mathematiques de la SMC, 15. Springer-Verlag, New York, second edition, Theory and Applications, 2003.

[12] J. V. Burke and M. C. Ferris, A Gauss-Newton method for convex composite optimization, *Math. Programming Ser A.*, 71, 179–194, 1995.

[13] J. E. Jr. Dennis and R. B. Schnabel, Numerical methods for unconstrained optimization and nonlinear equations, Classics in Appl. Math., 16, SIAM, Philadelphia, PA, 1996.

[14] O. P. Ferreira, M. L. N. Gonçalves and P. R. Oliveira, Local convergence analysis of the Gauss–Newton method under a majorant condition, *J. Complexity*, 27 (1), 111–125, 2011.

[15] O. P. Ferreira and B. F. Svaiter, Kantorovich's theorem on Newton's method in Riemmanian manifolds, *J. Complexity*, 18 (1), 304–329, 2002.

[16] O. P. Ferreira and B. F. Svaiter, Kantorovich's majorants principle for Newton's method, *Comput. Optim. Appl.*, 42, 213–229, 2009.

[17] O. P. Ferreira, M. L. N. Gonçalves and P. R. Oliveira, Convergence of the Gauss-Newton method for convex composite optimization under a majorant condition, *SIAM J. Optim.*, 23 (3), 1757–1783, 2013.

[18] W. M. Häussler, A Kantorovich-type convergence analysis for the Gauss-Newton method. *Numer. Math.*, 48, 119–125, 1986.

[19] J. B. Hiriart-Urruty and C. Lemaréchal, Convex analysis and minimization algorithms (two volumes). I. Fundamentals. II. Advanced theory and bundle methods, 305 and 306, Springer-Verlag, Berlin, 1993.

[20] L. V. Kantorovich and G. P. Akilov, Functional Analysis, Pergamon Press, Oxford, 1982.

[21] C. Li and K. F. Ng, Majorizing functions and convergence of the Gauss-Newton method for convex composite optimization, *SIAM J. Optim.*, 18 (2), 613–692, 2007.

[22] C. Li and X. H. Wang, On convergence of the Gauss-Newton method for convex composite optimization, *Math. Program. Ser A.*, 91, 349–356, 2002.

[23] A. Magrenán and I. K. Argyros, Expanding the applicability of the Gauss-Newton method for convex optimization under a majorant condition, *SeMA*, 65, 37–56, 2014.

[24] F. A. Potra, Sharp error bounds for a class of Newton-like methods, *Libertas Mathematica*, 5, 71–84, 1985.

[25] S. M. Robinson, Extension of Newton's method to nonlinear functions with values in a cone, *Numer. Math.*, 19, 341–347, 1972.

[26] R. T. Rockafellar, Monotone proccesses of convex and concave type, Memoirs of the American Mathematical Society, 77. American Mathematical Society, Providence, R.I., 1967.

[27] R. T. Rockafellar, Convex analysis, Princeton Mathematical Series, 28, Princeton University Press, Princeton, N.J., 1970.

[28] S. Smale, Newton's method estimates from data at one point. The merging of disciplines: new directions in pure, applied, and computational mathematics (Laramie, Wyo., 1985), 185–196, Springer, New York, 1986.

[29] X. H. Wang, Convergence of Newton's method and uniqueness of the solution of equations in Banach space, *IMA J. Numer. Anal.*, 20, 123–134, 2000.

Chapter 19

Directional Newton methods and restricted domains

Ioannis K. Argyros

Department of Mathematical Sciences, Cameron University, Lawton, OK 73505, USA, Email: iargyros@cameron.edu

Á. Alberto Magreñán

Universidad Internacional de La Rioja, Escuela Superior de Ingeniería y Tecnología, 26002 Logroño, La Rioja, Spain, Email: alberto.magrenan@unir.net

CONTENTS

19.1 Introduction

Let $F : D \subseteq \mathbb{R}^n \rightarrow \mathbb{R}$ be a differentiable function. In computer graphics, we often need to compute and display the intersection $\mathcal{C} = \mathcal{A} \cap \mathcal{B}$ of two surfaces $\mathcal{A}$ and $\mathcal{B}$ in $\mathbb{R}^3$ [5], [6]. If the two surfaces are explicitly given by

$$\mathcal{A} = \{(u, v, w)^T : w = F_1(u, v)\}$$

and

$$\mathcal{B} = \{(u,v,w)^T \ : w = F_2(u,v)\},$$

then the solution $x^\star = (u^\star, v^\star, w^\star)^T \in \mathcal{C}$ must satisfy the nonlinear equation

$$F_1(u^\star, v^\star) = F_2(u^\star, v^\star)$$

and

$$w^\star = F_1(u^\star, v^\star).$$

Therefore, we must solve a nonlinear equation in two variables $x = (u,v)^T$ of the form

$$F(x) = F_1(x) - F_2(x) = 0.$$

The marching method can be used to compute the intersection $\mathcal{C}$. In this method, we first need to compute a starting point $x_0 = (u_0, v_0, w_0)^T \in \mathcal{C}$, and then compute the succeeding intersection points by succesive updating.

In mathematical programming [9], for an equality–constraint optimization problem, e.g.,

$$\begin{array}{ll} \min & \psi(x) \\ s.t. & F(x) = 0 \end{array}$$

where, ψ, $F : \mathcal{D} \subseteq \mathbb{R}^n \longrightarrow \mathbb{R}$ are nonlinear functions, we need a feasible point to start a numerical algorithm. That is, we must compute a solution of equation $F(x) = 0$.

In the case of a system of nonlinear equations $G(x) = 0$, with $G : \mathcal{D} \subseteq \mathbb{R}^n \longrightarrow \mathbb{R}^n$, we may solve instead

$$\| G(x) \|_2 = 0,$$

if the zero of function G is isolated or locally isolated and if the rounding error is neglected [3], [7], [10], [11], [12].

We use the directional Newton method (DNM) [5] given by

$$x_{k+1} = x_k - \frac{F(x_k)}{\nabla F(x_k) \cdot d_k} \, d_k \quad (k \geq 0)$$

to generate a sequence $\{x_k\}$ converging to $x^\star$.

Now, we will explain DNM. First of all, we begin by choosing an initial guess $x_0 \in U_0$, where F is differentiable and a direction vector d_0.

Then, we restrict F on the line $\mathcal{A} = \{x_0 + \theta \, d_0, \ \theta \in \mathbb{R}\}$, where it is a function of one variable $f(\theta) = F(x_0 + \theta \, d_0)$.

Set $\theta_0 = 0$ to obtain the Newton iteration for f, that is the next point:

$$v_1 = -\frac{f(0)}{f'(0)}.$$

The iteration for F is

$$x_1 = x_0 - \frac{F(x_0)}{\nabla F(x_0) \cdot d_0} d_0.$$

Note that $f(0) = F(x_0)$ and $f'(0)$ is the directional derivative

$$f'(0) = F'(x_0, d_0) = \nabla F(x_0) \cdot d_0.$$

By repeating this process we arrive at DNM.

If $n = 1$, DNM reduces to the classical Newton method [1]– [3], [7].

In [5], Levin and Ben–Israel provide a semilocal convergence analysis for the DNM.

The quadratic convergence of the method was established for directions d_k sufficiently close to the gradients $\nabla F(x_k)$, and under standard Newton–Kantorovich–type hypotheses [1]– [3], [7].

In this chapter, motivated by the aforementioned [5], optimization considerations, and using the new idea of restricted convergence domains, we find more precise location where the iterates lie than in earlier studies such as [1,5]. This way, we provide a semilocal convergence analysis with the following advantages

1. Weaker hypotheses;
2. Larger convergence domain for DNM;
3. Finer error bounds on the distances $\| x_{k+1} - x_k \|$, $\| x_k - x^\star \|$ $(k \geq 0)$;
4. An at least as precise information on the location of the zero $x^\star$.

A numerical example where our results apply, but the earlier ones in [5] are violated is also provided at the end of this chapter.

Here after, we use the Euclidean inner product, the corresponding norm $\| x \|$, and the corresponding matrix norm $\| A \|$, except in Section 19.3 where the ∞–norm is used for vectors and matrices, denoted by $\| x \|_\infty$, and $\| A \|_\infty$, respectively.

19.2 Semilocal convergence analysis

First of all we present some needed lemma on majorizing sequences for DNM.

Lemma 19.2.1 *Suppose that there exist constants $L_0 > 0$, $L > 0$ and $\eta \geq 0$ such that:*

$$q_0 := \overline{L}\eta \leq \frac{1}{2} \tag{19.1}$$

where,

$$\overline{L} = \frac{1}{8}\left(4\,L_0 + \sqrt{L_0\,L} + \sqrt{8\,L_0^2 + L_0\,L}\right). \tag{19.2}$$

Then, sequence $\{t_k\}$ ($k \geq 0$) given by

$$t_0 = 0, \quad t_1 = \eta, \quad t_2 = t_1 + \frac{L_0\,(t_1 - t_0)^2}{2\,(1 - L_0\,t_1)}, \quad t_{k+1} = t_k + \frac{L\,(t_k - t_{k-1})^2}{2\,(1 - L_0\,t_k)} \quad (k \geq 2), \tag{19.3}$$

is well defined, nondecreasing, bounded above by $t^{\star\star}$, and converges to its unique least upper bound $t^\star \in [0, t^{\star\star}]$, where

$$t^{\star\star} = \left[1 + \frac{L_0\eta}{2(1-\delta)(1-L_0\eta)}\right]\eta, \tag{19.4}$$

and

$$\delta = \frac{2\,L}{L + \sqrt{L^2 + 8\,L_0\,L}}, \; L_0, L \neq 0. \tag{19.5}$$

The proof is shown in [3]. Here, $\angle$ denotes the angle between two vectors u and v, given by

$$\angle(u,v) = \arccos\frac{u \cdot v}{\| u \| \cdot \| v \|}, \quad u \neq 0, \quad v \neq 0.$$

Throughout the chapter $B(x,\rho)$, and $\overline{B}(x,\rho)$ stand, respectively for the open and closed ball in $\mathbb{R}^n$ with center $x \in \mathbb{R}^n$ and radius $\rho > 0$.

Now, we provide the main semilocal convergence theorem for DNM.

Theorem 19.2.2 *Let $F : \mathcal{D} \subseteq \mathbb{R}^n \longrightarrow \mathbb{R}$ be a differentiable function. Assume that (i) there exist a point $x_0 \in \mathcal{D}$, such that*

$$F(x_0) \neq 0, \qquad \nabla F(x_0) \neq 0.$$

Let $d_0 \in \mathbb{R}^n$ be such that $\| d_0 \| = 1$, and set

$$h_0 = -\frac{F(x_0)}{\nabla F(x_0) \cdot d_0}\, d_0,$$

$$x_1 = x_0 + h_0.$$

(ii) For $F \in \mathcal{C}^2[\mathcal{D}]$, *there exist constants* $M_0 > 0$ *and* $M > 0$ *such that*

$$\| \nabla F(x) - \nabla F(x_0) \| \leq M_0 \| x - x_0 \|, \qquad x \in \mathcal{D}, \tag{19.6}$$

$$\sup_{x \in \mathcal{D}_0} \| F''(x) \| = M, \mathcal{D}_0 = \mathcal{D} \cap B(x_0, 1/M_0), \tag{19.7}$$

$$p_0 = |F(x_0)| \, \overline{M} \, |\nabla F(x_0) \cdot d_0|^{-2} \leq \frac{1}{2}, \tag{19.8}$$

and

$$B_0 := B(x_0, t^\star) \subseteq \mathcal{D} \tag{19.9}$$

where

$$\overline{M} = \frac{1}{8}\left(4 M_0 + \sqrt{M_0 M} + \sqrt{8 M_0^2 + M_0 M}\right). \tag{19.10}$$

(iii) Sequence $\{x_k\}$ $(k \geq 0)$ *given by*

$$x_{k+1} = x_k + h_k, \tag{19.11}$$

where,

$$h_k = -\frac{F(x_k)}{\nabla F(x_k) \cdot d_k} \, d_k, \tag{19.12}$$

satisfies

$$\angle(d_k, \nabla F(x_k)) \leq \angle(d_0, \nabla F(x_0)) \qquad k \geq 0, \tag{19.13}$$

where, each $d_k \in \mathbb{R}^n$ *is such that* $\| d_k \| = 1$.

Then, sequence $\{x_k\}_{k \geq 0}$ *remains in* B_0 *and converges to a zero* $x^\star \in B_0$ *of function* F.

On the other hand, $\nabla F(x^\star) \neq 0$ *unless* $\| x^\star - x_0 \| = t^\star$.

Moreover, the following estimates hold for all $k \geq 0$:

$$\| x_{k+1} - x_k \| \leq t_{k+1} - t_k \tag{19.14}$$

and

$$\| x_k - x^\star \| \leq t^\star - t_k, \tag{19.15}$$

where, iteration $\{t_k\}$ *is given by (19.3), for*

$$L_0 = |\nabla F(x_0)|^{-1} M_0, \quad L = |\nabla F(x_0) \cdot d_0|^{-1} M, \quad \eta = |\nabla F(x_0) \cdot d_0|^{-1} \, |F(x_0)|.$$

Note that condition (19.13) is equivalent to

$$\frac{|\nabla F(x_k) \cdot d_k|}{\| \nabla F(x_k) \|} \geq \frac{|\nabla F(x_0) \cdot d_0|}{\| \nabla F(x_0) \|} \qquad k \geq 0. \tag{19.16}$$

Proof: The iterates $\{x_n\}$ are shown to be in D_0 which is more precise location than D used in [1,5], since $D_0 \subseteq D$ leading to advantages. We shall show using mathematical induction on $k \geq 0$:

$$\| x_{k+1} - x_k \| \leq t_{k+1} - t_k, \tag{19.17}$$

and

$$\overline{B}(x_{k+1}, t^\star - t_{k+1}) \subseteq \overline{B}(x_k, t^\star - t_k). \tag{19.18}$$

For each $z \in \overline{B}(x_1, t^\star - t_1)$, we have

$$\begin{aligned} \| z - x_0 \| &\leq \| z - x_1 \| + \| x_1 - x_0 \| \\ &\leq t^\star - t_1 + t_1 - t_0 = t^\star - t_0, \end{aligned}$$

which shows that $z \in \overline{B}(x_0, t^\star - t_0)$.

Since, also

$$\| x_1 - x_0 \| = \| h_0 \| \leq \eta = t_1 - t_0,$$

estimates (19.17) and (19.18) hold for $k = 0$.

Assume (19.17) and (19.18) hold for all $i \leq k$. Then we have:

$$\begin{aligned} \| x_{k+1} - x_0 \| &\leq \| x_{k+1} - x_k \| + \| x_k - x_{k-1} \| + \cdots + \| x_1 - x_0 \| \\ &\leq (t_{k+1} - t_k) + (t_k - t_{k-1}) + \cdots + (t_1 - t_0) = t_{k+1}, \end{aligned} \tag{19.19}$$

and

$$\| x_k + t\,(x_{k+1} - x_k) - x_0 \| \leq t_k + t\,(t_{k+1} - t_k) \leq t^\star, \quad t \in [0,1].$$

Using condition (19.6) for $x = x_k$, we obtain:

$$\begin{aligned} \| \nabla F(x_k) \| &\geq \| \nabla F(x_0) \| - \| \nabla F(x_k) - \nabla F(x_0) \| \\ &\geq \| \nabla F(x_0) \| - M_0 \; \| x_k - x_0 \| \\ &\geq \| \nabla F(x_0) \| - M_0 \; (t_k - t_0) \\ &\geq \| \nabla F(x_0) \| - M_0 \; t_k > 0 \qquad \text{(by (19.18), and Lemma 19.2.1)}. \end{aligned} \tag{19.20}$$

We also get

$$\begin{aligned} \int_{x_{k-1}}^{x_k} (x_k - x)\, F''(x)\, dx &= -(x_k - x_{k-1})\, \nabla F(x_{k-1}) + F(x_k) - F(x_{k-1}) \\ &= -h_{k-1}\, \nabla F(x_{k-1}) + F(x_k) - F(x_{k-1}) \\ &= \frac{F(x_{k-1})}{(F(x_{k-1}) \cdot d_{k-1})}\, (d_{k-1} \cdot \nabla F(x_{k-1})) + F(x_k) \\ &\quad - F(x_{k-1}) \\ &= F(x_k). \end{aligned} \tag{19.21}$$

Now, we introduce a change of variable given by $x = x_{k-1} + t\, h_{k-1}$, $t \in [0,1]$. We can write

$$x_k - x = x_k - x_{k-1} - t\, h_{k-1} = h_{k-1} - t\, h_{k-1} = (1-t)\, h_{k-1}, \qquad dx = h_{k-1}\, dt.$$

Then (19.21) can be written as:

$$F(x_k) = \int_0^1 (1-t)\, h_{k-1}\, F''(x_{k-1} + \theta\, h_{k-1})\, h_{k-1}\, d\theta. \tag{19.22}$$

Using (19.7), (19.11)–(19.16), we obtain

$$\begin{aligned}
\| x_{k+1} - x_k \| = \| h_k \| &= \frac{|F(x_k)|}{|\nabla F(x_k) \cdot d_k|} \\
&\leq \frac{|\int_0^1 (1-t)\, h_{k-1}\, F''(x_0 + \theta\, h_{k-1})\, h_{k-1}\, d\theta|}{|\nabla F(x_k) \cdot d_k|} \\
&\leq \frac{M\ \| h_{k-1} \|^2}{2\,|\nabla F(x_k) \cdot d_k|} \\
&\leq \frac{M\ \| h_{k-1} \|^2}{2\ \| \nabla F(x_k) \|}\ \frac{\| \nabla F(x_0) \|}{|\nabla F(x_0) \cdot d_0|} \\
&\leq \frac{M\ \| h_{k-1} \|^2\ \| \nabla F(x_0) \|}{2\,(\| \nabla F(x_0) \| - M_0\, t_k)\ |\nabla F(x_0) \cdot d_0|} \\
&\leq \frac{M\ \| h_{k-1} \|^2}{2\,(1 - \| \nabla F(x_0) \|^{-1}\ M_0\, t_k)\ |\nabla F(x_0) \cdot d_0|} \\
&\leq \frac{M\, (t_k - t_{k-1})^2}{2\,(1 - |\nabla F(x_0) \cdot d_0|^{-1}\, M_0\, t_k)\ |\nabla F(x_0) \cdot d_0|} \\
&= t_{k+1} - t_k,
\end{aligned}$$

which shows (19.17) for all $k \geq 0$.

Then, for each $w \in \overline{B}(x_{k+2}, t^\star - t_{k+2})$, we get

$$\begin{aligned}
\| w - x_{k+1} \| &\leq \| w - x_{k+2} \| + \| x_{k+2} - x_{k+1} \| \\
&\leq t^\star - t_{k+2} + t_{k+2} - t_{k+1} = t^\star - t_{k+1},
\end{aligned}$$

showing (19.18) for all $k \geq 0$.

Lemma 19.2.1 implies that $\{t_n\}$ is a Cauchy sequence. It then follows from (19.17), and (19.18) that $\{x_n\}$ is a Cauchy sequence too. Since B_0 is a closed set, it follows that $\{x_n\}$ converges to some $x^\star \in B_0$.

The point $x^\star$ is a zero of F, since

$$0 \leq |F(x_k)| \leq \frac{1}{2} M\,(t_k - t_{k-1})^2 \longrightarrow 0 \quad \text{as} \quad k \rightarrow \infty.$$

On the other hand, we prove $\nabla F(x^\star) \neq 0$, except if, $\| x^\star - x_0 \| = t^\star$.

Using (19.6) and the definition of constant L_0, we get

$$\begin{aligned} \| \nabla F(x) - \nabla F(x_0) \| &\leq M_0 \, \| x - x_0 \| \\ &\leq M\, t^\star \leq |\nabla F(x_0) \cdot d_0| \leq \| \nabla F(x_0) \| \end{aligned}$$

If $\| x - x_0 \| < t^\star$, then by (19.6), we obtain:

$$\| \nabla F(x) - \nabla F(x_0) \| \leq M_0 \, \| x - x_0 \| < M_0 t^\star \leq \| \nabla F(x_0) \|,$$

or

$$\| \nabla F(x_0) \| > \| \nabla F(x) - \nabla F(x_0) \|,$$

which shows $\nabla F(x) \neq 0$. □

Now, we present the following very useful remark.

Remark 19.2.3 **(a)** *$t^\star$ in (19.9) can be replaced by $t^{\star\star}$ given in closed form by (19.4).*

(b) *The proof followed the corresponding proof in [3], but the new sufficient convergence condition (19.8) is weaker than the one in [3].*

To show the second part of the previous remark we need the following definition.

Definition 19.2.4 *Let $F : \mathcal{D} \subseteq \mathbb{R}^n \longrightarrow \mathbb{R}$ be a differentiable function, where $\mathcal{D}$ is a convex region. Function ∇F is called Lipschitz continuous, if there exists a constant $M_1 \geq 0$, such that:*

$$\| \nabla F(x) - \nabla F(y) \| \leq M_1 \, \| x - y \| \quad \text{for all} \quad x, y \in \mathcal{D}. \tag{19.23}$$

If F is twice differentiable, then we can set

$$\sup_{x \in \mathcal{D}_0} \, \| F''(x) \| := M_1. \tag{19.24}$$

Notice that

$$M_0 \leq M_1 \tag{19.25}$$

and

$$M \leq M_1, (\text{ since } D_0 \subseteq D) \tag{19.26}$$

hold in general, and $\dfrac{M_1}{M_0}$ can be arbitrarily large [1]– [3].

Remark 19.2.5 *If F is twice differentiable, then, in view of the proof of Theorem 19.2.2, constant L_0 can be defined by the more precise:*

$$L_0 = \| \nabla F(x_0) \|^{-1} M_0. \tag{19.27}$$

Remark 19.2.6 *Our Theorem 19.2.2 improves Theorem 2.2 in [1].*

Theorem 2.2 in [5] uses condition

$$p = |F(x_0)|\, N\, |\nabla F(x_0) \cdot d_0|^{-2} \leq \frac{1}{2} \tag{19.28}$$

corresponding to our condition (19.8), where

$$N = 4M_0 + M_1 + \sqrt{M_1^2 + 8M_0M_1}.$$

But, we have

$$\overline{M} < N. \tag{19.29}$$

That is,

$$p \leq \frac{1}{2} \Longrightarrow p_0 \leq \frac{1}{2}, \tag{19.30}$$

but not necessarily vice verca (unless if $M_0 = M = M_1$).
Moreover corresponding error bounds as well as the information on the location of we improved if $M_0 < M$ or $M < M_1$.

Examples were $M_0 < M < M < M_1$ can be found in [1–4]. The rest of the results in [1] (and consequently the results in [5] can be improved along the same lines) It is worth noting that in the proof of Theorem 2.2 in [3], we used M_1 instead of M (i.e., D instead of D_0). However, the iterates x_n remain in D_0 which is more precise domain than D, since $D_0 \subseteq D$.

References

[1] I. K. Argyros, A semilocal convergence analysis for directional Newton methods, Math. Compt. AMS, 80, 327–343 (2011).

[2] I. K. Argyros, Convergence and applications of Newton-type iterations, Springer–Verlag, New York, 2008.

[3] I. K. Argyros and S. Hilout, Weaker conditions for the convergence of Newton's method, *J. Complexity*, 28, 364–387, 2012.

[4] A. Ben–Israel and Y. Levin, Maple programs for directional Newton methods, are available at ftp://rutcor.rutgers.edu/pub/bisrael/Newton–Dir.mws.

[5] Y. Levin and A. Ben–Israel, Directional Newton methods in n variables, *Math. Comput. AMS*, 71, 237, 251–262, 2002.

[6] G. Lukács, The generalized inverse matrix and the surface–surface intersection problem. Theory and practice of geometric modeling (Blaubeuren, 1988), 167–185, Springer, Berlin, 1989.

[7] J. M. Ortega and W. C. Rheinboldt, Iterative solution of nonlinear equations in several variables, Academic Press, New York, 1970.

[8] V. Pereyra, Iterative methods for solving nonlinear least square problems, *SIAM J. Numer. Anal.*, 4, 27–36, 1967.

[9] B. T. Polyak, Introduction to optimization. Translated from the Russian. With a foreword by D. P. Bertsekas. Translations series in Mathematics and engineering, Optimization Software, Inc., Publications Division, New York, 1987.

[10] A. Ralston and P. Rabinowitz, A first course in numerical analysis, 2nd Edition, Mc Graw-Hill, 1978.

[11] J. Stoer and K. Bulirsch, Introduction to Numerical Analysis, Springer–Verlag, 1976.

[12] H. F. Walker and L. T. Watson, Least–change Secant update methods, *SIAM J. Numer. Anal.*, 27, 1227–1262, 1990).

Chapter 20

Expanding the applicability of the Gauss-Newton method for convex optimization under restricted convergence domains and majorant conditions

Ioannis K. Argyros

Department of Mathematical Sciences, Cameron University, Lawton, OK 73505, USA, Email: iargyros@cameron.edu

Á. Alberto Magreñán

Universidad Internacional de La Rioja, Escuela Superior de Ingeniería y Tecnología, 26002 Logroño, La Rioja, Spain, Email: alberto.magrenan@unir.net

CONTENTS

20.1 Introduction

In this chapter we are concerned with the convex composite optimizations problem. This work is mainly motivated by the work in [17,23]. We present a convergence analysis of Gauss–Newton method (defined by Algorithm (GNA) in Sec. 20.2). The convergence of GNA is based on the majorant function in [17] (to be precised in Sec. 20.2). They follow the same formulation using the majorant function provided in [23] (see [21,23,28,29]). In [3,5,8], a convergence analysis in a Banach space setting was given for (GNM) defined by

$$x_{k+1} = x_k - \left[F'(x_k)^+F'(x_k)\right]^{-1} F'(x_k)^+F(x_k) \text{ for each } k = 0,1,2,\ldots,$$

where x_0 is an initial point and $F'(x)^+$ in the Moore-Penrose inverse [11–13, 19,26] of operator $F'(x)$ with $F:\mathbb{R}^n \to \mathbb{R}^m$ being continuously differentiable. In [23], a semilocal convergence analysis using a combination of a majorant and a center majorant function was given with the advantages $(\mathcal{A})$: tighter error estimates on the distances involved and the information on the location of the solution is at least as precise. These advantages were obtained (under the same computational cost) using same or weaker sufficient convergence hypotheses. Here, we extend the same advantages $(\mathcal{A})$ but to hold for GNA.

The chapter is organized as follows: Section 20.2 contains the definition of GNA. In order for us to make the chapter as self contained as possible, the notion of quasi-regularity is also re-introduced (see, e.g., [12, 17, 21]). The semilocal convergence analysis of GNA is presented in Section 20.3. Numerical examples and applications of our theoretical results and favorable comparisons to earlier studies (see, e.g., [12, 17, 18, 21, 22]) are presented in Section 20.4.

20.2 Gauss-Newton Algorithm and Quasi-Regularity condition

20.2.1 Gauss-Newton Algorithm GNA

Using the idea of restricted convergence domains, we study the convex composite optimization problem

$$\min_{x\in \mathbb{R}^n} p(x) := h(F(x)), \tag{20.1}$$

where $h:\mathbb{R}^m \longrightarrow \mathbb{R}$ is convex, $F:\mathbb{R}^n \longrightarrow \mathbb{R}^m$ is Fréchet-differentiable operator and $m,l \in \mathbb{N}^\star$. The importance of (20.1) can be found in [2, 10, 12, 19, 21–23,

25, 27]. We assume that the minimum h_{min} of the function h is attained. Problem (20.1) is related to

$$F(x) \in \mathcal{C}, \tag{20.2}$$

where

$$\mathcal{C} = \operatorname{argmin} h \tag{20.3}$$

is the set of all minimum points of h.

Let $\xi \in [1, \infty[$, $\Delta \in]0, \infty]$ and for each $x \in I\!\!R^n$, define $\mathcal{D}_\Delta(x)$ by

$$\begin{aligned} \mathcal{D}_\Delta(x) \quad = \quad & \{d \in I\!\!R^n : \| d \| \leq \Delta,\ h(F(x) + F'(x)d) \leq h(F(x) + F'(x)d') \\ & \text{for all } d' \in I\!\!R^n \text{ with } \| d' \| \leq \Delta\}. \end{aligned} \tag{20.4}$$

Let $x_0 \in I\!\!R^n$ be an initial point. The Gauss-Newton algorithm GNA associated with (ξ, Δ, x_0) as defined in [12] (see also [17]) is as follows:

Algorithm GNA : (ξ, Δ, x_0)

INICIALIZATION. Take $\xi \in [1, \infty)$, $\Delta \in (0, \infty]$ and $x_0 \in \mathbb{R}^n$, set $k = 0$.
STOP CRITERION. Compute $D_\Delta(x_k)$. If $0 \in \mathcal{D}_\Delta(x_k)$, STOP. Otherwise.
ITERATIVE STEP. Compute d_k satisfying $d_k \in D_\Delta(x_k)$, $\|d_k\| \leq \xi d(0, D\Delta(x_k))$,

Then, set $x_{k+1} = x_k + d_k$, $k = k + 1$ and GO TO STOP CRITERION.

Here, $d(x, W)$ denotes the distance from x to W in the finite dimensional Banach space containing W. Note that the set $\mathcal{D}_\Delta(x)$ $(x \in I\!\!R^n)$ is nonempty and is the solution of the following convex optimization problem

$$\min_{d \in I\!\!R^n,\, \|d\| \leq \Delta} \quad h(F(x) + F'(x)d), \tag{20.5}$$

which can be solved by well known methods such as the subgradient or cutting plane or bundle methods (see, e.g., [12, 19, 25–27]).

Let $U(x, r)$ denote the open ball in $\mathbb{R}^n$ (or $\mathbb{R}^m$) centered at x and of radius $r > 0$. By $\overline{U}(x, r)$ we denote its closure. Let W be a closed convex subset of $\mathbb{R}^n$ (or $\mathbb{R}^m$). The negative polar of W denoted by $W^\ominus$ is defined as

$$W^\ominus = \{z : < z, w > \leq 0 \quad \text{for each} \quad w \in W\}. \tag{20.6}$$

20.2.2 Quasi Regularity

In this section, we mention some concepts and results on regularities which can be found in [12] (see also, e.g., [17, 21–23, 25]). For a set-valued mapping $T : \mathbb{R}^n \rightrightarrows \mathbb{R}^m$ and for a set A in $\mathbb{R}^n$ or $\mathbb{R}^m$, we denote by

$$D(T) = \{x \in \mathbb{R}^n : Tx \neq \emptyset\}, \quad R(T) = \bigcup_{x \in D(T)} Tx, \tag{20.7}$$

$$T^{-1}y = \{x \in \mathbb{R}^n : y \in Tx\} \quad \text{and} \quad \| A \| = \inf_{a \in A} \| a \| .$$

Consider the inclusion

$$F(x) \in C, \tag{20.8}$$

where C is a closed convex set in $\mathbb{R}^m$. Let $x \in \mathbb{R}^n$ and

$$\mathcal{D}(x) = \{d \in \mathbb{R}^n : F(x) + F'(x)d \in C\}. \tag{20.9}$$

Definition 20.2.1 Let $x_0 \in \mathbb{R}^n$.

(a) x_0 is quasi-regular point of (20.8) if there exist $R_0 \in]0,+\infty[$ and an increasing positive function β on $[0,R_0[$ such that

$$\mathcal{D}(x) \neq \emptyset \text{ and } d(0,\mathcal{D}(x)) \leq \beta(\| x - x_0 \|)\, d(F(x),C) \text{ for all } x \in U(x_0,R_0). \tag{20.10}$$

$\beta(\| x - x_0 \|)$ is an "error bound" in determining how for the origin is away from the solution set of (20.8).

(b) x_0 is a regular point of (20.8) if

$$ker(F'(x_0)^T) \cap (C - F(x_0))^{\circ} = \{0\}. \tag{20.11}$$

Proposition 20.2.2 (see, e.g., [12,17,21,25]) *Let x_0 be a regular point of (20.8). Then, there are constants $R_0 > 0$ and $\beta > 0$ such that (20.10) holds for R_0 and $\beta(\cdot) = \beta$. Therefore, x_0 is a quasi-regular point with the quasi-regular radius $R_{x_0} \geq R_0$ and the quasi-regular bound function $\beta_{x_0} \leq \beta$ on $[0,R_0]$.*

Remark 20.2.3 (a) $\mathcal{D}(x)$ can be considered as the solution set of the linearized problem associated to (20.8)

$$F(x) + F'(x)d \in C. \tag{20.12}$$

(b) If C defined in (20.8) is the set of all minimum points of h and if there exists $d_0 \in \mathcal{D}(x)$ with $\| d_0 \| \leq \Delta$, then $d_0 \in \mathcal{D}_\Delta(x)$ and for each $d \in \mathbb{R}^n$, we have the following equivalence

$$d \in \mathcal{D}_\Delta(x) \Longleftrightarrow d \in \mathcal{D}(x) \Longleftrightarrow d \in \mathcal{D}_\infty(x). \tag{20.13}$$

(c) Let R_{x_0} denote the supremum of R_0 such that (20.10) holds for some function β defined in Definition 20.2.1. Let $R_0 \in [0,R_{x_0}]$ and $\mathcal{B}_R(x_0)$ denotes the set of function β defined on $[0,R_0)$ such that (20.10) holds. Define

$$\beta_{x_0}(t) = \inf\{\beta(t) : \beta \in \mathcal{B}_{R_{x_0}}(x_0)\} \quad \text{for each} \quad t \in [0,R_{x_0}). \tag{20.14}$$

All function $\beta \in \mathcal{B}_R(x_0)$ with $\lim_{t \to R^-} \beta(t) < +\infty$ can be extended to an element of $\mathcal{B}_{R_{x_0}}(x_0)$ and we have that

$$\beta_{x_0}(t) = \inf\{\beta(t) : \beta \in \mathcal{B}_R(x_0)\} \quad \text{for each} \quad t \in [0,R_0). \tag{20.15}$$

R_{x_0} and β_{x_0} are called the quasi-regular radius and the quasi-regular function of the quasi-regular point x_0, respectively.

Definition 20.2.4 (a) *A set-valued mapping* $T : \mathbb{R}^n \rightrightarrows \mathbb{R}^m$ *is convex if the following items hold*

(i) $Tx + Ty \subseteq T(x+y)$ *for all* $x, y \in \mathbb{R}^n$.

(ii) $T\lambda x = \lambda Tx$ *for all* $\lambda > 0$ *and* $x \in \mathbb{R}^n$.

(iii) $0 \in T0$.

20.3 Semi-local convergence

In this section we present the semi-local convergence of GNA. First, we study the convergence of majorizing sequences for GNA. Then, we study the convergence of GNA. We need the definition of the center-majorant function and the definition of the majorant function for F.

Definition 20.3.1 *Let* $R > 0$, $x_0 \in \mathbb{R}^n$ *and* $F : \mathbb{R}^n \to \mathbb{R}^m$ *be continuously Fréchet-differentiable. A twice-differentiable function* $f_0 : [0, R) \to \mathbb{R}$ *is called a center-majorant function for F on* $U(x_0, R)$, *if for each* $x \in U(x_0, R)$,

(h_0^0) $\|F'(x) - F'(x_0)\| \leq f_0'(\|x - x_0\|) - f_0'(0)$;

(h_1^0) $f_0(0) = 0$, $f_0'(0) = -1$;

and

(h_2^0) f_0' *is convex and strictly increasing.*

Definition 20.3.2 *[5,8,17] Let* $x_0 \in \mathbb{R}^n$ *and* $F : \mathbb{R}^n \to \mathbb{R}^m$ *be continuously differentiable. Define* $R_0 = \sup\{t \in [0, R) : f_0'(t) < 0\}$. *A twice-differentiable function* $f : [0, R_0) \to \mathbb{R}$ *is called a majorant function for F on* $U(x_0, R_0)$, *if for each* $x, y \in U(x_0, R_0)$, $\|x - x_0\| + \|y - x\| < R_0$,

(h_0) $\|F'(y) - F'(x)\| \leq f'(\|y - x\| + \|x - x_0\|) - f'(\|x - x_0\|)$;

(h_1) $f(0) = 0$, $f'(0) = -1$;

and

(h_2) f' *is convex and strictly increasing.*

Moreover, assume that

(h_3) $f_0(t) \leq f(t)$ and $f_0'(t) \leq f'(t)$ for each $t \in [0, R_0)$.

Remark 20.3.3 *Suppose that* $R_0 < R$. *If* $R_0 \geq R$, *then we do not need to introduce Definition 20.3.2.*

In Section 20.4, we present examples where hypothesis (h_3) is satisfied. Let $\xi > 0$ and $\alpha > 0$ be fixed and define auxiliary function $\varphi : [0, R_0) \to \mathbb{R}$ by

$$\varphi(t) = \xi + (\alpha - 1)t + \alpha f(t). \tag{20.16}$$

We shall use the following hypotheses

(h_4) there exists $s^* \in (0, R)$ such that for each $t \in (0, s^*)$, $\varphi(t) > 0$ and $\varphi(s^*) = 0$;

(h_5) $\varphi(s^*) < 0$.

From now on we assume the hypotheses $(h_0) - (h_4)$ and $(h_0^0) - (h_2^0)$ which will be called the hypotheses (H).

Next, we present the main semi-local convergence result of the Gauss-Newton method generated by the Algorithm GNA for solving (20.1).

Theorem 20.3.4 *Suppose that the (H) conditions are satisfied. Then,*

(i) *sequence $\{s_k\}$ generated by the Gauss-Newton method for $s_0 = 0, s_{k+1} = s_k - \frac{\varphi(s_k)}{\varphi'(s_k)}$ for solving equation $\psi(t) = 0$ is: well defined; strictly increasing; remains in $[0, s^*)$ and converges Q-linearly to s^*.*

Let $\eta \in [1, \infty]$, $\Delta \in (0, \infty]$ and $h : \mathbb{R}^m \to \mathbb{R}$ be real-valued convex with minimizer set C such that $C \neq \emptyset$.

(ii) *Suppose that $x_0 \in \mathbb{R}^n$ is a quasi-regular point of the inclusion*

$$F(x) \in C,$$

with the quasi-regular radius r_{x_0} and the quasi-regular bound function β_{x_0} defined by (20.14) and (20.15), respectively. If $d(F(x_0), C) > 0$, $s^* \leq r_{x_0}$, $\Delta \geq \xi \geq \eta\beta_{x_0}(0)d(F(x_0), C)$,

$$\alpha \geq \sup\left\{\frac{\eta\beta_{x_0}(t)}{\eta\beta_{x_0}(t)(1 + f'(t)) + 1} : \xi \leq t < s^*\right\}$$

then, sequence $\{x_k\}$ generated by GNA is well defined, remains in $U(x_0, s^*)$ for each $k = 0, 1, 2, \ldots$, such that

$$F(x_k) + F'(x_k)(x_{k+1} - x_k) \in C \text{ for each } k = 0, 1, 2 \ldots. \tag{20.17}$$

Moreover, the following estimates hold

$$\|x_{k+1} - x_k\| \leq s_{k+1} - s_k, \tag{20.18}$$

$$\|x_{k+1} - x_k\| \leq \frac{s_{k+1} - s_k}{(s_k - s_{k-1})^2}\|x_k - x_{k-1}\|^2, \tag{20.19}$$

for each $k = 0, 1, 2 \ldots$, and $k = 1, 2, \ldots$, respectively and converges to a point $x^* \in U(x_0, s^*)$ satisfying $F(x^*) \in C$ and

$$\|x^* - x_k\| \leq t^* - s_k \text{ for each } k = 0, 1, 2, \ldots. \tag{20.20}$$

The convergence is R-linear. If hypothesis (h_5) holds, then the sequences $\{s_k\}$ and $\{x_k\}$ converge Q-quadratically and R-quadratically to s^* and x^*, respectively. Furthermore, if

$$\alpha > \overline{\alpha} := \sup\left\{\frac{\eta_{\beta_{x_0}}(t)}{\eta_{\beta_{x_0}}(t)(1+f'(t))+1} : \xi \leq t < s^*\right\},$$

then, the sequence $\{x_k\}$ converges R-quadratically to x^*.

Proof: Simply replace function g in [23] (see also [17]) by function f in the proof, where g is a majorant function for F on $U(x_0, R)$. That is we have instead of (h_0):
(h_0')

$$\|F'(y) - F'(x)\| \leq g'(\|y-x\| + \|x-x_0\|) - g'(\|x-x_0\|) \tag{20.21}$$

for each $x, y \in U(x_0, R)$ with $\|x - x_0\| + \|y - x\| < R$. The iterates x_n lie in $U(x_0, R_0)$ which is a more precise location than $U(x_0, R)$.

Remark 20.3.5 *(a) If $f(t) = g(t) = f_0(t)$ for each $t \in [0, R_0)$ and $R_0 = R$, then Theorem 2.1 reduces to the corresponding Theorem in [17]. Moreover, if $f_0(t) \leq f(t) = g(t)$ we obtain the results in [23]. Notice that, we have that*

$$f_0'(t) \leq g'(t) \text{ for each } t \in [0, R) \tag{20.22}$$

and

$$f'(t) \leq g'(t) \text{ for each } t \in [0, R_0). \tag{20.23}$$

Therefore, if

$$f_0'(t) \leq f'(t) < g'(t) \text{ for each } t \in [0, R_0), \tag{20.24}$$

then the following advantages denoted by $(\mathcal{A})$ are obtained: weaker sufficient convergence criteria, tighter error bounds on the distances $\|x_n - x^*\|$, $\|x_{n+1} - x_n\|$ and an at least as precise information on the location of the solution x^*. These advantages are obtained using less computational cost, since in practice the computation of function g requires the computation of functions f_0 and f as special cases. It is also worth noticing that under (h_0^0) function f_0' is defined and therefore R_0 which is at least as small as R. Therefore the majorant function to satisfy (i.e., f') is at least as small as the majorant function satisfying (h_0') (i.e., g') leading to the advantages *of the new approach over the approach in [17] or [23]. Indeed, we have that if function* ψ *has a solution* t^**, then, since* $\varphi(t^*) \leq \psi(t^*) = 0$ *and* $\varphi(0) = \psi(0) = \xi > 0$*, we get that function* φ *has a solution* r^* *such that*

$$r^* \leq t^*. \tag{20.25}$$

but not necessarily vice versa. If also follows from (20.25) that the new information about the location of the solution x^* *is at least as precise as the one given in [18].*

(b) Let us specialize conditions (20.8)–(20.10) even further in the case when L_0, K *and* L *are constant functions and* $\alpha = 1$*. Then, function corresponding to (20.16) reduce to*

$$\psi(t) = \frac{L}{2}t^2 - t + \xi \tag{20.26}$$

[17, 23] and

$$\varphi(t) = \frac{K}{2}t^2 - t + \xi, \tag{20.27}$$

respectively. In this case the convergence criteria become, respectively

$$h = L\xi \leq \frac{1}{2} \tag{20.28}$$

and

$$h_1 = K\xi \leq \frac{1}{2}. \tag{20.29}$$

Notice that

$$h \leq \frac{1}{2} \Longrightarrow h_1 \leq \frac{1}{2} \tag{20.30}$$

but not necessarily vice versa. Unless, if $K = L$*. Criterion (20.28) is the famous for its simplicity and clarity Kantorovich hypotheses for the semilocal convergence of Newton's method to a solution* x^* *of equation* $F(x) = 0$ *[20]. In the case of Wang's condition [29] we have*

$$\varphi(t) = \frac{\gamma t^2}{1 - \gamma t} - t + \xi,$$

$$\psi(t) = \frac{\beta t^2}{1 - \beta t} - t + \xi,$$

$$L(u) = \frac{2\gamma}{(1 - \gamma u)^3}, \ \gamma > 0, \ 0 \leq t \leq \frac{1}{\gamma}$$

and

$$K(u) = \frac{2\beta}{(1 - \beta u)^3}, \ \beta > 0, \ 0 \leq t \leq \frac{1}{\beta}$$

with convergence criteria, given respectively by

$$H = \gamma\xi \leq 3 - 2\sqrt{2} \tag{20.31}$$

$$H_1 = \beta\xi \leq 3 - 2\sqrt{2}. \tag{20.32}$$

Then, again we have that

$$H \leq 3 - 2\sqrt{2} \Longrightarrow H_1 \leq 3 - 2\sqrt{2}$$

but not necessarily vice versa, unless, if $\beta = \gamma$.

(c) *Concerning the error bounds and the limit of majorizing sequence, let us define majorizing sequence* $\{r_{\alpha,k}\}$ *by*

$$r_{\alpha,0} = 0; r_{\alpha,k+1} = r_{\alpha,k} - \frac{\varphi(r_{\alpha,k})}{\varphi'_{\alpha,0}(r_{\alpha,k})}$$

for each $k = 0, 1, 2, \ldots,$ *where*

$$\varphi_{\alpha,0}(t) = \xi - t + \alpha \int_0^t L_0(t)(t-u)du.$$

Suppose that

$$-\frac{\varphi(r)}{\varphi'_{\alpha,0}(r)} \leq -\frac{\varphi(s)}{\varphi'(s)}$$

for each $r, s \in [0, R_0]$ *with* $r \leq s$. *According to the proof of Theorem 20.3.1 sequence* $\{r_{\alpha,k}\}$ *is also a majorizing sequence for GNA.*

Moreover, a simple inductive argument shows that

$$r_k \leq s_k,$$

$$r_{k+1} - r_k \leq s_{k+1} - s_k$$

and

$$r^* = \lim_{k \longrightarrow \infty} r_k \leq s^*.$$

Furthermore, the first two preceding inequality are strict for $n \geq 2$, *if*

$$L_0(u) < K(u) \text{ for each } u \in [0, R_0].$$

Similarly, suppose that

$$-\frac{\varphi(s)}{\varphi'(s)} \leq -\frac{\psi(t)}{\psi'(t)} \tag{20.33}$$

for each $s, t \in [0, R_0]$ *with* $s \leq t$. *Then, we have that*

$$s_{\alpha,k} \leq t_{\alpha,k}$$

$$s_{\alpha,k+1} - s_{\alpha,k} \leq t_{\alpha,k+1} - t_{\alpha,k}$$

and

$$s^* \leq t^*.$$

The first two preceding inequalities are also strict for $k \geq 2$, if strict inequality holds in (20.33).

20.4 Numerical examples

Specializations of Theorem 20.3.3 to some interesting cases such as Smale's $\alpha-$theory (see also Wang's $\gamma-$theory) and Kantorovich theory have been reported in [17, 23, 25], if $f_0'(t) = f'(t) = g'(t)$ for each $t \in [0, R)$ with $R_0 = R$ and in [23], if $f_0'(t) < f'(t) = g'(t)$ for each $t \in [0, R_0)$. Next, we present examples where $f_0'(t) < f'(t) < g'(t)$ for each $t \in [0, R_0)$ to show the advantages of the new approach over the ones in [17, 24, 26]. We choose for simplicity $m = n = \alpha = 1$.

Example 20.4.1 *Let $x_0 = 1, D = U(1, 1-q)$, $q \in [0, \frac{1}{2})$ and define function F on D by*

$$F(x) = x^3 - q. \tag{20.34}$$

Then, we have that $\xi = \frac{1}{3}(1-q), L_0 = 3-q, L = 2(2-q)$ and $K = 2(1+\frac{1}{L_0})$. The Newton-Kantorovich condition (20.28) is not satisfied, since

$$h > \frac{1}{2} \text{ for each } q \in [0, \frac{1}{2}). \tag{20.35}$$

Hence, there is no guarantee by the Newton-Kantorovich Theorem [17] that Newton's method (20.1) converges to a zero of operator F. Let us see what gives: We have by (20.29) that

$$h_1 \leq \frac{1}{2}, \text{ if } 0.461983163 < q < \frac{1}{2} \tag{20.36}$$

Hence, we have demonstrated the improvements using this example.

References

[1] S. Amat, S. Busquier and J. M. Gutiérrez, Geometric constructions of iterative functions to solve nonlinear equations, *J. Comput. Appl. Math.*, 157, 197–205, 2003.

[2] I. K. Argyros, Computational theory of iterative methods, Studies in Computational Mathematics, 15, Editors: K. Chui and L. Wuytack. Elsevier, New York, U.S.A., 2007.

[3] I. K. Argyros, Concerning the convergence of Newton's method and quadratic majorants, *J. Appl. Math. Comput.*, 29, 391–400, 2009.

[4] I. K. Argyros and S. Hilout, On the Gauss-Newton method, *J. Appl. Math.*, 1–14, 2010.

[5] I. K. Argyros and S. Hilout, Extending the applicability of the Gauss-Newton method under average Lipschitz-conditions, *Numer. Algorithms*, 58, 23–52, 2011.

[6] I. K. Argyros, A semilocal convergence analysis for directional Newton methods, *Math. Comput., AMS*, 80, 327–343, 2011.

[7] I. K. Argyros, Y. J. Cho and S. Hilout, Numerical methods for equations and its applications, CRC Press/Taylor and Francis Group, New York, 2012.

[8] I. K. Argyros and S. Hilout, Improved local convergence of Newton's method under weaker majorant condition, *J. Comput. Appl. Math.*, 236 (7), 1892–1902, 2012.

[9] I. K. Argyros and S. Hilout, Weaker conditions for the convergence of Newton's method, *J. Complexity*, 28, 364–387, 2012.

[10] I. K. Argyros and S. Hilout, Computational Methods in Nonlinear Analysis, World Scientific Publ. Comp., New Jersey, 2013.

[11] A. Ben-Israel and T. N. E. Greville, Generalized inverses. CMS Books in Mathematics/Ouvrages de Mathematiques de la SMC, 15. Springer-Verlag, New York, second edition, Theory and Applications, 2003.

[12] J. V. Burke and M. C. Ferris, A Gauss-Newton method for convex composite optimization, *Math. Programming Ser A.*, 71, 179–194, 1995.

[13] J. E. Jr. Dennis and R. B. Schnabel, Numerical methods for unconstrained optimization and nonlinear equations, Classics in Appl. Math., 16, SIAM, Philadelphia, PA, 1996.

[14] O. P. Ferreira, M. L. N. Gonçalves and P. R. Oliveira, Local convergence analysis of the Gauss–Newton method under a majorant condition, *J. Complexity*, 27 (1), 111–125, 2011.

[15] O. P. Ferreira and B. F. Svaiter, Kantorovich's theorem on Newton's method in Riemmanian manifolds, *J. Complexity*, 18 (1), 304–329, 2002.

[16] O. P. Ferreira and B. F. Svaiter, Kantorovich's majorants principle for Newton's method, *Comput. Optim. Appl.*, 42, 213–229, 2009.

[17] O. P. Ferreira, M. L. N. Gonçalves and P. R. Oliveira, Convergence of the Gauss-Newton method for convex composite optimization under a majorant condition, *SIAM J. Optim.*, 23 (3), 1757–1783, 2013.

[18] W. M. Häussler, A Kantorovich-type convergence analysis for the Gauss-Newton method. *Numer. Math.*, 48, 119–125 , 1986.

[19] J. B. Hiriart-Urruty and C. Lemaréchal, Convex analysis and minimization algorithms (two volumes). I. Fundamentals. II. Advanced theory and bundle methods, 305 and 306, Springer–Verlag, Berlin, 1993.

[20] L. V. Kantorovich and G. P. Akilov, Functional Analysis, Pergamon Press, Oxford, 1982.

[21] C. Li and K. F. Ng, Majorizing functions and convergence of the Gauss-Newton method for convex composite optimization, SIAM J. Optim., 18, 2, 613–692 (2007).

[22] C. Li and X. H. Wang, On convergence of the Gauss-Newton method for convex composite optimization, *Math. Program. Ser A.*, 91, 349–356, 2002.

[23] A. Magrenán and I. K. Argyros, Expanding the applicability of the Gauss-Newton method for convex optimization under a majorant condition, *SeMA*, 65, 37–56, 2014.

[24] F. A. Potra, Sharp error bounds for a class of Newton-like methods, *Libertas Mathematica*, 5, 71–84, 1985.

[25] S. M. Robinson, Extension of Newton's method to nonlinear functions with values in a cone, *Numer. Math.*, 19, 341–347, 1972.

[26] R. T. Rockafellar, Monotone proccesses of convex and concave type, Memoirs of the American Mathematical Society, 77. American Mathematical Society, Providence, R.I., 1967.

[27] R. T. Rockafellar, Convex analysis, Princeton Mathematical Series, 28, Princeton University Press, Princeton, N.J., 1970.

[28] S. Smale, Newton's method estimates from data at one point. The merging of disciplines: new directions in pure, applied, and computational mathematics (Laramie, Wyo., 1985), 185–196, Springer, New York, 1986.

[29] X. H. Wang, Convergence of Newton's method and uniqueness of the solution of equations in Banach space, *IMA J. Numer. Anal.*, 20, 123–134, 2000.

Chapter 21

Ball Convergence for eighth order method

Ioannis K. Argyros

Department of Mathematical Sciences, Cameron University, Lawton, OK 73505, USA, Email: iargyros@cameron.edu

Á. Alberto Magreñán

Universidad Internacional de La Rioja, Escuela Superior de Ingeniería y Tecnología, 26002 Logroño, La Rioja, Spain, Email: alberto.magrenan@unir.net

CONTENTS

21.1 Introduction

Consider the problem of approximating a locally unique solution x^* of the nonlinear equation

$$F(x) = 0, \tag{21.1}$$

where F is a Fréchet-differentiable operator defined on a convex subset D of a Banach space X with values in a Banach space Y. The equation (21.1) covers wide range of problems in classical analysis and applications [1–30]. Closed form solutions of these nonlinear equations exist only for few special cases which may not be of much practical value. Therefore solutions of these nonlinear equations (21.1) are approximated by iterative methods.

Newton-like methods are undoubtedly the most popular method for approximating a locally unique solution x^* provided that the initial point is close enough to the solution [1–30]. The number of function evaluations per step increases with the order of convergence. In the scalar case the efficiency index [3, 6, 21] $EI = p^{\frac{1}{m}}$ provides a measure of balance where p is the order of the method and m is the number of function evaluations.

It is well known that according to the Kung-Traub conjuncture the convergence of any multi-point method without memory cannot exceed the upper bound 2^{m-1} [21] (called the optimal order). Hence the optimal order for a method with three function evaluations per step is 4. The corresponding efficiency index is $EI = 4^{\frac{1}{3}} = 1.58740...$ which is better than Newtons method which is $EI = 2^{\frac{1}{2}} = 1.414....$ Therefore, the study of new optimal methods of order four is important.

In the present paper, we consider the following method:

$$\begin{aligned} y_n &= x_n - \frac{1}{2}F'(x_n)^{-1}F(x_n) \\ z_n &= \frac{1}{3}(4y_n - x_n) \\ u_n &= y_n + (F'(x_n) - 3F'(z_n))^{-1}F(x_n) \\ v_n &= u_n + 2(F'(x_n) - 3F'(z_n))^{-1}F(u_n) \\ x_{n+1} &= v_n + 2(F'(x_n) - 3F'(z_n))^{-1}F(v_n). \end{aligned} \tag{21.2}$$

Using hypotheses up to the first derivative of function F and contractions on a Banach space setting we proved the convergence of the method. Moreover we avoid Taylor expansions and use instead Lipschitz parameters. We do not have to use higher order derivatives to show the convergence of method (21.2). This way we expand the applicability of method (21.2).

The rest of the paper is organized as follows. In Section 21.2 we present the local convergence analysis. We also provide a radius of convergence, computable error bounds and uniqueness result not given in the earlier studies using Taylor expansions [13, 18, 20]. Special cases and numerical examples are presented in the concluding Section 21.3.

21.2 Local convergence analysis

The local convergence analysis of method (21.2) that follows is based on some scalar functions and parameters. Let v, w_0, w be continuous and non-decreasing functions defined on $[0,+\infty)$ with values in $[0,+\infty)$ satisfying $w_0(0) = w(0) = 0$. Let r_0 be the smallest positive solution of the equation

$$w_0(t) = 1. \tag{21.3}$$

Define functions g_1, g_2, p, h_p, on the interval $[0,r_0)$ by,

$$g_1(t) = \frac{\int_0^1 w((1-\theta)t)d\theta + \frac{1}{2}\int_0^1 v(\theta t)d\theta}{1-w_0(t)},$$

$$g_2(t) = \frac{1}{2}(1+4g_1(t)),$$

$$p(t) = \frac{1}{2}(w_0(t)+3w_0(g_2(t)t).$$

$$h_p(t) = p(t) - 1.$$

We have that $h_p(0) = -1 < 0$ and $h_p(t) \to +\infty$ as $t \to r_0^-$. It follows again from the intermediate value theorem that function h_p has zeros in the interval $(0, r_0)$. Denote by r_p the smallest such zero. Moreover define functions g_i, h_i, $i = 3,4,5$ on the interval $[0, r_0)$ by

$$g_3(t) = \frac{\int_0^1 w((1-\theta)t)d\theta}{1-w_0(t)} + \frac{3}{4}\frac{[w_0(t)+w_0(g_2(t)t)]\int_0^1 v(\theta t)d\theta}{(1-w_0(t))(1-p(t))},$$

$$g_4(t) = (1+\frac{2}{1-p(t)})g_3(t),$$

$$g_5(t) = (1+\frac{2}{1-p(t)})g_4(t)$$

and

$$h_i(t) = g_i(t) - 1.$$

Then, we get that $h_i(0) = -1 < 0$ and $h_i(t) \to +\infty$ as $t \to r_p^-$. Denote by r_i the smallest zeros of functions h_i on the interval $(0,r_p)$. Furthermore, define the radius of convergence r by

$$r = \min\{r_i\}. \tag{21.4}$$

Finally define functions r_1 and r_2 by $r_1 = g_1(r)r$ and $r_2 = g_2(r)r$. Then, we have that for each $t \in [0, r)$

$$0 \leq g_i(t) < 1, \quad i = 3,4,5. \tag{21.5}$$

Let $U(z,\rho)$, $\overline{U}(z,\rho)$ stand for the open and closed balls in X, respectively with center $z \in X$ and of radius $\rho > 0$. Next, we present the local convergence analysis of method (21.2) using the preceding notation.

Theorem 21.2.1 *Let $F : D \subset X \to Y$ be a Fréchet differentiable operator. Suppose:*

(a) *There exists $x^* \in D$, such that $F(x^*) = 0$, $F'(x^*)^{-1} \in L(Y,X)$;*

(b) There exists function $w_0 : [0,+\infty) \to [0,+\infty)$ *continuous, non-decreasing with* $w_0(0) = 0$ *such that for each* $x \in D$,

$$\|F'(x^*)^{-1}(F'(x) - F'(x^*)\| \leq w_0(\|x - x^*\|).$$

(c) Let $R = \max\{r_1, r_2, r\}$. *For* r_0 *defined by (21.3), set* $D_0 = D \cap U(x^*, \bar{r}_0)$ *where* $\bar{r}_0 = \max\{r_1, r_2, r_0\}$. *There exists functions* $w, v : [0,+\infty) \to [0,+\infty)$ *continuous, non-decreasing with* $w(0) = 0$ *such that for each* $x, y \in D_0$,

$$\|F'(x^*)^{-1}(F'(x) - F'(y)\| \leq w(\|x - y\|)$$

and

$$\|F'(x^*)^{-1}F'(x)\| \leq v(\|x - x^*\|).$$

(d) $U(x^*, R) \subseteq D$.

Then, the sequence $\{x_n\}$ *generated for* $x_0 \in U(x^*, r) - \{x^*\}$ *is well defined for each* $n = 0, 1, 2, \ldots$ *and converges to* x^*. *Moreover, the following estimates hold for each* $n = 0, 1, \cdots$

$$\|y_n - x^*\| \leq g_1(\|x_n - x^*\|)\|x_n - x^*\| \leq g_1(r)r = r_1, \tag{21.6}$$

$$\|z_n - x^*\| \leq g_2(\|x_n - x^*\|)\|x_n - x^*\| \leq g_2(r)r = r_2, \tag{21.7}$$

$$\|u_n - x^*\| \leq g_3(\|x_n - x^*\|)\|x_n - x^*\| \leq \|x_n - x^*\| < r \tag{21.8}$$

$$\|v_n - x^*\| \leq g_4(\|x_n - x^*\|)\|x_n - x^*\| \leq \|x_n - x^*\| \tag{21.9}$$

and

$$\|x_{n+1} - x^*\| \leq g_5(\|x_n - x^*\|)\|x_n - x^*\| \leq \|x_n - x^*\| \tag{21.10}$$

where the g_i, $i = 1, 2, 3, 4, 5$ *functions are defined above the Theorem. Furthermore, if there exists* $r^* > r$ *such that*

$$\int_0^1 w_0(\theta r^*)d\theta < 1 \tag{21.11}$$

then the point x^* *is the only solution of equation* $F(x) = 0$ *in* $D_1 = D \cap U(x^*, r^*)$.

Proof: We shall show using mathematical induction that sequence $\{x_n\}$ is well defined, converges to x^* and estimates (21.6)-(21.10) are satisfied. Let $x \in U(x^*, r)$. Using (21.3), (21.4) and (b) we have in turn that

$$\|F'(x^*)^{-1}(F'(x) - F'(x^*)\| \leq w_0(\|x - x^*\|) \leq w_0(r) < 1. \tag{21.12}$$

It follows from (21.12) and the Banach Lemma on invertible operators [4,5,28] that $F'(x)^{-1} \in L(Y,X)$ and

$$\|F'(x)^{-1}F'(x^*)\| \leq \frac{1}{1-w_0(\|x-x^*\|)}. \tag{21.13}$$

In particular (21.13) holds for $x = x_0$, since by hypotheses $x_0 \in U(x^*,r) - \{x^*\}$, y_0, z_0 are well defined by the first and second substeps of method (21.2) for $n=0$. We can write by (a),

$$y_0 - x^* = x_0 - x^* - F'(x_0)^{-1}F(x_0) + \frac{1}{2}F'(x_0)^{-1}F(x_0) \tag{21.14}$$

$$F(x_0) = F(x_0) - F(x^*) = \int_0^1 F'(x^* + \theta(x_0 - x^*))(x_0 - x^*)d\theta. \tag{21.15}$$

Notice that $\|x^* + \theta(x_0 - x^*) - x^*\| = \theta\|x_0 - x^*\| < r$, for each $\theta \in [0,1]$, so $x^* + \theta(x_0 - x^*) \in U(x^*,r)$.

Then, using (21.4), (21.5) (for i = 3), (c), (21.13), (21.14) and (21.15), we get in turn that

$$\begin{aligned} \|y_0 - x^*\| &\leq \|F'(x_0)^{-1}F'(x^*)\| \| \textstyle\int_0^1 F'(x^*)^{-1}[F'(x^* + \theta(x_0 - x^*)) \\ &\quad -F'(x_0)]\|d\theta\|x_0 - x^*\| + \tfrac{1}{2}\|F'(x_0)^{-1}F(x^*)\|\|F'(x^*)^{-1}F(x_0)\| \\ &\leq \frac{\int_0^1 w((1-\theta)\|x_0-x^*\|)d\theta + \frac{1}{2}\int_0^1 v(\theta\|x_0-x^*\|)d\theta}{1-w_0(\|x_0-x^*\|)}, \\ &\leq g_1(r)r = r_1, \end{aligned} \tag{21.16}$$

which shows (21.6) for $n = 0$ and $y_0 \in U(x^*,r_1)$.

Then, from the second sub-step of method (21.2) and (21.16) we also get that

$$\begin{aligned} \|z_0 - x^*\| &= \tfrac{1}{3}\|4(y_0 - x^*) + (x_0 - x^*)\| \\ &\leq \tfrac{1}{3}[4\|(y_0 - x^*) + (x_0 - x^*)\|] \\ &\leq \tfrac{1}{3}[4g_1\|(x_0 - x^*) + 1]\|(x_0 - x^*)\| \\ &\leq g_2(r)r = r_2, \end{aligned} \tag{21.17}$$

which shows (21.17) for $n = 0$ and $z_0 \in U(x^*,r_2)$. Next, we must show that $A_0 = (F'(x_0) - 3F'(z_0))^{-1} \in L(Y,X)$. Using (b), (21.4) and (21.17), we obtain in turn that

$$\|(-2F'(x^*))^{-1}[A_0 - F'(x^*)]\|$$

$$\le \tfrac{1}{2}[\|F'(x^*)^{-1}[F'(x_0) - F'(x^*)]\| + 3\|F'(x^*)^{-1}(F'(z_0) - F'(x^*))\|]$$

$$\le \tfrac{1}{2}[w_0(\|x^* - x_0\|) + 3w_0(\|z_0 - x^*\|)$$

$$\le \tfrac{1}{2}[w_0(\|x^* - x_0\|) + 3w_0(g_2(\|x^* - x_0\|)\|x^* - x_0\|)] = p(\|x^* - x_0\|) \le p(r) < 1, \tag{21.18}$$

so,

$$\|A_0^{-1}F'(x^*)\| \le \frac{1}{2(1 - p(\|x_0 - x^*\|))}. \tag{21.19}$$

Hence, u_0, v_0 and x_1 are well defined. Then, using (21.4), (21.5), (21.13), (21.16), (21.17), (c) and (21.19), we obtain in turn that

$$\begin{aligned} \|u_0 - x^*\| &= \|(x_0 - x^* - F'(x_0)^{-1}F(x_0)) + \tfrac{3}{2}[(F'(x_0) - F'(x^*)) \\ &\quad +(F'(x^*) - F'(z_0)]A_0^{-1}F(x_0)\| \\ &\le \frac{\int_0^1 w((1-\theta)(\|x_0-x^*\|))d\theta\|x_0-x^*\|}{1-w_0(\|x_0-x^*\|)} \\ &\quad +\frac{3}{4}\frac{[w_0(\|x_0-x^*\|)+w_0(g_2(\|x_0-x^*\|)\|x_0-x^*\|)\int_0^1 v(\theta\|x_0-x^*\|)d\theta]\|x_0-x^*\|}{(1-w_0(\|x_0-x^*\|))(1-p(\|x_0-x^*\|))} \\ &= g_3(\|x_0 - x^*\|)\|x_0 - x^*\| \le \|x_0 - x^*\| < r, \end{aligned} \tag{21.20}$$

$$\begin{aligned} \|v_0 - x^*\| &\le \|u_0 - x^*\| + 2\frac{\int_0^1 v(\theta\|x_0-x^*\|)d\theta\|x_0-x^*\|}{1-p(\|x_0-x^*\|)} \\ &\le (1 + \frac{2}{1-p(\|x_0-x^*\|)})g_3(\|x_0 - x^*\|)\|x_0 - x^*\| \qquad (21.21) \\ &= g_4(\|x_0 - x^*\|)\|x_0 - x^*\| \le \|x_0 - x^*\| < r \end{aligned}$$

and

$$\begin{aligned} \|x_1 - x^*\| &\le \|v_0 - x^*\| + \frac{2\|v_0-x^*\|}{1-p(\|x_0-x^*\|)} \\ &\le (1 + \frac{2}{1-p(\|x_0-x^*\|)})g_4(\|x_0 - x^*\|)\|x_0 - x^*\| \qquad (21.22) \\ &= g_5(\|x_0 - x^*\|)\|x_0 - x^*\| \le \|x_0 - x^*\| < r, \end{aligned}$$

which show (21.8)-(21.10) for $n = 0$ and $u_0, v_0, x_1 \in U(x^*, r)$. By simply replacing $x_0, y_0, z_0, u_0, v_0, x_1$ by $x_k, y_k, z_k, u_k, v_k, x_{k+1}$ in the preceding estimates we arrive at (21.6)-(21.10). Then, in view of the estimate

$$\|x_{k+1} - x^*\| \leq c\|x_k - x^*\| < r, \tag{21.23}$$

where $c = g_5(\|x_0 - x^*\|) \in [0, 1)$, we conclude that $\lim_{k\to\infty} x_k = x^*$ and $x_{k+1} \in U(x^*, r)$. Let $y^* \in D_1$ with $F(y^*) = 0$ and set $Q = \int_0^1 F'(x^* + \theta(y^* - x^*))d\theta$. Using (b) and (21.11), we get that

$$\begin{aligned} \|F'(x^*)^{-1}(Q - F'(x^*))\| &\leq \int_0^1 w_0(\theta\|x^* - y^*\|)d\theta \\ &\leq \int_0^1 w_0(\theta r^*)d\theta < 1, \end{aligned} \tag{21.24}$$

so, $Q^{-1} \in L(Y, X)$. Then, from the identity $0 = F(y^*) - F(x^*) = Q(y^* - x^*)$, we deduce that $x^* = y^*$. □

Remark 21.2.2 *(a) The radius r_1 was obtained by Argyros in [4] as the convergence radius for Newton's method under condition (21.9)-(21.11). Notice that the convergence radius for Newton's method given independently by Rheinboldt [19] and Traub [21] is given by*

$$\rho = \frac{2}{3L} < r_1.$$

As an example, let us consider the function $f(x) = e^x - 1$. Then $x^* = 0$. Set $\Omega = U(0, 1)$. Then, we have that $L_0 = e - 1 < l = e$, so $\rho = 0.24252961 < r_1 = 0.324947231$.

Moreover, the new error bounds [2, 14] are:

$$\|x_{n+1} - x^*\| \leq \frac{L}{1 - L_0\|x_n - x^*\|}\|x_n - x^*\|^2,$$

whereas the old ones [26, 28]

$$\|x_{n+1} - x^*\| \leq \frac{L}{1 - L\|x_n - x^*\|}\|x_n - x^*\|^2.$$

Clearly, the new error bounds are more precise, if $L_0 < L$. Clearly, we do not expect the radius of convergence of method (21.2) given by r to be larger than r_1 (see (21.5)).

(b) The local results can be used for projection methods such as Arnoldi's method, the generalized minimum residual method (GMREM), the generalized conjugate method (GCM) for combined Newton/finite projection methods and in connection to the mesh independence principle in order to develop the cheapest and most efficient mesh refinement strategy [4–7].

(c) *The results can be also be used to solve equations where the operator F' satisfies the autonomous differential equation [5, 7, 15, 17]:*

$$F'(x) = P(F(x)),$$

where $P: Y \to Y$ is a known continuous operator. Since $F'(x^*) = P(F(x^*)) = P(0)$, we can apply the results without actually knowing the solution x^*. Let as an example $F(x) = e^x - 1$. Then, we can choose $P(x) = x+1$ and $x^* = 0$.

(d) *It is worth noticing that method (21.2) are not changing if we use the new instead of the old conditions [23]. Moreover, for the error bounds in practice we can use the computational order of convergence (COC)*

$$\xi = \frac{ln\frac{\|x_{n+2}-x_{n+1}\|}{\|x_{n+1}-x_n\|}}{ln\frac{\|x_{n+1}-x_n\|}{\|x_n-x_{n-1}\|}}, \qquad \text{for each } n = 1, 2, \ldots$$

or the approximate computational order of convergence (ACOC)

$$\xi^* = \frac{ln\frac{\|x_{n+2}-x^*\|}{\|x_{n+1}-x^*\|}}{ln\frac{\|x_{n+1}-x^*\|}{\|x_n-x^*\|}}, \qquad \text{for each } \ n = 0, 1, 2, \ldots$$

instead of the error bounds obtained in Theorem 21.2.1.

(e) *In view of (21.10) and the estimate*

$$\begin{aligned} \|F'(x^*)^{-1}F'(x)\| &= \|F'(x^*)^{-1}(F'(x) - F'(x^*)) + I\| \\ &\leq 1 + \|F'(x^*)^{-1}(F'(x) - F'(x^*))\| \leq 1 + L_0\|x - x^*\| \end{aligned}$$

condition (21.12) can be dropped and M can be replaced by

$$M(t) = 1 + L_0 t$$

or

$$M(t) = M = 2,$$

since $t \in [0, \frac{1}{L_0})$.

21.3 Numerical examples

The numerical examples are presented in this section.

Example 21.3.1 *Let $X = Y = \mathbb{R}^3, D = \bar{U}(0,1), x^* = (0,0,0)^T$. Define function F on D for $w = (x,y,z)^T$ by*

$$F(w) = (e^x - 1, \frac{e-1}{2}y^2 + y, z)^T.$$

Then, the Fréchet-derivative is given by

$$F'(v) = \begin{bmatrix} e^x & 0 & 0 \\ 0 & (e-1)y+1 & 0 \\ 0 & 0 & 1 \end{bmatrix}.$$

Notice that using the (21.11) conditions, we get $w_0(t) = L_0 t, w(t) = Lt, v(t) = L$, $L_0 = e-1, L = e^{\frac{1}{L_0}}$. *The parameters are*

$r_p = 0.1165$, $r_3 = 0.0577$, $r_4 = 0.0261$, $r_5 = 0.0107 = r$, $r_1 = 0.0098$, $r_2 = 0.0250$.

Example 21.3.2 *Let* $X = Y = C[0,1]$, *the space of continuous functions defined on* $[0,1]$ *and be equipped with the max norm. Let* $D = \overline{U}(0,1)$. *Define function F on D by*

$$F(\varphi)(x) = \varphi(x) - 5\int_0^1 x\theta\varphi(\theta)^3 d\theta. \tag{21.25}$$

We have that

$$F'(\varphi(\xi))(x) = \xi(x) - 15\int_0^1 x\theta\varphi(\theta)^2\xi(\theta)d\theta, \text{ for each } \xi \in D.$$

Then, we get that $x^* = 0$, $w_0(t) = L_0 t, w(t) = Lt, v(t) = 2$, $L_0 = 7.5$, $L = 15$. *The parameters for method are*

$r_p = 0.0054$, $r_3 = 0.0022$, $r_4 = 0.0229$, $r_5 = 0.0115 = r$, $r_1 = 0.0949$, $r_2 = 0.02889$.

References

[1] M. F. Abad, A. Cordero and J. R. Torregrosa, Fourth and Fifth-order methods for solving nonlinear systems of equations: An application to the global positioning system, *Abstract and Applied Analysis* Volume 2013, Article ID 586708, 10 pages.

[2] S. Amat, M. A. Hernández and N. Romero, Semilocal convergence of a sixth order iterative method for quadratic equations, *Appl. Numer. Math.*, 62, 833–841, 2012.

[3] I. K. Argyros, Computational theory of iterative methods. Series: Studies in Computational Mathematics, 15, Editors: C. K. Chui and L. Wuytack, Elsevier Publ. Co. New York, U.S.A., 2007.

[4] I. K. Argyros, A semilocal convergence analysis for directional Newton methods, *Math. Comput.*, 80, 327–343, 2011.

[5] I. K. Argyros and S. Hilout, Weaker conditions for the convergence of Newton's method, *J. Complexity*, 28, 364–387, 2012.

[6] I. K. Argyros and Said Hilout, Computational methods in nonlinear analysis. Efficient algorithms, fixed point theory and applications, World Scientific, 2013.

[7] I. K. Aryros and H. Ren, Improved local analysis for certain class of iterative methods with cubic convergence, *Numer. Algorithms*, 59, 505–521, 2012.

[8] V. Candela and A. Marquina, Recurrence relations for rational cubic methods I: The Halley method, *Computing*, 44, 169–184, 1990.

[9] V. Candela and A. Marquina, Recurrence relations for rational cubic methods II: The Chebyshev method, *Computing*, 45 (4), 355–367, 1990.

[10] C. Chun, P. Stănică and B. Neta, Third-order family of methods in Banach spaces, *Comput. Appl. Math. with Appl.*, 61, 1665–1675, 2011.

[11] A. Cordero, F. Martinez and J. R. Torregrosa, Iterative methods of order four and five for systems of nonlinear equations, *J. Comput. Appl. Math.*, 231, 541–551, 2009.

[12] A. Cordero, J. Hueso, E. Martinez and J. R. Torregrosa, A modified Newton-Jarratt's composition, *Numer. Algorithms*, 55, 87–99, 2010.

[13] A. Cordero, J. R. Torregrosa and M. P. Vasileva, Increasing the order of convergence of iterative schemes for solving nonlinear systems, *J. Comput. Appl. Math.*, 252, 86–94, 2013.

[14] R. Ezzati and E. Azandegan, A simple iterative method with fifth order convergence by using Potra and Ptak's method, *Math. Sci.*, 3, 2, 191–200, 2009.

[15] J. M. Gutiérrez, Á. A. Magreñán and N. Romero, On the semi-local convergence of Newton-Kantorovich method under center-Lipschitz conditions, *Appl. Math. Comput.*, 221, 79–88, 2013.

[16] V. I. Hasanov, I. G. Ivanov and F. Nebzhibov, A new modification of Newton's method, *Appl. Math. Eng.*, 27, 278–286, 2002.

[17] M. A. Hernández and M. A. Salanova, Modification of the Kantorovich assumptions for semi-local convergence of the Chebyshev method, *J. Comput. Appl. Math.*, 126, 131–143, 2000.

[18] J. P. Jaiswal, Semilocal convergence of an eighth-order method in Banach spaces and its computational efficiency, *Numer. Algorithms*, 71, 933–951, 2016.

[19] L. V. Kantorovich and G. P. Akilov, Functional Analysis, Pergamon Press, Oxford, 1982.

[20] J. S. Kou, Y. T. Li and X. H. Wang, A modification of Newton method with third-order convergence, *Appl. Math. Comput.*, 181, 1106–1111, 2006.

[21] Á. A. Magreñán, Different anomalies in a Jarratt family of iterative root finding methods, *Appl. Math. Comput.*, 233, 29–38, 2014.

[22] Á. A. Magreñán, A new tool to study real dynamics: The convergence plane, *Appl. Math. Comput.*, 248, 29–38, 2014.

[23] M. S. Petkovic, B. Neta, L. Petkovic and J. Džunič, Multipoint methods for solving nonlinear equations, Elsevier, 2013.

[24] F. A. Potra and V. Pták, Nondiscrete Induction and Iterative Processes, in: Research Notes in Mathematics, Vol. 103, Pitman, Boston, 1984.

[25] H. Ren and Q. Wu, Convergence ball and error analysis of a family of iterative methods with cubic convergence, *Appl. Math. Comput.*, 209, 369–378, 2009.

[26] W. C. Rheinboldt, An adaptive continuation process for solving systems of nonlinear equations, In: Mathematical models and numerical methods (A. N. Tikhonov et al. eds.) pub. 3, 129–142 Banach Center, Warsaw Poland, 1977.

[27] J. R. Sharma, P. K. Guha and R. Sharma, An efficient fourth order weighted-Newton method for systems of nonlinear equations, *Numer. Algorithms*, 62 (2), 307–323, 2013.

[28] J. F. Traub, Iterative methods for the solution of equations, AMS Chelsea Publishing, 1982.

[29] S. Weerakoon and T. G. I. Fernando, A variant of Newton's method with accelerated third-order convergence, *Appl. Math. Let.*, 13, 87–93, 2000.

[30] T. Zhanlar, O. Chulunbaatar and G. Ankhbayar, On Newton-type methods with fourth and fifth-order convergence, *Bulletin of PFUR Series Mathematics. Information Sciences, Physics*, 2 (2), 29–34, 2010.

Chapter 22

Expanding Kantorovich's theorem for solving generalized equations

Ioannis K. Argyros

Department of Mathematical Sciences, Cameron University, Lawton, OK 73505, USA, Email: iargyros@cameron.edu

Á. Alberto Magreñán

Universidad Internacional de La Rioja, Escuela Superior de Ingeniería y Tecnología, 26002 Logroño, La Rioja, Spain, Email: alberto.magrenan@unir.net

CONTENTS

22.1 Introduction

In [18], G. S. Silva considered the problem of approximating the solution of the generalized equation

$$F(x)+Q(x) \ni 0, \tag{22.1}$$

where $F : D \longrightarrow H$ is a Fréchet differentiable function, H is a Hilbert space with inner product $\langle .,. \rangle$ and corresponding norm $\|.\|$, $D \subseteq H$ an open set and T :

$H \rightrightarrows H$ is set-valued and maximal monotone. It is well known that the system of nonlinear equations and abstract inequality system can be modeled as equation of the form (22.1) [17]. If $\psi : H \longrightarrow (-\infty, +\infty]$ is a proper lower semi continuous convex function and

$$Q(x) = \partial \psi(x) = \{u \in H : \psi(y) \geq \psi(x) + \langle u, y - x \rangle\}, \text{ for all } y \in H \tag{22.2}$$

then (22.1) becomes the variational inequality problem

$$F(x) + \partial \psi(x) \ni 0,$$

including linear and nonlinear complementary problems. Newton's method for solving (22.1) for an initial guess x_0 is defined by

$$F(x_k) + F'(x_k)(x_{k+1} - x_k) + Q(x_{k+1}) \ni 0, \ k = 0, 1, 2 \ldots \tag{22.3}$$

has been studied by several authors [1]– [24]. In [13], Kantorovich obtained a convergence result for Newton's method for solving the equation $F(x) = 0$ under some assumptions on the derivative $F'(x_0)$ and $\|F'(x_0)^{-1}F(x_0)\|$. Kantorovich, used the majorization principle to prove his results. Later in [16], Robinson considered generalization of the Kantorovich theorem of the type $F(x) \in K$, where K is a nonempty closed and convex cone, and obtained convergence results and error bounds for this method. Josephy [12], considered a semilocal Newton's method of the kind (22.3) in order to solving (22.1) with $F = N_C$ the normal cone mapping of a convex set $C \subset \mathbb{R}^2$.

The rest of this paper is organized as follows. Preliminaries are given in Section 22.2 and the main results are presented in the concluding Section 22.3.

22.2 Preliminaries

Let $U(x, \rho)$ and $\bar{U}(x, \rho)$ stand respectively for open and closed balls in H with center $x \in H$ and radius $\rho > 0$. The following Definitions and Lemmas are used for proving our results. These items are stated briefly here in order to make the paper as self contains as possible. More details can be found in [18].

Definition 22.2.1 *Let $D \subseteq H$ be an open nonempty subset of H, $h : D \longrightarrow H$ be a Fréchet differentiable function with Fréchet derivative h' and $Q : H \rightrightarrows H$ be a set mapping. The partial linearization of the mapping $h + Q$ at $x \in H$ is the set-valued mapping $L_h(x, .) : H \rightrightarrows H$ given by*

$$L_h(x, y) := h(x) + h'(x)(y - x) + Q(y). \tag{22.4}$$

For each $x \in H$, the inverse $L_h(x, .)^{-1} : H \rightrightarrows H$ of the mapping $L_h(x, .)$ at $z \in H$ is defined by

$$L_h(x, .)^{-1} := \{y \in H : z \in h(x) + h'(x)(y - x) + Q(y)\}. \tag{22.5}$$

Definition 22.2.2 *Let $Q : H \rightrightarrows H$ be a set-valued operator. Q is said to be monotone if for any $x, y \in domQ$ and $u \in Q(y), v \in Q(x)$ implies that the following inequality holds:*

$$\langle u - v, y - x \rangle \geq 0.$$

A subset of $H \times H$ is monotone if it is the graph of a monotone operator. If $\varphi : H \longrightarrow (-\infty, +\infty]$ is a proper function then the subgradient of φ is monotone.

Definition 22.2.3 *Let $Q : H \rightrightarrows H$ be monotone. Then Q is maximal monotone if the following implication holds for all $x, u \in H$:*

$$\langle u - v, y - x \rangle \geq 0 \text{ for each } y \in domQ \text{ and } v \in Q(y) \Rightarrow x \in domQ \text{ and } v \in Q(y). \tag{22.6}$$

Lemma 22.2.4 *([22]) Let G be a positive operator (i.e., $\langle G(x), x \rangle \geq 0$). The following statements about G hold:*

- $\|G^2\| = \|G\|^2$;
- *If G^{-1} exists, then G^{-1} is a positive operator.*

Lemma 22.2.5 *([21, Lemma 2.2]) Let G be a positive operator. Suppose G^{-1} exists, then for each $x \in H$ we have*

$$\langle G(x), x \rangle \geq \frac{\|x\|^2}{\|G^{-1}\|}.$$

22.3 Semilocal convergence

We present the semilocal convergence analysis of generalized Newton's method using some more flexible scalar majorant functions than in [18].

Theorem 22.3.1 *Let $F : D \subseteq H \longrightarrow H$ be continuous with Fréchet derivative F' continuous on D. Let also $Q : H \rightrightarrows H$ be a set-valued operator. Suppose that there exists $x_0 \in D$ such that $F'(x_0)$ is a positive operator and $\hat{F}'(x_0)^{-1}$ exists. Let $R > 0$ and $\rho := \sup\{t \in [0, R) : U(x_0, t) \subseteq D\}$. Suppose that there exists $f_0 : [0, R) \longrightarrow \mathbb{R}$ twice continuously differentiable such that for each $x \in U(x_0, \rho)$*

$$\|\hat{F}'(x_0)^{-1}\| \|F'(x) - F'(x_0)\| \leq f_0'(\|x - x_0\|) - f_0'(0). \tag{22.7}$$

Moreover, suppose that

(h_1^0) $f_0(0) > 0$ *and* $f_0'(0) = -1$.

(h_2^0) f_0' *is convex and strictly increasing.*

(h_3^0) $f_0(t) = 0$ *for some* $t \in (0, R)$.

Then, sequence $\{t_n^0\}$ *generated by* $t_0^0 = 0$,

$$t_{n+1}^0 = t_n^0 - \frac{f_0(t_n^0)}{f_0'(t_n^0)}, \quad n = 0, 1, 2, \ldots$$

is strictly increasing, remains in $(0, t_0^*)$ *and converges to* t_0^*, *where* t_0^* *is the smallest zero of function* f_0 *in* $(0, R)$. *Furthermore, suppose that for each* $x, y \in D_1 := \bar{U}(x_0, \rho) \cap U(x_0, t_0^*)$ *there exists* $f_1 : [0, \rho_1) \longrightarrow \mathbb{R}$, $\rho_1 = \min\{\rho, t_0^*\}$ *such that*

$$\|\hat{F}'(x_0)^{-1}\| \|F'(y) - F'(x)\| \leq f_1'(\|x - y\| + \|x - x_0\|) - f_1'(\|x - x_0\|) \tag{22.8}$$

$$\text{and } \|x_1 - x_0\| \leq f_1'(0)$$

Moreover, suppose that

(h_1^1) $f_1(0) > 0$ *and* $f_1'(0) = -1$.

(h_2^1) f_1' *is convex and strictly increasing.*

(h_3^1) $f_1(t) = 0$ *for some* $t \in (0, \rho_1)$.

(h_4^1) $f_0(t) \leq f_1(t)$ *and* $f_0'(t) \leq f_1'(t)$ *for each* $t \in [0, \rho_1)$.

Then, f_1 *has a smallest zero* $t_1^* \in (0, \rho_1)$, *the sequences generated by generalized Newton's method for solving the generalized equation* $F(x) + Q(x) \ni 0$ *and the scalar equation* $f_1(0) = 0$, *with initial point* x_0 *and* t_0^1 *(or* $s_0 = 0$*), respectively,*

$$0 \in F(x_n) + F'(x_n)(x_{n+1} - x_n) + Q(x_{n+1}),$$

$$t_{n+1}^1 = t_n^1 - \frac{f_1(t_n^1)}{f_1'(t_n^1)} \; (\text{or } s_{n+1} = s_n - \frac{f_1(s_n)}{f_0'(s_n)})$$

are well defined, $\{t_n^1\}$ *(or* s_n*) is strictly increasing, remains in* $(0, t_1^*)$ *and converges to* t_1^*. *Moreover, sequence* $\{x_n\}$ *generated by generalized Newton's method (22.3) is well defined, remains in* $U(x_0, t_1^*)$ *and converges to a point* $x^* \in \bar{U}(x_0, t_1^*)$, *which is the unique solution of generalized equation* $F(x) + Q(x) \ni 0$ *in* $\bar{U}(x_0, t_1^*)$. *Furthermore, the following estimates hold:*

$$\|x_n - x^*\| \leq t_1^* - t_n^1, \quad \|x_n - x^*\| \leq t_1^* - s_n,$$

$$\|x_{n+1} - x^*\| \leq \frac{t_1^* - t_{n+1}^1}{(t_1^* - t_n^1)^2} \|x_n - x^*\|^2$$

$$\|x_{n+1} - x^*\| \leq \frac{t_1^* - s_{n+1}^1}{(t_1^* - s_n^1)^2} \|x_n - x^*\|^2$$

$$s_{n+1} - s_n \leq t_{n+1}^1 - t_n^1,$$

and sequences $\{t_n^1\}$, $\{s_n\}$ *and* $\{x_n\}$ *converge* $Q-$*linearly as follows:*

$$\|x_{n+1} - x^*\| \leq \frac{1}{2}\|x_n - x^*\|,$$

$$t_1^* - t_{n+1}^1 \leq \frac{1}{2}(t_1^* - t_n),$$

and

$$t_1^* - s_{n+1}^1 \leq \frac{1}{2}(t_1^* - s_n).$$

Finally, if
(h_5^1) $f_1'(t_1^*) < 0$,
then the sequences $\{t_n^1\}$, $\{s_n\}$ *and* $\{x_n\}$ *converge* $Q-$*quadratically as follows:*

$$\|x_{n+1} - x^*\| \leq \frac{D^- f_1'(t_1^*)}{-2f_1'(t_1^*)}\|x_n - x^*\|^2,$$

$$\|x_{n+1} - x^*\| \leq \frac{D^- f_1'(t_1^*)}{-2f_0'(t_1^*)}\|x_n - x^*\|^2,$$

$$t_1^* - t_{n+1} \leq \frac{D^- f_1'(t_1^*)}{-2f_1'(t_1^*)}(t_1^* - t_n)^2,$$

and

$$t_1^* - s_{n+1} \leq \frac{D^- f_1'(t_1^*)}{-2f_0'(t_1^*)}(t_1^* - s_n)^2,$$

where D^- *stands for the left directional derivative of function* f_1.

Remark 22.3.2 *(a) Suppose that there exists* $f : [0,R) \longrightarrow \mathbb{R}$ *twice continuously differentiable such that for each* $x, y \in U(x_0, \rho)$

$$\|\hat{F}'(x_0)^{-1}\|\|F'(y) - F'(x)\| \leq f'(\|x-y\| + \|x - x_0\|) - f'(\|x - x_0\|). \tag{22.9}$$

If $f_1(t) = f_0(t) = f(t)$ *for each* $t \in [0,R)$, *then Theorem 22.3.1 specializes to Theorem 4 in [18]. Otherwise, i.e., if*

$$f_0(t) \leq f_1(t) \leq f(t) \text{ for each } t \in [0, \rho_1), \tag{22.10}$$

then, our Theorem is an improvement of Theorem 4 under the same computational cost, since in practice the computation of function f *requires the computation of functions* f_0 *or* f_1 *as special cases. Moreover, we have that for each* $t \in [0, \rho_1)$

$$f_0(t) \leq f(t) \tag{22.11}$$

and

$$f_1(t) \leq f(t) \tag{22.12}$$

leading to $t_n^1 \leq t_n$, $s_n \leq t_n$,

$$t_{n+1}^1 - t_n^1 \leq t_{n+1} - t_n \tag{22.13}$$

$$s_{n+1}^1 - s_n^1 \leq s_{n+1} - s_n \tag{22.14}$$

and

$$t_1^* \leq t^*, \tag{22.15}$$

where $\{t_n\}$ *is defined by*

$$t_0 = 0,\, t_{n+1} = t_n - \frac{f(t_n)}{f'(t_n)},$$

$t^* = \lim_{n \longrightarrow \infty} t_n$ *and* t^* *is the smallest zero of function* f *in* $(0,R)$ *(provided that the "h" conditions hold for function* f *replacing* f_1 *and* f_0*). If*

$$-\frac{f_1(t)}{f_1'(t)} \leq -\frac{f(s)}{f'(s)} \tag{22.16}$$

or

$$-\frac{f_1(t)}{f_0'(t)} \leq -\frac{f(s)}{f'(s)}, \tag{22.17}$$

respectively for each $t \leq s$*. Estimates (22.13) and (22.14) can be strict if (22.16) and (22.17) hold as strict inequalities.*

(b) *So far we have improved the error bounds and the location of the solution* x^* *but not necessarily the convergence domain of the generalized Newton's method (22.3). We can also show that convergence domain can be improved in some interesting special cases. Let* $F \equiv \{0\}$,

$$f(t) = \frac{L}{2}t^2 - t + \eta,$$

$$f_0(t) = \frac{L_0}{2}t^2 - t + \eta$$

and

$$f_1(t) = \frac{L_1}{2}t^2 - t + \eta,$$

where $\|x_1 - x_0\| \leq \eta$ *and* L, L_0 *and* L_1 *are Lipschitz constants satisfying:*

$$\|\hat{F}'(x_0)^{-1}\|\|F'(y) - F'(x)\| \leq L\|y - x\|$$

$$\|\hat{F}'(x_0)^{-1}\|\|F'(x) - F'(x_0)\| \leq L_0\|x - x_0\|$$

and

$$\|\hat{F}'(x_0)^{-1}\|\|F'(y) - F'(x)\| \leq L_1\|y - x\|,$$

on the corresponding balls. Then, we have that

$$L_0 \leq L$$

and

$$L_1 \leq L.$$

The corresponding majorizing sequences are

$$t_0 = 0,\, t_1 = \eta,\, t_{n+1} = t_n - \frac{f(t_n)}{f'(t_n)} = t_n + \frac{L(t_n - t_{n-1})^2}{2(1 - Lt_n)}, \quad n = 1,2,\ldots$$

$$t_0^1 = 0,\, t_1^1 = \eta,\, t_{n+1}^1 = t_n^1 - \frac{f_1(t_n^1)}{f_1'(t_n^1)} = t_n^1 + \frac{L_1(t_n^1 - t_{n-1}^1)^2}{2(1 - Lt_n^1)}, \quad n = 1,2,\ldots$$

$$\begin{aligned} s_0 = 0,\, s_1 = \eta,\, s_{n+1} &= s_n - \frac{f_1(s_n) - f_1(s_{n-1}) - f_1'(s_{n-1})(s_n - s_{n-1})}{f_0'(s_n)} \\ &= s_n + \frac{L_1(s_n - s_{n-1})^2}{2(1 - L_0 s_n)}, \quad n = 1,2,\ldots. \end{aligned}$$

Then, sequences converge provided, respectively that

$$q = L\eta \leq \frac{1}{2} \tag{22.18}$$

and for the last two

$$q_1 = L_1\eta \leq \frac{1}{2},$$

so

$$q \leq \frac{1}{2} \Longrightarrow q_1 \leq \frac{1}{2}.$$

It turns out from the proof of Theorem 22.3.1 that sequence $\{r_n\}$ *defined by [6]*

$$r_0 = 0,\, r_1 = \eta,\, r_2 = r_1 + \frac{L_0(r_1 - r_0)^2}{2(1 - L_0 r_1)},$$

$$r_{n+2} = r_{n+1} + \frac{L_1(r_{n+1} - r_n)^2}{2(1 - L_0 r_{n+1})}$$

is also a tighter majorizing sequence than the preceding ones for $\{x_n\}$. *The sufficient convergence condition for* $\{r_n\}$ *is given by [6]:*

$$q_2 = K\eta \leq \frac{1}{2},$$

where

$$K = \frac{1}{8}(4L_0 + \sqrt{L_1 L_0 + 8L_0^2} + \sqrt{L_0 L_1}).$$

Then, we have that

$$q_1 \leq \frac{1}{2} \Longrightarrow q_2 \leq \frac{1}{2}.$$

Hence, the old results in [18] have been improved. Similar improvements can follow for the Smale's alpha theory [2, 6] or Wang's γ−theory [18, 22, 24]. Examples where $L_0 < L$ or $L_1 < L$ or $L_0 < L_1$ can be found in [6]. It is worth noticing that (22.18) is the famous for its simplicity and clarity Newton-Kantorovich hypothesis for solving nonlinear equations using Newton's method [13] employed as a sufficient convergence condition in all earlier studies other than ours.

(c) *The introduction of (22.8) depends on (22.7) (i.e., f_1 depends on f_0). Such an introduction was not possible before (i.e., when f was used instead of f_1).*

Proof of Theorem 22.3.1 Simply notice that the iterates $\{x_n\} \in D_1$ which is a more precise location than $\bar{U}(x_0,\rho)$ used in [18], since $D_1 \subseteq \bar{U}(x_0,\rho)$. Then, the definition of function f_1 becomes possible and replaces f in the proof [18], whereas for the computation on the upper bounds $\|\hat{F}'(x)^{-1}\|$ we use the more precise f_0 than f as it is shown in the next perturbation Banach lemma [13].

□

Lemma 22.3.3 *Let $x_0 \in D$ be such that $\bar{F}'(x_0)$ is a positive operator and $\bar{F}'(x_0)^{-1}$ exists. If $\|x-x_0\| \leq t < t^*$, then $\bar{F}'(x)$ is a positive operator and $\bar{F}'(x)^{-1}$ exists. Moreover,*

$$\|\bar{F}'(x)^{-1}\| \leq \frac{\bar{F}'(x_0)^{-1}\|}{f_0'(t)}. \tag{22.19}$$

Proof: Observe that

$$\|\bar{F}'(x)-\bar{F}'(x_0)\| \leq \frac{1}{2}\|\bar{F}'(x)-\bar{F}'(x_0)\| + \frac{1}{2}\|(\bar{F}'(x)-\bar{F}'(x_0))^*\| = \|\bar{F}'(x)-\bar{F}'(x_0)\|. \tag{22.20}$$

Let $x \in \bar{U}(x_0,t)$, $0 \leq t < t^*$. Thus $f'(t) < 0$. Using, (h_1^1) and (h_2^1), we obtain that

$$\begin{aligned} \|\bar{F}'(x_0)^{-1}\|\|\bar{F}'(x)-\bar{F}'(x_0)\| &\leq \|\bar{F}'(x_0)^{-1}\|\|\bar{F}'(x)-\bar{F}'(x_0)\| \\ &\leq f_0'(\|x-x_0\|)-f_0(0) \\ &< f_0'(t)+1<1. \end{aligned} \tag{22.21}$$

So by Banach's Lemma on invertible operators, we have $\bar{F}'(x)^{-1}$ exists. Moreover by above inequality,

$$\|\bar{F}'(x)^{-1}\| \leq \frac{\|\bar{F}'(x_0)^{-1}\|}{1-\|\bar{F}'(x_0)^{-1}\|\|F'(x)-F'(x_0)\|} \leq \frac{\|\bar{F}'(x_0)^{-1}\|}{1-(f_0'(t)+1)} = -\frac{\|F'(x_0)^{-1}\|}{f_0'(t)}.$$

Using (22.21) we have

$$\|\bar{F}'(x)-\bar{F}'(x_0)\| \leq \frac{1}{\|\bar{F}'(x_0)^{-1}\|}. \tag{22.22}$$

Thus, we have

$$\langle(\bar{F}'(x_0)-\bar{F}'(x))y,y\rangle \leq \|\bar{F}'(x_0)-\bar{F}'(x)\|\|y\|^2 \leq \frac{\|y\|^2}{\|\bar{F}'(x_0)^{-1}\|},$$

which implies,

$$\langle\bar{F}'(x_0)y,y\rangle - \frac{\|y\|^2}{\|\bar{F}'(x_0)^{-1}\|} \leq \langle\bar{F}'(x)y,y\rangle. \tag{22.23}$$

Now since $\bar{F}'(x_0)$ is a positive operator and $\bar{F}'(x_0)^{-1}$ exists by assumption, we obtain that

$$\langle\bar{F}'(x_0)y,y\rangle \geq \frac{\|y\|^2}{\|\bar{F}'(x_0)^{-1}\|}. \tag{22.24}$$

The result now follows from (22.23) and (22.24).

□

Remark 22.3.4 *This result improves the corresponding one in [18, Lemma 8] (using function f instead of f_0 or f_1) leading to more precise estimates on the distances $\|x_{n+1}-x^*\|$ which together with idea of restricted convergence domains lead to the aforementioned advantages stated in Remark 22.3.2.*

Next, we present an academic example to show that: (22.11), (22.12) hold as strict inequalities; the old convergence criteria (22.18) does not hold but the new involving q_1 and q_2 do hold and finally the new majorizing sequences $\{s_n\}$ and $\{r_n\}$ are tighter than the old majorizing sequence $\{t_n\}$. Hence, the advantages of our approach (as already stated in the abstract of this study) over the corresponding ones such as the ones in [18] follow.

Example 22.3.5 *Let $F \equiv \{0\}, H=\mathbb{R}, D=U(x_0,1-p), x_0=1, p\in(0,\frac{1}{2})$ and define function F on D by*

$$F(x)=x^3-p. \tag{22.25}$$

Then, we get using (22.25) that $L=2(2-p), L_0=3-p, L_1=2(1+\frac{1}{L_0})$ and $\eta=\frac{1}{3}(1-p)$, so we have

$$L_0<L_1<L$$

and

$$f_0(t)<f_1(t)<f(t) \text{ for each } p\in(0,\frac{1}{2}).$$

In particular the sufficient Kantorovich convergence criterion (22.18) used in [18] is not satisfied, since

$$q=L\eta>\frac{1}{2} \text{ for each } p\in(0,\frac{1}{2}). \tag{22.26}$$

Hence, there is no guarantee under the results in [18] that Newton's method converges to the solution $x^* = \sqrt[3]{p}$. However, our convergence criteria are satisfied:

$$q_1 = L_1\eta \leq \frac{1}{2} \text{ for each } p \in (0.4619831630, \frac{1}{2}). \qquad (22.27)$$

and

$$q_2 = K\eta \leq \frac{1}{2} \text{ for each } p \in (0.2756943786, \frac{1}{2}). \qquad (22.28)$$

In order for us to compare the majorizing sequences $\{t_n\}, \{s_n\}, \{r_n\}$, let $p = 0.6$. Then, old condition (22.26) as well as new conditions (22.27) and (22.28) are satisfied. Then, we present the following table showing that the new error bounds are tighter. It is worth noticing that these advantages are obtained under the same computational cost since the evaluation of the old Lipschitz constants L requires the computation of the new constants L_0, L_1 and K as special cases.

Table 22.1: Error bound comparison.

n	$t_{n+1} - t_n$	$s_{n+1} - s_n$	$r_{n+1} - r_n$
2	0.0397	0.0370	0.0314
3	0.0043	0.0033	0.0023
4	5.1039e-05	2.6253e-05	1.2571e-05
5	7.2457e-09	1.6743e-09	3.7365e-10

References

[1] I. K. Argyros, Concerning the convergence of Newton's method and quadratic majorants, *J. Appl. Math. Comput.*, 29, 391–400, 2009.

[2] I. K. Argyros, A Kantorovich-type convergence analysis of the Newton-Josephy method for solving variational inequalities, *Numer. Algorithms*, 55, 447–466, 2010.

[3] I. K. Argyros, Variational inequalities problems and fixed point problems, *Comput. Math. with Appl.*, 60, 2292–2301, 2010.

[4] I. K. Argyros, Improved local convergence of Newton's method under weak majorant condition, J. Comput. Appl. Math., 236, 1892–1902, 2012.

[5] I. K. Argyros, Improved local converge analysis of inexact Gauss-Newton like methods under the majorant condition, *J. Franklin Inst.*, 350 (6), 1531–1544, 2013.

[6] I. K. Argyros and S. Hilout, Weaker conditions for the convergence of Newton's method, *J. Complexity*, 28, 364–387, 2012.

[7] A. I. Dontchev and R. T. Rockafellar, Implicit functions and solution mappings, Springer Monographs in Mathematics, Springer, Dordrecht, 2009.

[8] O. Ferreira, A robust semi-local convergence analysis of Newtons method for cone inclusion problems in Banach spaces under affine invariant majorant condition, *J. Comput. Appl. Math.*, 279, 318–335, 2015.

[9] O. P. Ferreira, M. L. N. Goncalves and P. R. Oliveria, Convergence of the Gauss-Newton method for convex composite optimization under a majorant condition, *SIAM J. Optim.*, 23 (3), 1757–1783, 2013.

[10] O. P. Ferreira and G. N. Silva, Inexact Newton's method to nonlinear functions with values in a cone, arXiv: 1510.01947, 2015.

[11] O. P. Ferreira and B. F. Svaiter, Kantorovich's majorants principle for Newton's method, *Comput. Optim. Appl.*, 42 (2), 213–229, 2009.

[12] N. Josephy, Newton's method for generalized equations and the PIES energy model, University of Wisconsin-Madison, 1979.

[13] L. V. Kantorovič, On Newton's method for functional equations, *Doklady Akad Nauk SSSR (N.S.)*, 59, 1237–1240, 1948.

[14] A. Pietrus and C. Jean-Alexis, Newton-secant method for functions with values in a cone, *Serdica Math. J.*, 39 (3-4), 271–286, 2013.

[15] F. A. Potra, The Kantorivich theorem and interior point methods, *Math. Programming*, 102 (1), 47–70, 2005.

[16] S. M. Robinson, Strongly regular generalized equations, *Math. Oper. Res.*, 5 (1), 43–62, 1980.

[17] R. T. Rochafellar, Convex analysis, Princeton Mathematical Series, No. 28, Princeton University Press, Princeton, N. J., 1970.

[18] G. N. Silva, On the Kantorovich's theorem for Newton's method for solving generalized equations under the majorant condition, *Appl. Math. Comput.*, 286, 178–188, 2016.

[19] S. Smale, Newtons method estimates from data at one point. In R. Ewing, K. Gross and C. Martin, editors, The Merging of Disciplines: New Directions in pure, Applied and Computational Mathematics, pages 185–196, Springer New York, 1986.

[20] J. F. Traub and H. Woźniakowski, Convergence and complexity of Newton iteration for operator equations, *J. Assoc. Comput. Mach.*, 26 (2), 250–258, 1979.

[21] L. U. Uko and I. K. Argyros, Generalized equation, variational inequalities and a weak Kantorivich theorem, *Numer. Algorithms*, 52 (3), 321–333, 2009.

[22] J. Wang, Convergence ball of Newton's method for generalized equation and uniqueness of the solution, *J. Nonlinear Convex Anal.*, 16 (9), 1847–1859, 2015.

[23] P. P. Zabrejko and D. F. Nguen, The majorant method in the theory of Newton-Kantorivich approximations and the Pták error estimates, *Numer. Funct. Anal. Optim.*, 9 (5-6), 671–684, 1987.

[24] Y. Zhang, J. Wang and S. M. Gau, Convergence criteria of the generalized Newton method and uniqueness of solution for generalized equations, *J. Nonlinear Convex. Anal.*, 16 (7), 1485–1499, 2015.

Index

U

V

W